Lecture Notes in Civil Engineering

Volume 449

Lecture Notes in Civil Engineering (LNCE) publishes the latest developments in Civil Engineering—quickly, informally and in top quality. Though original research reported in proceedings and post-proceedings represents the core of LNCE, edited volumes of exceptionally high quality and interest may also be considered for publication. Volumes published in LNCE embrace all aspects and subfields of, as well as new challenges in, Civil Engineering. Topics in the series include:

- Construction and Structural Mechanics
- Building Materials
- Concrete, Steel and Timber Structures
- Geotechnical Engineering
- Earthquake Engineering
- Coastal Engineering
- Ocean and Offshore Engineering; Ships and Floating Structures
- Hydraulics, Hydrology and Water Resources Engineering
- Environmental Engineering and Sustainability
- Structural Health and Monitoring
- Surveying and Geographical Information Systems
- Indoor Environments
- Transportation and Traffic
- Risk Analysis
- Safety and Security

To submit a proposal or request further information, please contact the appropriate Springer Editor:

- Pierpaolo Riva at pierpaolo.riva@springer.com (Europe and Americas);
- Swati Meherishi at swati.meherishi@springer.com (Asia—except China, Australia, and New Zealand);
- Wayne Hu at wayne.hu@springer.com (China).

All books in the series now indexed by Scopus and EI Compendex database!

Sreevalsa Kolathayar · N. Vinod Chandra Menon ·
K. S. Sreekeshava
Editors

Best Practices in Geotechnical and Pavement Engineering

Select Proceedings of IACESD 2023

Springer

Editors
Sreevalsa Kolathayar
Department of Civil Engineering
National Institute of Technology
Mangaluru, Karnataka, India

N. Vinod Chandra Menon
Amrita Vishwa Vidyapeetham
Kollam, Kerala, India

K. S. Sreekeshava
Department of Civil Engineering
Jyothy Institute of Technology
Bengaluru, Karnataka, India

ISSN 2366-2557 ISSN 2366-2565 (electronic)
Lecture Notes in Civil Engineering
ISBN 978-981-99-8504-3 ISBN 978-981-99-8505-0 (eBook)
https://doi.org/10.1007/978-981-99-8505-0

This Springer imprint is published by the registered company Springer Nature Singapore Pte Ltd.
The registered company address is: 152 Beach Road, #21-01/04 Gateway East, Singapore 189721, Singapore

Paper in this product is recyclable.

Preface

The complexities of the modern built environment demand continuous innovation, adaptation and the sharing of knowledge to ensure that our foundations remain strong and our pathways smooth. Geotechnical and pavement engineering stand as integral pillars, upholding the safety, functionality and sustainability of our infrastructure. There is need for innovative solutions and eco-friendly practices that are reshaping these disciplines.

This volume presents the select proceedings of the G20 C20 International Conference on Interdisciplinary Approaches in Civil Engineering for Sustainable Development (IACESD-2023) which showcase the remarkable diversity of thought and expertise that defines the field of civil engineering today. From soil mechanics and foundation design to asphalt technology and road maintenance strategies, the contents of this volume reflect the multifaceted nature of these fields. The integration of digital tools, remote sensing and data analytics is revolutionizing the way geotechnical and pavement engineering projects are planned, executed and monitored.

We extend our heartfelt gratitude to all the authors, presenters, participants and reviewers who contributed to the success of the G20 C20 International Conference on Interdisciplinary Approaches in Civil Engineering.

We thank all the staff of Springer for their full support and cooperation at all the stages of the publication. We hope that this book shall be beneficial to students, academicians, professionals and researchers.

Surathkal, India — Dr. Sreevalsa Kolathayar
Kollam, India — Prof. N. Vinod Chandra Menon
Bengaluru, India — Dr. K. S. Sreekeshava

Contents

About the Editors

Dr. Sreevalsa Kolathayar pursued M.Tech. from IIT Kanpur, Ph.D. from IISc, and served as International Research Staff at UPC Barcelona Tech Spain. He is an Associate Professor in the Department of Civil Engineering, National Institute of Technology Karnataka (NITK) Surathkal, India. He has authored five books, edited twelve books, and published over 100 research articles. He is on the Editorial Board of several International Journals. In 2017, The New Indian Express honoured him with South India's Most Inspiring Young Teachers Award. He received the ISET DK Paul Research Award from the Indian Society of Earthquake Technology for the best Ph.D. thesis on Earthquake Risk Reduction in India. He received the "IEI Young Engineers Award" from The Institution of Engineers (India) in 2019. He is on the roster of two technical committees of ASCE Geo-Institute and is a Member of the Working Groups of BIS CED 39 for three IS codes.

Prof. N. Vinod Chandra Menon has worked as Professor in the Centre for Disaster Management at the Yeshwantrao Chavan Academy of Development Administration (YASHADA), Pune. He has worked in charge of Emergency Preparedness and Response in UNICEF India Country Office in New Delhi. He is currently Adjunct Professor at Amrita Vishwa Vidyapeetham, India; President of RedR India, and Regional Director Asia of The International Emergency Management Society (TIEMS) Oslo. He has over 37 years of working experience, of which more than a quarter century has been in the fields of disaster risk reduction, climate change adaptation and public policy analysis. In 2005, he was nominated by the Prime Minister of India as one of the Founder Members of the National Disaster Management Authority (NDMA), Government of India, with the status of a Union Minister of State in the Government of India.

Dr. K. S. Sreekeshava is currently working as Associate Professor and Head of the Department of Civil Engineering, Jyothy Institute of Technology, Bengaluru. He obtained Ph.D. from BMS College of Engineering, Visvesvaraya Technological University, Belagavi, in the year 2020. His research interests are in the field of masonry structures, bio-composites, and structural design. He has published more than 30 research articles in reputed journals, Elsevier and Springer publishing houses. He has successfully coordinated funded AICTE-ISTE faculty refresher Programme and is also working as Student Project Proposal (SPP) Coordinator. He has the honours of life member of ISET, ISTE, ICI, INSC, NICEE, and IAENG. He has successfully completed the research funding grant under competitive research funding scheme initiated under scheme of TEQIP by Visvesvaraya Technological University, Belagavi. He has guided several undergraduate and postgraduate projects sponsored by Karnataka State Council for Science and Technology.

Best Practices in Geotechnical and Pavement Engineering—An Introduction

Sreevalsa Kolathayar, N. Vinod Chandra Menon, K. S. Sreekeshava, Poonam Shekhawat, and C. Bhargavi

1 Introduction

Geotechnical investigation is the process of collecting information about the physical, chemical, and engineering properties of soil, rock, and groundwater beneath the ground surface. It involves various techniques, such as drilling, sampling, testing, and analysis, to evaluate the subsurface conditions and to determine the suitability of the site for a particular construction project. Recently, increase in infrastructure development for rapid growth of population is demanding the construction on soft compressible ground. A tendency exists in geotechnical practice for rapid construction to meet the stringent construction schedules. Construction on soft ground develops large settlements and deforms laterally resulting in instability. Hence, constructing anything on top of it takes a lot of work. To transform this soil into one that can be used for construction, it will need to be stabilized using one of many different processes. Either the use of mechanical manipulation or some kind of chemical transformation may be used to successfully stabilize a soft soil. These ground improvement techniques should be sustainable considering the present climate change scenario.

Furthermore, the pavement should be durable and it should involve sustainable construction practices. The ingredients of pavement construction should be locally available. The pavement should withstand high temperature and effect of rain. Hence, it should be adopted to all weather conditions. The stabilized pavement materials are emphasized more in the recent research as they are typically incorporated into the

S. Kolathayar
National Institute of Technology-Karnataka, Surathkal, Karnataka, India
e-mail: sreevalsa@nitk.edu.in

N. Vinod Chandra Menon
Amrita Vishwa Vidyapeetham, Amritapuri, Kollam, India

K. S. Sreekeshava (✉) · P. Shekhawat · C. Bhargavi
Jyothy Institute of Technology, Bengaluru, Karnataka, India
e-mail: sreekeshava.ks@jyothyit.ac.in

S. Kolathayar et al. (eds.), *Best Practices in Geotechnical and Pavement Engineering*,
Lecture Notes in Civil Engineering 449, https://doi.org/10.1007/978-981-99-8505-0_1

pavement structure as base course and sub-base course. The use of marginal and waste materials reduces the use of virgin aggregates, which have become a scarce resource in recent decades.

The chapters in this book volume present diverse insights on sustainable ground improvement techniques, geosynthetics, geophysics, and pavement construction. This article is an attempt to summarize the contents of the book volume on best practices in geotechnical and pavement engineering. All the chapters in this volume are segregated in five clusters, e.g., soil–foundation interaction, sustainable ground improvement techniques, behavior of geosynthetic implemented soil, advanced geophysics and rock mechanics practices, recent advancement in pavement construction.

2 Foundation Engineering

Sumanth et al. analyzed a site consisting of 3 m of compacted silty sand-engineered fill to find out the optimal type and size of the foundation. Inconsistent standard penetration resistance values and non-uniform compaction were reported across different borehole locations on the site. Proper compaction control and engineered fill construction were adopted at the site to arrive at optimal foundation dimensions. Moreover, enhancing the assessment of the energy transfer ratio, utilizing appropriate tools, and implementing high-quality procedures within the geotechnical investigation program contribute significantly to advancing sustainability in construction practices. Dixit et al. present a nonlinear site response study of shallow foundation isolated on gravel tire crumb mixture layer on shallow bedrock sites. It was found that the addition of gravel tire crumb mixture layer has improved soil properties like shear strength, energy absorption capacity, and ductility. Mandeel et al. present a parametric study with the use of finite element method to elaborate the behavior of full-scale foundation resting on the layered soil within vertical concentric loading that simulates the field conditions more accurately. For the two layer of clay soil with different parameters, it was noticed that bearing capacity depends upon the cohesion of the first layer [1, 2]. Yesakova et al. presented the results of field surveys of the settlements of adjacent multi-story buildings on large-sized slab foundations, the soil base of which is composed of weak water-saturated sandy soils on top. The analysis of the results and comparison of the theoretical and actual values of the settlements of the building show the possibility of using of the model of the soil base as a linear elastic layer of finite width has been confirmed for forecasting adequate deformations of sectional buildings on large-sized slab foundations which are based on weak water-saturated soils on top. Saha et al. investigated the improvement potential of using geotextile on a sandy soil bed. The study's findings demonstrate that reinforcement configuration has a high potential to improve the load-bearing characteristics of the geotextile-reinforced solution, where experimental results depicted a good agreement with numerical simulations. Guruprasad presented a practical case of retrofit of a slab-type reinforced concrete raft foundation, that supports a mercantile building founded on medium sand.

3 Sustainable Ground Improvement Techniques

Sushma et al. analyzed the stability of various embankments constructed using Matsuo stability charts for different embankment loadings. Results of the present study truly confirm that the Matsuo chart predicts the stability of slopes under varying conditions of the embankments to higher degree accuracy. Khan et al. examined the physiochemical properties of clay soil samples obtained from a construction site. The authors found that soil faces continuous changes with extended curing time in strength and permeability due to the addition of metakaolin, which is supported by the test results showing that permeability decreases and strength increases with increasing time of curing of treated soil samples with metakaolin. Sagari et al. explained about the terrazyme, a liquid enzyme, stabilized soil with cement. The study found that the optimum dosage of 4.5% along with cement gave maximum strength for terrazyme 5x chemical, 5% along with cement gave maximum strength for terrazyme 11x chemical, and hence, bio-enzymes' chemical like terrazyme can be used for soil stabilization in problematic soil. Usmani et al. discussed and outlined geotechnical knowledge required for nurturing sustainable engineering practices and their integration with existing education and training of geotechnical engineers. Deepa et al. explored the effectiveness of magnesium-rich olivine or forsterite in reducing the hydraulic conductivity of silt. Test results indicate that the hydraulic conductivity values for both the olivine-treated silt samples satisfy the minimum recommended specification of hydraulic conductivity (less than 1×10^{-7} cm/s) as per the Environmental Protection Agency (EPA) guidelines. Shekhawat et al. discussed about stabilization of illite and kaolin-based clay at ambient temperature using flyash-based geopolymer. The results demonstrate that 30% geopolymer concentration in clay significantly increased compressive strength, tensile strength, and California bearing ratio. Finally, the geopolymer-treated clay mixtures found in the investigation have possible applications in pavements and embankments as sustainable building materials prepared from industrial and domestic wastes [3, 4]. Reddy et al. evaluated the influence of montmorillonite content on swelling behavior in relation to its plasticity. Bhatia et al. examined numerical simulations utilizing PLAXIS 3D, utilizing a FEM-based program, stone column behavior. Naik et al. applied the procedure for the estimation of uncertainty measurements based on the guidelines prescribed by National Accreditation Board for Testing and Calibration Laboratories (NABL) 141 (2016). Sharma and Das analyzed the slope stability for hill-based construction in NE India. Nagendra et al. investigated the effect of soil type on the rheological parameters of a reengineered cohesive soil. Koley et al. mitigated the harmful effect of plastic on the environment by performing a laboratory 3D consolidation test to investigate the impact of waste plastic fibers on the consolidation behavior of the soil. Kantesaria et al. examined the geotechnical properties of pond ash-bentonite mixtures as potential liner materials. The results show that the addition of bentonite significantly increases the plasticity index and thus the water-holding capacity of pond ash. Das et al. presented the deterministic 2D slope stability analysis for a specific slope in Agumbe Ghat at Karnataka and another landslide prone region in Dhobighat at Hyderabad. It has been found that the slope at Dhobighat where mostly the well-graded sand is available is more affected due to the change in the design parameters as compared to Agumbe Ghat slope where the soil is mostly stiff clay. Endait et al. reviewed the sustainable usage of calcium carbide residue, which exhibits pozzolanic properties for improving the soil, especially clayey soil. Turkane et al. presented the strength properties of local soil enhanced by mean of alkaline activated sawdust ash.

4 Behavior of Reinforced Soil

Azim and Sengupta describe about slope stability of a 8 m high road embankment resting on soft soil under saturated conditions under the action of seismic activity analogous to 0.38 g acceleration caused in seismic zone IV using equivalent linear dynamic model in Quake/W and Slope/W. Since the simulation yielded unsafe safety factors, based on the susceptible liquefaction zone, layers of geotextile were applied to improve the stability and effectiveness in terms of total and vertical displacements, q/p' ratio, induced accelerations, effect on total stress, and excess pore water pressures were analyzed. Geotextiles have proved to be effective in stabilizing the concerned simulation yielding factor of 2.813 and 1.793 pre- and post-earthquake events and declining displacements and acceleration by 33.3% and hence proved to be an effective reinforcement in minimizing seismic induced dynamic forces and enhancing the overall seismic performance of the simulation model. Utilizing geosynthetics in the lengths of currently weak formations is an alternate strategy to reduce the number of stress. Singh et al. investigated the importance of using geosynthetics (geogrids) for existing sub-grade. Using the finite element software PLAXIS 3D, the vertical deformations and stresses of a railway embankment (with and without geogrid) are calculated under a moving train load of 90 kN. The speed of the moving train is taken as 180 km/h. The results indicated that including geosynthetics helped in reducing deformations and stresses to a certain level. Yadav et al. examined the effect of inclusion of pond ash and randomly distributed polypropylene fibers on the strength behavior of cement-stabilized clayey soil. Based on favorable test results obtained, it can be concluded that partial replacement of cement-stabilized clayey soil by pond ash with the inclusion of polypropylene fibers causes a significant increase in strength, decrease in the stiffness, and rate of loss of post-peak strength. The fiber inclusion changed the brittle behavior of the composite to ductile [5, 6]. Kumar et al. evaluated the effect of fiber reinforcement (coir and jute fiber) obtained from Jodhpur, Rajasthan, in improvement of geotechnical behavior of the natural clayey soil mixed with sand along with comparison of two natural fibers in improving soil strength. Among coir and jute fiber content/percentage, jute seems to have shown improvement accompanied by enhanced resistance to the cracking. Jatin et al. investigated the strengths of subgrades reinforced with jute textile and polypropylene (PP) geotextile. According to the test results, jute textile (natural fiber) increased the soil sample's shear strength. Jayakrishnan et al. studied the effects of coir geocell reinforcement on the strength improvement of geocell-reinforced sand through laminar box shake table tests. The results have shown that the use of coir geocell, irrespective of its height, has improved the liquefaction resistance of coastal sand. Debnath et al. implemented a laboratory experimental program through a triaxial test and a numerical analysis using PLAXIS 2D finite element model to study the effect of jute geotextile on sub-grade soil modification. The experimental and numerical result shows that including jute geotextile on sub-grade soil significantly improves the soil strength by lowering the stress, strain, and displacement at the top of the sub-grade soil, increasing the shear strength and bearing capacity of the soil and by reducing the quantity of instant settlement.

5 Advanced Geophysics and Rock Mechanics

Rana et al. discusses the role of ground penetrating radar (GPR) in infrastructure projects, including its benefits, applications, and limitations. The paper also presents the simulation approach using forward modeling tools to visualize the response of GPR for various subsurface conditions. In summary, GPR is a valuable asset for infrastructure project professionals, providing accurate and reliable data for geotechnical engineers and civil engineers to evaluate structural performances of the building and avoid any dangers or risks. Rana et al. in another study explain the key applications of geophysics in the structural and geotechnical fields, focusing on how it has revolutionized site characterization, foundation design, and hazard assessment. Geophysics provides a non-intrusive method to investigate the subsurface without the need for drilling or excavation. Geophysics can play a vital role in hazard assessment by helping to identify potential hazards and allowing for more effective mitigation strategies. Arora et al. discussed about the tunnel-boring machine (TBM) excavation and support installation process in squeezing clay-rich rocks through a physical model. The observations are made to understand better the interaction between the support and the squeezing ground. Based on the test results and case studies, a recommendation to optimize the support requirement for tunnels in squeezing ground is proposed [6, 7]. Shradha et al. reviewed case studies of application of diaphragm walls in marine conditions as well as for underground structures like metros which is done. Purnanand et al. analyzed twin tunnels in granitic rock for Karwar–Kundapura highway.

6 Pavement Geotechnics

Goud et al. analyzed the effect of waste plastic bottles as fiber reinforcement in the pavement concrete. The test results of each trial were compared with the conventional mix. There was no considerable increase in compression strength of concrete with PET fibers when compared with control mix; however, about 5% increase in flexural strength was observed. Bhanusuresh et al. studied waste material-based fillers for patching potholes using warm mix asphalt (WMA). Marshall Stability tests were conducted, and it was found that the WMA with 25% of all the fillers showed a Marshal Stability 26.93% higher than that of conventional WMA. Kumar et al. studied the cohesionless pavement geomaterial reinforced with multi-layers of georeinforcement as jute fibers. The test results demonstrate that including jute fiber in single, double, and triple layers increases the stiffness capacity of the soil at the optimum depth of *D*/4. Bhonsle and Kandolkar investigated the stability of vertical slope/cut which is existing at Moira village, Goa, India. The stability of slope was found to be greater than 1. Prasad et al. addressed methodology of designing cement-treated base (CTB) for flexible pavements. The research findings were utilized to assess how well the CTB performed in terms of fatigue under repeated wheel loads. Kumar and Prakash aimed to replace the natural coarse aggregates using construction and demolition (C&D) waste. Girme et al. a cold mix technology is adopted for the preparation of road construction material using granite aggregate. Based on the experimental results, it is concluded that the use of granite aggregates gives improved results over basalt aggregates. Waghchaure et al. presented the feasibility study of using the construction and demolition wastes in the low volume road construction in rural areas. The coarse aggregates in concrete are replaced by the recycled aggregate (RCA) (0, 50, and 100%) obtained from construction and demolition waste. The numerical results of 7 and 28 days compressive strength of concrete using RCA are presented in the paper. Consoli et al. evaluated the impact of sodium chloride (NaCl) on the compressive strength and the accumulated loss of mass (ALM) after wetting–drying cycles on a mixture of crushed reclaimed asphalt pavement (RAP) with fly ash (FA) and carbide lime (L) (both industrial residues). Stabilization was carried out considering blends of 75% RAP and 25% FA. The results indicated that adding 1.0% NaCl to the crushed RAP-FA-lime blends almost doubles strength for a 7-day curing period and reduces the ALM by an average of 50% in durability tests [8, 9].

7 Summary

This chapter summarized the contents of the book volume on Best Practices in Geotechnical and Pavement Engineering. All the chapters in this volume are segregated in five sections such as soil–foundation interaction, sustainable ground improvement techniques, behavior of geosynthetic implemented soil, advanced geophysics and rock mechanics practices, and recent advancement in pavement construction.

References

1. Al-Ameri AF, Hussein SA, Mekkiyah H (2020) Estimate the bearing capacity of full-scale model shallow foundations on layered-soil using PLAXIS. Solid State Technol 63:1775–1787
2. Bariker P, Kolathayar S (2022) Dynamic soil structure interaction of a high-rise building resting over a finned pile mat. Infrastructures 7:142. https://doi.org/10.3390/infrastructures7100142
3. Shekhawat P, Sharma G, Singh RM (2023) Strength characteristics of hazardous wastes flyash and eggshell powder-based geopolymer-stabilized soft soil cured at ambient temperature. Arab J Geosci 16:1–12. https://doi.org/10.1007/s12517-023-11234-2
4. Puppala AJ, Pedarla A (2017) Innovative ground improvement techniques for expansive soils. Innov Infrastruct Solut 2:24. https://doi.org/10.1007/s41062-017-0079-2
5. Yadav JS, Tiwari SK, Shekhwat P (2018) Strength behaviour of clayey soil mixed with pond ash, cement and randomly distributed fibres. Transp Infrastruct Geotechnol 5. https://doi.org/10.1007/s40515-018-0056-z
6. Yang K-H, Zornberg JG, Liu C-N, Lin H-D (2012) Stress distribution and development within geosynthetic-reinforced soil slopes. Geosynth Int 19(1):62–78. Thomas Telford Ltd. https://doi.org/10.1680/gein.2012.19.1.62
7. Arora K, Gutierrez M, Hedayat A (2022) Physical model simulation of rock-support interaction for the tunnel in squeezing ground. J Rock Mech Geotech Eng 14:82–92. https://doi.org/10.1016/j.jrmge.2021.08.016
8. Consoli NC, Giese DN, Leon HB et al (2018) Sodium chloride as a catalyser for crushed reclaimed asphalt pavement—fly ash—carbide lime blends. Transp Geotech. https://doi.org/10.1016/j.trgeo.2018.02.001
9. Shukla S, Das B, Chandrasekaran S, Kolathayar S (2021) Smart technologies for sustainable development. https://doi.org/10.1007/978-981-15-5001-0

Foundation Engineering

Sustainable Implementation of Standard Geotechnical Practices to Optimize the Foundation Design of a Residential Building: A Case Study

H. C. Sumanth, A. R. Nandhagopal, C. Ashitha, C. R. Parthasarathy, and T. Prashanth

1 Introduction

When a suitable foundation has to be designed for a superstructure, the foundation engineer typically follows a decision-making process in selecting the optimum type of foundation. Geotechnical investigation is a primary and key activity that will provide the basic information to optimize the foundation design. Geotechnical investigation is the process of collecting information about the physical, chemical, and engineering properties of soil, rock, and groundwater beneath the ground surface. It involves various techniques, such as drilling, sampling, testing, and analysis, to evaluate the subsurface conditions and to determine the suitability of the site for a particular construction project. Neglecting the primary step of geotechnical investigation might lead the construction contractor to either underestimate or overestimate the soil strength. In case of underestimation of the strength of the soil shall mean the design of the structures is constructed with a higher factor of safety incurring significantly higher costs. Whereas in the case of overestimation of soil strength, the structure shall not be designed with a sufficient factor of safety and might lead to a potential disaster. The advances made in the solution of geotechnical problems are undeniable. However, the question must be asked "What are our problem-solving abilities worth if we are solving the wrong problems and the solutions either are not or cannot be implemented in practice? [1]. Understanding the scope of work, the benefit of adopting standard practices, and the incorporation of suitable investigation methods with respect to codal provisions is paramount for any construction activity.

H. C. Sumanth (✉) · A. R. Nandhagopal · C. Ashitha · C. R. Parthasarathy · T. Prashanth
Sarathy Geotech and Engineering Services Pvt Ltd, Bengaluru, India
e-mail: sumanth@sarathygeotech.com

S. Kolathayar et al. (eds.), *Best Practices in Geotechnical and Pavement Engineering*,
Lecture Notes in Civil Engineering 449, https://doi.org/10.1007/978-981-99-8505-0_2

In the case of seismic vulnerable zones, the occurrence of an earthquake may lead to differential settlements induced by liquefaction. A proper geotechnical investigation shall help the engineer designs that can withstand catastrophes due to such natural calamities. Liquefiable deposits can be identified by conducting a proper geotechnical investigation and ground improvement methods like Rapid Impact Compaction (RIC) which can be adopted [2].

1.1 Ground Improvement

One must consider applicable site-specific techniques for the improvement of soft ground conditions, before resorting to deep foundations [3]. There are possibly three different methods of ground Improvement.

1. Compaction: techniques that typically are used to compact or densify soil in situ. **Methods**: Dynamic Compaction, Vibro-compaction, Surcharging with PVDs, Compaction Grouting, Blast Densification, and Vacuum-Induced Consolidation.
2. Reinforcement: techniques that typically construct a reinforcing element within the soil mass without necessarily changing the soil properties. The performance of the soil mass is improved by the inclusion of reinforcing elements.
   Methods: Stone columns, Vibro-concrete columns, Soil Nailing, Micro-piles, Fracture grouting, and Geosynthetics.
3. 3. Fixation: techniques that fix or bind the soil particles together thereby increasing the soil's strength and decreasing its compressibility and permeability.
   Methods: Permeation grouting, Jet grouting, Deep soil mixing, freezing, and vitrification.

1.2 Standard Penetration Test

The standard penetration test (SPT) is a commonly used geotechnical investigation method in land-based and nearshore projects for determining the strength of soil and estimating its bearing capacity [4]. SPT involves driving a standard sampler with a 38 mm diameter and to a depth of 45 cm from the depth of testing. The split spoon sampler is attached to a string of rods till the required depth in a clean borehole of 100 or 150 mm diameter using a 63.5 N hammer falling freely through the height of 75 cm. The number of blows required to drive the sampler the first 15 cm (N1) is recorded but is disregarded as a seating drive and the number of blows required to drive it the next two consecutive 15 cm (N2 and N3). The N value of that particular SPT is reported as the sum of N2 and N3. If the penetration is < 30 cm, then the number of blows and the depth penetrated shall be logged onto the field bore log. The corrected N value or the N value corresponding to a 60% energy transfer ratio shall be estimated by applying the multiple correction factors. The advantages of SPT are that it is quick and inexpensive and there are multiple correlations to estimate the engineering properties from the N value. On the other hand, the disadvantages of SPT are slightly inaccurate measurements in gravelly soils or soils with high water content and the collection of undisturbed samples is also of utmost importance so as to estimate the shear strength parameters of the sub-soil. Other methods like Auger boring, cone penetration test (CPT), Pressure meter, and Dilatometer are used based on the scope and site feasibility. CPT is generally used for offshore geotechnical investigations or as a confirmatory investigation method for improved ground or reclamation sites. Geophysical methods can be effectively adopted to identify or map the sinkholes, cavities, weak zones of earth in a site. IS 1892 [5] presents the theory and suitability of all such methods.

1.3 Engineered Fill

Engineered fill is a type of fill material that has been meticulously designed and placed to meet specific engineering specifications. This material is frequently used in construction projects such as foundations, roads, and embankments. Engineered fill is typically created by selecting and processing specific types of soil and rock materials, which are then tested to ensure that they meet the desired engineering properties, such as strength, compaction, and permeability. The material is then laid out and compacted in layers, with each layer meticulously monitored and tested to ensure that it meets the required specifications. Engineered fill is frequently used when natural soils are unsuitable for construction or when additional support is required. If a building is being built on a site with poor soil conditions, for example, the engineered fill can be used to provide a stable foundation. Similarly, if a road or bridge is being constructed over a soft or unstable substrate, the engineered fill can be used to add support and prevent settling or failure.

Proper compaction with uniform lift thickness has to be performed at the site to ensure the uniform density of the earth laterally across the intended region. Overall, the use of engineered fill enables construction projects to be built on a broader range of sites while maintaining greater control over the materials and their properties. This can lead to more efficient and cost-effective construction processes, as well as increased product safety and durability.

2 Geotechnical Investigation and Ground Profile

The current study aims at performing a geotechnical investigation for a proposed G + 1 residential building in Bengaluru, India. An outline of the site location is shown in Fig. 1. Drilling consists of thirteen (13) exploratory boreholes by rotary drilling rig up to a maximum depth of 10.50 m from the existing ground level. Standard penetration test (SPT) for each borehole location was conducted at regular intervals. The groundwater table was encountered around 7 m across 13 boreholes in the project site. The proposed area of construction is filled up with external borrow material 1.5–2.0 m thick. Geotechnical investigation consisting of exploring subsurface conditions by advancing, sampling, and logging thirteen (13) nos. of exploratory boreholes, viz., BH-01 to BH-13, by rotary drilling method was carried out from the top of the fill layer. Soil stratification across the site consists of sandy clay or clayey sand with an increasing trend of stiffness along the vertical depth as shown in Fig. 2.

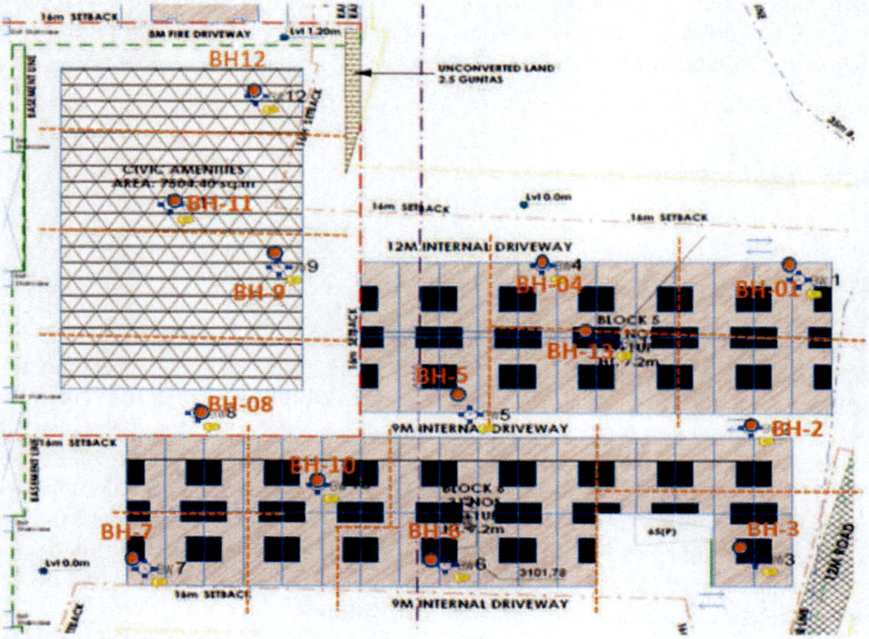

Fig. 1 Outline of site location

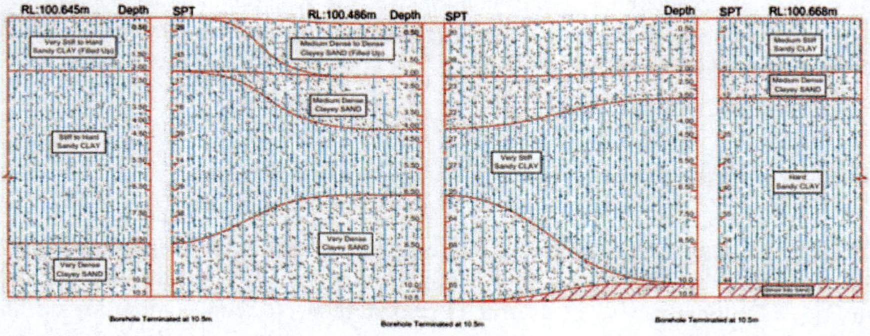

Fig. 2 Soil stratification across the site

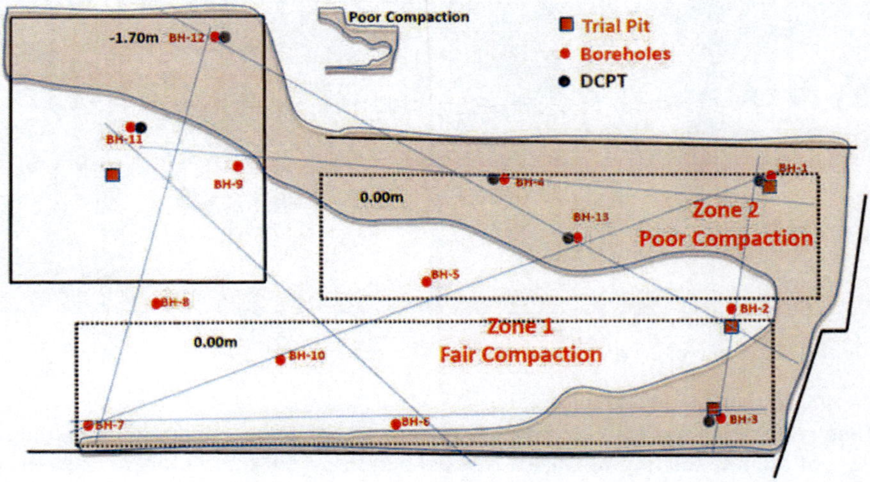

Fig. 3 Confirmatory geotechnical investigation layout

No rock was encountered in these 13 boreholes. The soil samples obtained from the boreholes were inspected in the field immediately upon retrieval for type, texture, and color. Soil samples were then sealed in plastic bags and transported to the laboratory for further testing.

2.1 *Problem Statement*

Apprehensions due to variability in the *N* values during the geotechnical investigation campaign across the site have paved the way for additional examinations through different methods. Dynamic cone penetration tests (DCPTs) were conducted at six (06) locations near BH-01, BH-03, BH-04, BH-11, BH-12, and BH-13. Additionally, four numbers of trial pits (Pit 1—near BH-2, Pit 2—near BH-1, Pit 3—near BH-3, and Pit 4—near BH-11) were inspected as shown in Fig. 3. Furthermore, undisturbed/representative samples were collected and subjected to additional laboratory tests.

3 Site Visit and In Situ Field Density

The proposed area of construction is filled up with external borrow material 1.5–2.0 m thick. Geotechnical investigation consisting of exploring subsurface conditions by advancing, sampling, and logging was carried out from the top of the fill layer. Inspection of in situ sounding test results (i.e., standard penetration test and dynamic

Fig. 4 Photographs of the site visit

cone penetration test) reveals the different degrees of compaction within the thickness of the fill. A site visit was carried out and inspected four numbers of the trial pits (Pit 1—near BH-2, Pit 2—near BH-1, Pit 3—near BH-3, and Pit 4—near BH-11) collected representative samples using the core cutter method. These samples were subjected to additional laboratory tests and were classified as per IS 1498 [6]. Figure 4 presents the sample photographs of the site visit. The round periphery of the compacted fill has breached due to lack of confinement, and in many places, the piping phenomenon is also noticed (Fig. 3). This can be attributed to the several pockets of poor compaction.

4 Analysis of Results

Core cutter method was performed in four trial pits to know the in situ density at the site as shown in Fig. 5. During driving the core cutter in Pit 1, it took about 30 blows with the dolly and hammer, while it was 2–3 blows in Pits 1 and 3, respectively. The compaction test on the representative samples determines the maximum dry density (MDD) of 1.91–1.99 g/cc, while the optimum moisture content (OMC) is 11.2–14.8%. The in situ dry density varied between 1.71 and 1.83 g/cc with the placement moisture content of about 8–14%. The degree of compaction with respect to dry density is 89–91%, while the placement moisture content varied from 69 to 101% of the respective OMC values. The cohesion of samples in Pit 2 was 75 kPa, while it was 350 kPa in Pit 1. The oedometer test results suggest higher compressibility of samples in Pit 3 vis-à-vis Pits 1 and 4. This confirms the different degrees of compaction within the compacted thickness of 1.5–2.0 m.

Based on in situ sounding tests, field visits, and further laboratory test results (IS 13030) [7], the compacted area can be demarcated into two zones (Zone 1—fair compaction and Zone 2—poor compaction) as shown in Fig. 3 and Table 1. The *N* values from standard penetration test and dynamic cone penetration test for both zones are presented in Figs. 6 and 7.

Fig. 5 Field density using the core cutter method

Table 1 Recommended design inputs from field and laboratory testing: foundation on top of the fill (IS 6403 [8] and IS 12070 [9])

Degree of compaction	Allowable bearing pressure of raft for 75 mm settlement (KN/m^2)			Modulus of sub-grade reaction (KN/m^3)		
	Based on SPT	Based on oedometer	Recommended	Based on SPT	Based on oedometer	Recommended
Zone 1 fair compaction	330	170	170	4410	2266	3300
Zone 2 poor compaction	110	130	110	1470	1733	1600

Under the circumstances, the foundation placed on top of the fill will undergo different degrees of differential settlement due to varied degrees of compaction. Thus, a lower bound N value of 14 and 7, respectively, is derived for allowable bearing pressure computation of Zones 1 and 2. The allowable bearing pressure is also computed for both zones based on oedometer test results as shown in Fig. 8.

Since the proposed construction was a residential villa of G + 1 floor, the foundation can be placed on top of the fill. It is recommended to adopt a raft foundation and the allowable pressure computed as per IS 6403 and IS 12070 is listed in Table 1. Alternatively, the foundation shall be designed with the allowable bearing pressure as shown below in Table 2 with the depth of foundation placed on the natural ground available at 1.5–2.0 m.

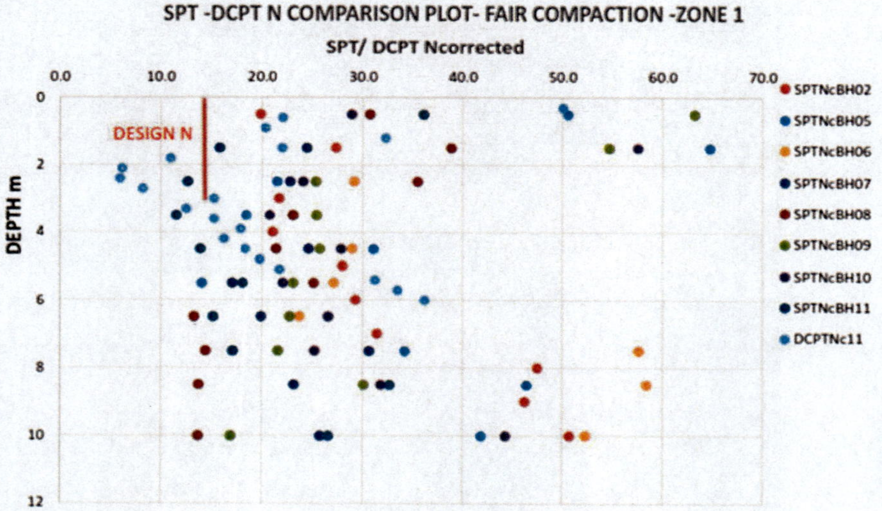

Fig. 6 Zone 1 fair compaction

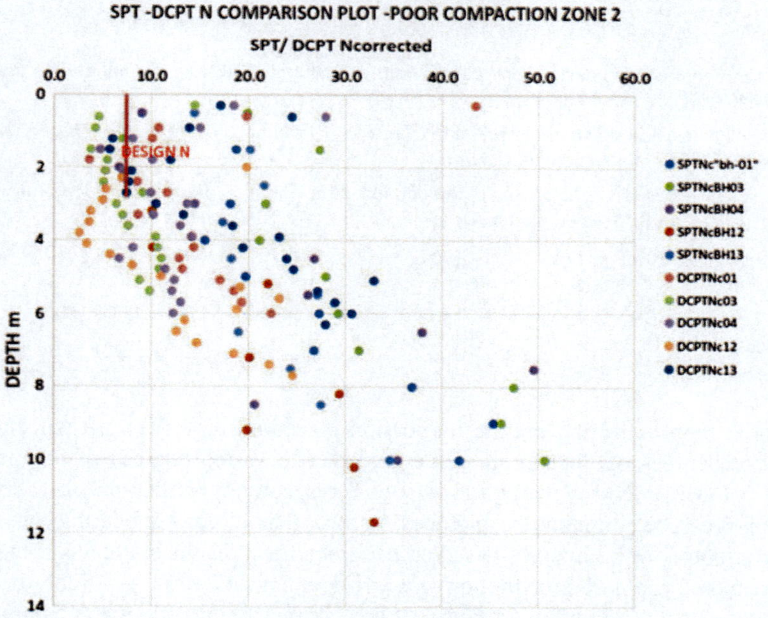

Fig. 7 Zone 2 poor compaction

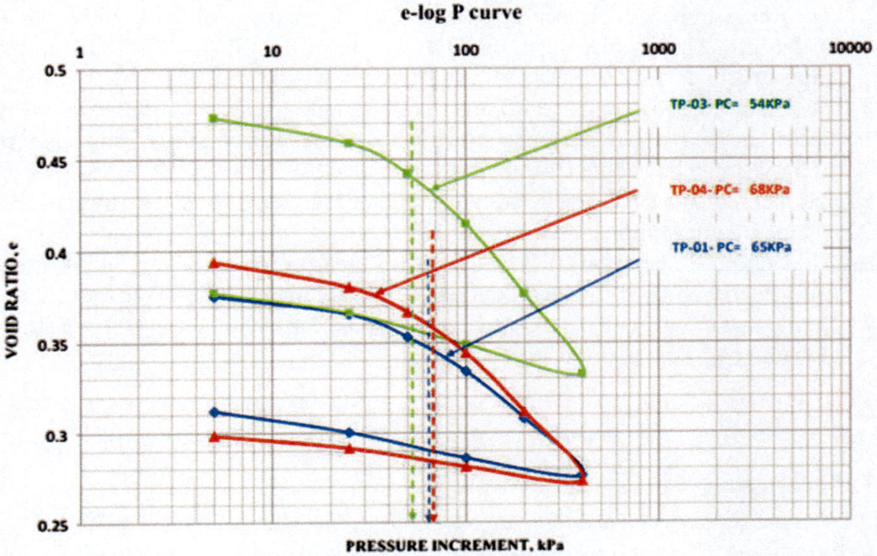

Fig. 8 One-dimensional consolidation results

Table 2 Recommended design inputs from field and laboratory testing: foundation at 1.5–2.0 m from NGL

Borehole No.	Depth of foundation from the top of the fill	Allowable bearing pressure (kN/m^2)		Modulus of sub-grade for the design of raft (kN/m^3)
		Isolated (for 25 mm maximum settlement)	Raft (for 50 mm maximum settlement)	
BH-01, 02, 03, 04, 05 and 07	1.5–2.0 m below natural ground level	200	400	8000
BH-08, 09 and 10		250	450	9000
BH-11 and 12		120	250	5000
BH-13		150	300	6000

5 Conclusions and Recommendations

1. The general layout of the site is raised by an external fill of 1.5–2.0 m thick (except near BH-12 locations). The compacted fill is assessed to have achieved different degrees of compaction, and thus, the area is demarcated as Zone 1 (fair compaction) and Zone 2 (poor compaction).
2. The proposed construction is a residential villa of G + 1 floor and it is understood that the foundation is planned to be placed on top of the fill. It is recommended to adopt a raft foundation and the allowable pressure as shown in Table 1.

3. The foundation shall be ensured to have a minimum depth of 0.5 m embedment. If the fill is further increased, then the weight of the fill shall be added to the superstructure load.
4. The Zone 2 (poor compaction) area can be improved by re-compacting or by heavy tamping (dynamic compaction method) and the allowable bearing pressure of Zone 1 can be adopted.
5. It is recommended to provide lateral confinement at the site boundary by geocells, gabion, or conventional raining walls.
6. The bottom of the excavation shall be inspected for loose pockets and if found any shall be replaced with lean concrete before laying the PCC for the foundation.
7. Settlement monitoring for the structures shall be carried out during the construction and post-construction stages.

References

1. Peter Day (2017) Challenges and shortcomings in geotechnical engineering practice in the context of a developing country. In: Proceedings of the 19th international conference on soil mechanics and geotechnical engineering, Seoul
2. Kristiansen H, Davies M (2004) Ground improvement using rapid impact compaction. In: Proceedings of the 3rd world conference on earthquake engineering
3. Hussin JD (2006) Methods of soft ground improvement. How to improve soft ground 5 professional development hours (PDH) or continuing education hours (CE) online PDH or CE course methods of soft ground improvement
4. IS 2131–1981 (Reaffirmed 2002): method of standard penetration test for soil
5. IS 1892–2021: code of practice for subsurface investigation for foundation
6. IS 1498–1970 (Reaffirmed 2007): classification and identification of soils for general engineering purpose
7. IS 13030–1991 (Reaffirmed 2010): method of test for laboratory determination of water content, porosity, density and related properties of rock material
8. IS 6403–1971: code of practice for determining shallow foundation's bearing capacity
9. IS 12070–1987 Code of practice for determining the bearing capacity of shallow foundation on rock

Nonlinear Behaviour of Foundation on Gravel Tire Crumb Mixture for Shallow Bedrock Sites

P. S. Naresh Dixit, G. Indudhara, A. Jithendra Kumar, Aniket I. Hiremath, and Goutham B. Reddy

1 Introduction

The unique properties of tire crumb rubbers have garnered attention for their use in engineering applications. These materials have a low density compared to soil and rock, making them ideal for lightweight fill applications. Their high porosity and elasticity also result in high compressibility, making them more suitable than gravel and other soils. Additionally, rubbers are excellent for sound and vibration absorption, which makes them useful in applications that require damping [1]. The use of tire crumbs not only provides a cost-effective method for repurposing waste materials but also helps address the growing problem of post-consumer tire disposal. With the increasing number of vehicles on the roads in developed countries, there is a large amount of space required for landfilling. Using tire crumbs in engineering applications can help to achieve a cleaner environment [2].

In seismic areas, the use of tire crumbs in construction can help save lives and minimize economic losses in the event of earthquakes. The high durability of tire crumbs under normal foundation conditions makes them a reliable choice for seismic applications. Overall, the properties of tire crumbs make them a versatile and cost-effective material for various engineering applications [3]. The construction industry can benefit from several applications of waste tire rubber. For instance, tire rubber can be used as a component of asphalt mixtures, which can improve pavement performance and durability. Rubber particles in the mixture can absorb shock and reduce noise while increasing resistance to cracking and deformation [4].

Waste tire rubber can also be used as a component of concrete to improve its resilience, durability, and impact resistance. It can also reduce the weight of concrete,

G. Indudhara · A. J. Kumar · A. I. Hiremath · G. B. Reddy
Department of Civil Engineering, BMS College of Engineering, Bengaluru, India

P. S. N. Dixit (✉)
Assistant Professor, Chanakya University, Bengaluru, India
e-mail: nareshdixit82@gmail.com

S. Kolathayar et al. (eds.), *Best Practices in Geotechnical and Pavement Engineering*, Lecture Notes in Civil Engineering 449, https://doi.org/10.1007/978-981-99-8505-0_3

making it easier to transport and handle. In addition, waste tire rubber can be used as a filler or reinforcement in various construction materials, such as plastic composites, insulation materials, and roofing products. Using waste tyre rubber in these products reduces the amount of virgin materials required and provides a new market for recycled materials [4]. Overall, using waste tire rubber in the construction industry can bring many benefits, such as improved material performance and durability, reduced environmental impact, and the creation of new markets for recycled materials. Numerous studies have explored the potential of waste tire rubber as an aggregate in concrete and evaluated its mechanical and durability properties. For instance, examined the strength and durability characteristics of self-compacting rubberised concrete with and without the addition of fibres. The inclusion of steel fibres could offset the reduction in compressive strength resulting from the use of scrap rubber in self-compacting concrete. Additionally, rubberised concrete with fibres demonstrated superior resistance against abrasion. Therefore, self-compacting rubberised concrete could prove to be a more durable cementitious composite than conventional self-compacting concrete.

However, concluded that concrete with rubber aggregate contents > 10% by mass would be unacceptable for primary structural elements [5]. Nevertheless, there are numerous structural applications with medium-to-low strength requirements for which blocks or other precast concrete units have the advantage of a lower unit weight compared to usual concrete mixes.

There is a shift in the current design practice to achieve better seismic performance by reuse of waste tires. Sustainable options used are granulated tire rubber [GTR] which is expected to reduce the accelerations and consequent seismic inertial forces by 40%, steel-fibre-reinforced- rubberized concrete [SFRRuC], and recycled rubber particles [RRP's] which is used as a practice substitution of mineral aggregates in concrete which give a resultant material [6].

The results of UU triaxial and compression tests on the sand–tire crumb (S-TC) mixtures, where the content of tire crumbs varies from 0 to 40% by volume showed that, higher the content of tire crumbs in the mixture, the larger the reduction in the shear strength. The peak shear strength and the corresponding axial strain of the mixture increase with the increase in the confining pressure. All mixtures with a percentage of tire crumbs equal to or > 10% showed a Brittleness Index (BI) of at least one third of the value of pure sand, indicating a ductile material [7].

The direct shear and unconsolidated undrained triaxial tests on STCM's with varying compositions and confining pressures were conducted. The coarse grain particle content in the soil was observed to affect the STCM's performance. The addition of tire crumbs improved the soil's shear strength and energy absorption capacity, with the increase being significant at times. The study showed a reduction in the brittleness index and an increase in the energy absorption capacity of the soil on higher percentages of tire crumbs and at higher confining pressures [8].

The direct shear and triaxial tests on sand-tire crumb mixtures (STCM) to investigate the influence of tire crumb size, on strength characteristics, energy absorption, brittleness index, and stiffness characteristics were conducted. The study found that the addition of waste tire crumbs to sand can increase the energy absorption capacity and ductility of STCM by about 40–70% [9]. In this paper, an attempt has been made to check the nonlinear site response of gravel rubber mixture layer as a base isolation material for low-cost housing.

2 Methodology

The selection of the site was determined based on the borehole data collected from various locations, where the values of *N* were obtained through SPT tests. Four sites were shortlisted, namely Mysore Road (Site-1), Whitefield (Site-2), Shivajinagar (Site- 3), and Jayanagar (Site-4), which had shallow bedrock. Once the necessary tests were completed and the corresponding data were obtained, it was analyzed using the DEEPSOIL software.

Direct shear test was conducted on the GRM mixtures where the samples were created with 5–40% rubber based on volume content with increment of 5%. The 0.150 mm × 150 mm × 75 mm direct shear box apparatus is used to evaluate the stress–strain-volumetric behaviour of the two reference materials (gravel and rubber) and their selected mixes. The testing was done under stresses of 100, 150, 200 kPa as per IS 2720 (Part 39)-1977 [10]. The specimen was then prepared in four layers inside the shear box. After thoroughly mixing gravel and rubber particles, uniform rigid-soft granular mixtures are attained by the dry tamping method. In the process, segregation is avoided by carefully placing the mixes in the testing apparatus and gently tamping dry specimens until achieving 50% relative density. The stress–strain curves were further used to find the total energy absorption capacity of GRM sample. The energy absorption capacity was determined by finding the area under the stress versus strain curves. The area under the stress–strain curve up to a given value of strain is the total mechanical energy per unit volume, which the material consumed while straining to the value, and it is given by:

$$\mathrm{EA} = \int \sigma(\varepsilon)\mathrm{d}\varepsilon, \tag{1}$$

where σ (ε) is stress as a function of strain and EA is the energy absorption capacity.

3 DEEPSOIL Analysis

After conduction of required tests in lab the results obtained were used in analysis using Deepsoil software. Using this software, both the linear and the nonlinear analyses can be performed in both the time and the frequency domains. Post-analysis, the output can be exported to the EXCEL file for our further reference. For our study, nonlinear time-domain analysis was used, and these options can be selected in the

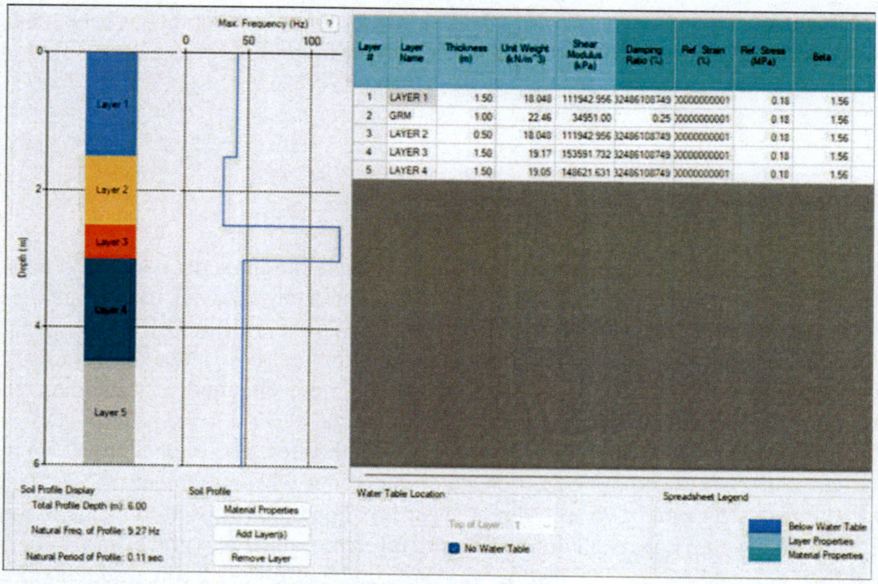

Fig. 1 DEEPSOIL interface showing the soil profile and GRM layer

first page of the DEEPSOIL software. The main window of the DEEPSOIL software is shown in Fig. 1.

Looking into the borehole data that were previously collected, soil profiles were created. In DEEPSOIL software, the soil layers having different soil properties can be created in the next page, i.e. the analysis tab in the software. When the layers are created, the thickness/depth of each layer is entered in to the software and its respective soil properties such as the unit weight, shear modulus, and the damping ratio were entered for each layer considered. Here, the condition of NO ground water was selected. The unit weight for each layer was obtained from the borehole data collected. The shear modulus for each layer was determined using the relation 15.09N0.74 [11] and the bulk density was calculated using the formula 1.232N0.141 [12] (for all types of soil). Rest of the parameters are according to the G/Gmax, and damping ratio curve is calculated by curve fitting procedure by the software. Rest of the input for analysis by software is done with input motion as within, elastic half space and damping ratio of 24%, where *N* is the SPT '*N*' value. The damping ratio for the GRM layer was taken as 25% (0.25).

The input motion given to the elastic half space was the Koyna earthquake, which had occurred at Koynanagar, Maharashtra, on December 11, 1967. The earthquake had a magnitude of 6.6 and had an intensity of VIII which led to the death of 180 people. This earthquake in the Deccan Plateau had an intensity which has an occurrence period of 100 years. The Peak Ground Acceleration (PGA) of the input motions used was 0.31 g. In context to the elastic half space, bedrock properties of Bengaluru bedrock were used for analysis, where the shear modulus was kept as 4.2×10^7 kPa having unit weight of 32 kN/m^3 with a damping ratio of 24%. The analysis was carried out nonlinearly in the time domain. The analysis was carried out for all the soils, varying the depth starting from 0.1 to 1 m thickness with increment of 0.1 m. The analysis was carried out, the output file was exported into Microsoft Excel for our future reference, it contains details like the PGA, strain, stress–strain ratio, maximum displacements for each layer that was analyzed, the required data was considered, and the results were concluded.

4 Results and Discussion

4.1 Direct Shear Test

Observation of direct shear test plots its clear that there is a sudden drop of shear till 10% of GRM, and then, the reduction in value of shear stress is less. After the percentage of tire in mix exceeds 30%, the change in shear strength is similar to the behavior of tire crumbs where large strains are taken with lesser change in shear strength and the curve becomes almost parallel to X axis. This is also one of the findings from the paper, Recycling of ELT's for sustainable geotechnical applications: A New Zealand perspective. From the experimental results, it is observed that as the GRM percentage increases, there is a reduction in the shear strength value from 35 to 43%.

4.2 Brittleness Index and Energy Absorption Capacity

With a compaction ratio of 1, as the confining pressure applied increases for a particular layer, we can observe that the energy absorption capacity also increases. Now, this can be correlated to the GRM mixture percentage used from different layers from 0 to 40% where the energy absorption capacity increases with increase in its confining pressure. As the GRM mixture increases every 5%, there is a good amount of improvement in its energy absorption capacity with doubling the confining pressure. The energy absorption increases in the range of 90–98% for various percentages of GRM under particular pressure.

4.3 Nonlinear Analysis

Nonlinear analysis was carried out on the soil profiles in the time domain using DEEPSOIL software. The effective stress, PGA, maximum displacement and stress ratios are shown for various sites for the given input motions. After the analysis, it is observed that effective stress has increased significantly in each layer, as the thickness of GRM layer increases by 0.1 m.

The use of gravel rubber mixture (GRM) layer as a soil improvement technique was analyzed for Site-1, and the results are shown in Table 1. The analysis was carried out using DEEPSOIL software. The study revealed a decrease in Peak Ground Acceleration (PGA) by 16.8 and 9.05% at the surface level and a depth of 1.5 m, respectively, with the use of GRM layer of varying depths from 0.1 to 1 m. The decrease in stress ratio and displacement further supports the improvement in the soil performance due to the GRM layer. Based on the analysis, the optimum depth of GRM layer for the site was found to be 1 m. The implications of this study are relevant to geotechnical engineers and researchers involved in soil improvement techniques using waste materials such as gravel rubber mixtures.

According to Table 2, the use of gravel rubber mixture (GRM) layers with a thickness ranging from 0.1 to 1 m resulted in a reduction in Peak Ground Acceleration (PGA) of 13.77 and 9.54% at the surface level and at a depth of 1.5 m, respectively, for Site-2. Based on the overall improvement in site response behavior, a GRM layer thickness of 1 m is considered the optimum for the given input motions. This implies that incorporating a GRM layer of the appropriate thickness can enhance the seismic performance of the soil. The results of this study align with previous research that has demonstrated the effectiveness of using waste tire rubber in earthquake engineering applications.

Table 1 DEEPSOIL nonlinear analysis PGA output for Site-1

PGA (in terms of g)		Layer 1	Layer 2	Layer 3	Layer 4	Layer 5
Thickness of GRM (m)	0	0.5618	0.0000	0.5581	0.4840	0.3176
	0.1	0.6145	0.6252	0.5507	0.4873	0.3207
	0.2	0.6161	0.6363	0.5660	0.4858	0.3187
	0.3	0.6255	0.6187	0.5966	0.4625	0.3075
	0.4	0.6559	0.5755	0.5995	0.4384	0.3113
	0.5	0.6430	0.5535	0.5076	0.4074	0.3054
	0.6	0.5991	0.5612	0.4680	0.3666	0.3214
	0.7	0.5170	0.5937	0.3835	0.3707	0.3214
	0.8	0.4740	0.5759	0.4089	0.3733	0.3161
	0.9	0.4740	0.5759	0.4089	0.3733	0.3161
	1	0.4672	0.5076	0.3987	0.4278	0.3127
	1.1	0.4767	0.4847	0.3976	0.3934	0.3142
	1.2	0.4795	0.4576	0.3923	0.3713	0.3202
	1.3	0.4646	0.4452	0.3843	0.3798	0.3140
	1.4	0.4490	0.4394	0.3745	0.3769	0.3141
	1.5	0.4301	0.4368	0.0000	0.3587	0.3184

Table 2 DEEPSOIL nonlinear analysis PGA output for Site-2

PGA (in terms of g)		Layer 1	Layer 2	Layer 3	Layer 4	Layer 5
Thickness of GRM (m)	0	0.5705	0.0000	0.5632	0.4848	0.3201
	0.1	0.6250	0.6505	0.5581	0.4839	0.3112
	0.2	0.6565	0.6261	0.5882	0.4596	0.3142
	0.3	0.6687	0.5964	0.5966	0.4436	0.3130
	0.4	0.6405	0.5837	0.5248	0.4673	0.3146
	0.5	0.6514	0.5564	0.4715	0.4495	0.3159
	0.6	0.5530	0.6119	0.4209	0.4243	0.3187
	0.7	0.5098	0.5994	0.3582	0.3915	0.3179
	0.8	0.5258	0.5621	0.3553	0.3764	0.3149
	0.9	0.4912	0.5304	0.4344	0.3920	0.3152
	1	0.4920	0.5094	0.4223	0.3782	0.3171
	1.1	0.5037	0.4850	0.4001	0.3679	0.3117
	1.2	0.4853	0.4607	0.3832	0.3870	0.3140
	1.3	0.4744	0.4445	0.3873	0.3769	0.3143
	1.4	0.4585	0.4377	0.3931	0.3855	0.3186
	1.5	0.4307	0.4415	0.0000	0.3814	0.3197

The study observed a significant reduction in Peak Ground Acceleration (PGA) at Site-3 by 4.04% at surface level and 12.96% at a depth of 1.5 m with GRM layer thickness varying from 0.1 to 1 m. These findings suggest that incorporating GRM layer in soil could improve soil response to seismic input. The optimal thickness of the GRM layer was determined to be 1 m based on the observed reduction in PGA. The DEEPSOIL analysis provided additional data, which are summarized in Table 3. These findings have implications for seismic design and construction, where the use of GRM layer in soil could potentially improve the seismic resistance of structures.

The DEEPSOIL analysis for Site-4 with varying thicknesses of gravel rubber mixture (GRM) layer ranging from 0.1 to 1 m showed that the Peak Ground Acceleration (PGA) is decreased by 19.4 and 23.8% at the surface level and at a depth of 1.5 m, respectively. The overall improvement in soil performance to the provided input motions indicated that the optimum thickness of the GRM layer is 1 m. The results of the PGA output from the DEEP SOIL analysis are presented in Table 4.

The reduction in PGA achieved by using a GRM layer can have significant implications for earthquake-resistant design and construction of structures. GRM is a low-cost, environmentally-friendly material that can be used to mitigate the effects of earthquakes on structures. The results of this study suggest that a 1 m GRM layer can effectively reduce the PGA and improve the seismic performance of the soil. This study provides valuable insights for engineers and designers in selecting appropriate materials and thickness for improving the seismic performance of the soil.

The two Figs. 2 and 3 presented above depict the variation of Peak Ground Acceleration (PGA) with the thickness of the gravel rubber mixture (GRM) layer at surface

Table 3 DEEPSOIL nonlinear analysis PGA output for Site-3

PGA (in terms of g)		Layer 1	Layer 2	Layer 3	Layer 4
Thickness of GRM (m)	0	0.6185	0.0000	0.6206	0.3829
	0.1	0.6417	0.6136	0.5943	0.3907
	0.2	0.6741	0.6202	0.5816	0.3664
	0.3	0.6489	0.6384	0.5571	0.3481
	0.4	0.6445	0.6292	0.5113	0.3367
	0.5	0.6629	0.5881	0.4703	0.3252
	0.6	0.6335	0.5779	0.4470	0.3303
	0.7	0.6184	0.5742	0.4471	0.3198
	0.8	0.6094	0.5906	0.4457	0.3406
	0.9	0.6405	0.5430	0.4374	0.3243
	1	0.5936	0.5401	0.4196	0.3403
	1.1	0.5297	0.5541	0.3777	0.3519
	1.2	0.5132	0.5236	0.3886	0.3455
	1.3	0.5299	0.5143	0.4069	0.3274
	1.4	0.5404	0.5033	0.3915	0.3382
	1.5	0.4836	0.4882	0.4109	0.3476

Table 4 DEEPSOIL nonlinear analysis PGA output for Site-4

PGA (in terms of g)		Layer 1	Layer 2	Layer 3	Layer 4	Layer 5
Thickness of GRM (m)	0	0.688	0.000	0.659	0.519	0.443
	0.1	0.706	0.672	0.580	0.512	0.450
	0.2	0.675	0.654	0.563	0.517	0.443
	0.3	0.624	0.655	0.511	0.533	0.407
	0.4	0.589	0.693	0.469	0.522	0.374
	0.5	0.592	0.658	0.430	0.483	0.363
	0.6	0.611	0.607	0.412	0.436	0.325
	0.7	0.626	0.551	0.396	0.391	0.311
	0.8	0.624	0.510	0.421	0.411	0.318
	0.9	0.574	0.505	0.439	0.392	0.333
	1	0.554	0.501	0.415	0.383	0.313
	1.1	0.537	0.499	0.409	0.380	0.323
	1.2	0.529	0.472	0.395	0.381	0.333
	1.3	0.480	0.463	0.394	0.374	0.320
	1.4	0.472	0.459	0.387	0.384	0.333
	1.5	0.446	0.456	0.000	0.395	0.319

level and at a depth of 1.5 m for various sites. The results indicate a significant decrease in the PGA values with the increase in the thickness of the GRM layer. The initial increase in PGA value as the thickness of GRM increases could be due to the added mass of the layer, which initially increases the seismic forces acting on the layer. However, as the thickness of the GRM layer increases, the energy dissipation and damping characteristics of the GRM layer come into play, resulting in a significant decrease in the PGA values.

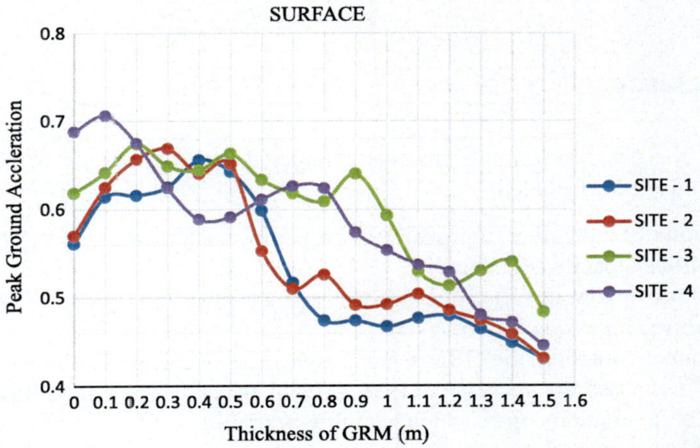

Fig. 2 Figure shows the variation of PGA at surface for varying depth of GRM

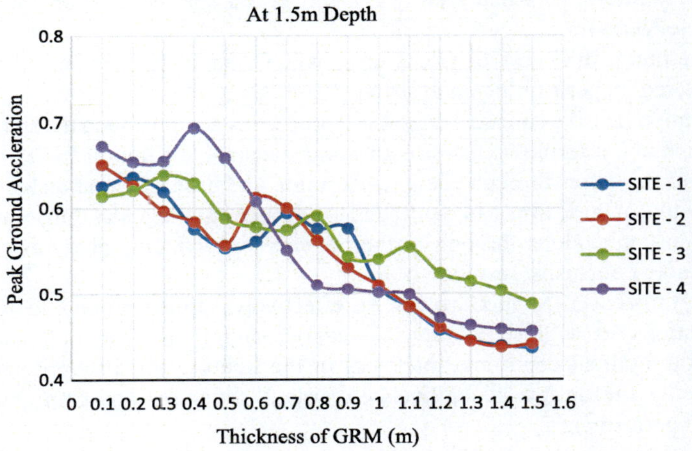

Fig. 3 Figure shows the variation of PGA at depth of 1.5 m for varying depth of GRM

The findings have significant implications in the field of earthquake engineering, where the use of GRM layers can help to reduce the seismic vulnerability of structures. The damping properties of the rubber particles in the GRM layer help absorb and dissipate the seismic energy, thereby reducing the seismic forces transmitted to the foundation and the structure. The use of GRM layers has been shown to be effective in reducing the seismic response of structures in previous studies, and the current study further reinforces the effectiveness of the GRM layer in reducing the seismic vulnerability of structures [13, 14].

5 Conclusion

Based on the tests performed, analysis conducted, and the results obtained, the following may be concluded:

1. Addition of GRM layer improved soil properties such as shear strength, energy absorption capacity, and ductility
2. The improvement in soil properties was site-specific, and different GRM layer thicknesses improved different soils.
3. The rubber content in the GRM used is 10% mass of the total volume.
4. There was a definite decrease in PGA (g) and effective stress (kPa) upon using GRM at foundation depth, which is highly desirable.
5. The PGA(g) value decreased by 16.8 and 9.05% at surface level and at depth of 1.5 m, respectively, for Site-1, 13.77 and 9.54% at surface level and at depth of 1.5 m, respectively, for Site-2, by 4.04% at surface level and 12.96% at depth of 1.5 m for Site-3, 19.4 and 23.8%, respectively, at surface level and at depth of 1.5 m for Site-4.
6. The optimum thickness of GRM layer was found to be 1 m for all the sites considered on performing analysis on DEEPSOIL.
7. The comprehensive enhancement of soil properties demonstrated the significance of this study, offering a cost-effective and sustainable solution for earthquake-resistant structures through energy absorption and base isolation techniques.
8. The results of these tests can be utilized to advance our understanding of vibration isolation applications through further studies and dynamic analyses that may yield more conclusive results.
9. The optimized GRM mixture can be effectively applied around footings and foundations to achieve vibration isolation, and future research can determine the optimal depth and thickness of the mixture for different soil types, enhancing the design of earthquake-resistant structures and improving their overall performance.

References

1. Naresh Dixit PS, Zubair M (2022) Linear and nonlinear site response study of soil-tire crumb mixture. In: Earthquake geotechnics: select proceedings of 7th ICRAGEE 2021. Springer, Singapore, pp 81–97
2. Tasalloti A, Chiaro G, Murali A, Banasiak L, Palermo A, Granello G (2021) Recycling of end-of-life tires (ELTs) for sustainable geotechnical applications: a New Zealand perspective. Appl Sci 11(17):7824
3. Banasiak L, Chiaro G, Palermo A, Granello G (2021) Environmental implications of the recycling of end-of-life tires in seismic isolation foundation systems. Advances in sustainable construction and resource management. Springer, Singapore, pp 43–52
4. Choudhary V, Choudhary A (2017) Use of tyre waste in concrete: a review. Int Res J Eng Technol 4(10):1924–1936
5. Alfayez SA (2018) Eco-efficient preplaced recycled aggregate concrete incorporating recycled tire waste rubber granules and steel wire fibre reinforcement (Doctoral dissertation, The University of Western Ontario (Canada))
6. Chiaro G, Tasalloti A, Palermo A, Granello G, Banasiak L (2021) Reuse of waste tires to develop eco-rubber seismic-isolation foundation systems: preliminary results. Advances in sustainable construction and resource management. Springer, Singapore, pp 159–169
7. Neaz Sheikh M, Mashiri MS, Vinod JS, Tsang HH (2013) Shear and compressibility behavior of sand–tire crumb mixtures. J Mater Civil Eng 25(10):1366–1374
8. Naresh Dixit PS, Zubair M (2021) Shear strength characteristics and energy absorption capacity of soil-tire crumb mixtures. J Earthq Sci Soil Dyn Eng 4(1):148
9. IS:2720 (Part 39) (1977) Methods of test for soils, Part 39: direct shear test for soils containing gravel, Indian Standards
10. Anbazhagan P, Manohar DR (2015) Energy absorption capacity and shear strength characteristics of waste tire crumbs and sand mixtures. Int J Geotech Earthq Eng 6(1):28–49
11. Anbazhagan P, Parihar A, Rashmi HN (2012) Review of correlations between SPT N and shear modulus: a new correlation applicable to any region. Soil Dyn Earthq Eng 36:52–69
12. Anbazhagan P, Uday A, Moustafa SS, Al-Arifi NS (2016) Correlation of densities with shear wave velocities and SPT N values. J Geophys Eng 13(3):320–341
13. Narendran KV, Rajagopal K (2020) Effectiveness of gravel rubber mixture (GRM) layerin reducing seismic vulnerability of structures. Soil Dyn Earthq Eng 128:106047
14. Sohail HM, Bhatti MA, Asif M (2020) Experimental study on seismic behavior of gravel-rubber mixture. Geomech Eng 18(4):277–286

Experience in Designing and Calculating Slab Foundations of a Two-Section Multi-story Building on Weak Water-Saturated Soils

O. V. Samorodov, S. V. Tabachnikov, S. V. Yesakova, and O. V. Krotov

1 Introduction

In the field of construction of multi-story and high-rise buildings with significant loads on the base, especially in the presence of weak water-saturated soils on top, pile-raft foundations [6, 10] are generally used to meet the requirements of the regulatory document [9] for maximum permissible deformations. The possibility of using slab foundations in this case must be confirmed by calculations with a scientific and technical justification of the accepted soil base model and its parameters.

The purpose of this paper is to justify the model of the soil base [4, 7] and its parameters based on the surveys of the settlements of the adjacent multi-story buildings on slab foundations [1, 2] during construction and compare the actual values of the deformations of the foundations with the calculated ones.

2 Research Results

The object of the research is the settlement of the soil base of the slab foundations of the adjacent sixteen-story residential buildings with underground and technical floors. The construction site is located in Ukraine, Kharkiv, Osnovyansky district, 2Б, Lysavetynska Street. The general view of the arrangement of the buildings is shown in Fig. 1 (experimental building 7, Sections 1 and 2). The sections of the building are similar, and either section is a mirror image of the adjacent one in plan. The buildings use a frameless structural system. The load-bearing elements of the buildings are as follows:

O. V. Samorodov · S. V. Tabachnikov · S. V. Yesakova (✉) · O. V. Krotov
O.M. Beketov National University of Urban Economy in Kharkiv, Kharkiv, Ukraine
e-mail: esakova.svetlana@gmail.com

S. Kolathayar et al. (eds.), *Best Practices in Geotechnical and Pavement Engineering*, Lecture Notes in Civil Engineering 449, https://doi.org/10.1007/978-981-99-8505-0_4

Fig. 1 Arrangement of the experimental building 7 (Sections 1 and 2) during erection

- Brick walls made of silicate brick with a thickness of 510 and 380 mm.
- Precast reinforced concrete floor slabs are made of hollow-core reinforced concrete slabs with a thickness of 220 mm.
- The basement walls are precast of concrete blocks with a thickness of 500 mm.

The layout of the boreholes is shown in Fig. 2.

The consistent level of groundwater for the study period (June 2020) was recorded at the depths of 1.8–3.3 m (at the absolute elevations of 101.51–102.92 m). The site is waterlogged. During the flooding period and intense atmospheric precipitation, the level of underground water can reach the marks close to the ground surface.

The description and physical and mechanical properties of the soils are given in Tables 1 and 2 (the specific gravity of soil, specific gravity of water-saturated soil, modulus of deformation, angle of internal friction, and specific soil adhesion are given as calculated at a confidence probability of 0.85. In addition, the stamp

Fig. 2 Layout of the boreholes

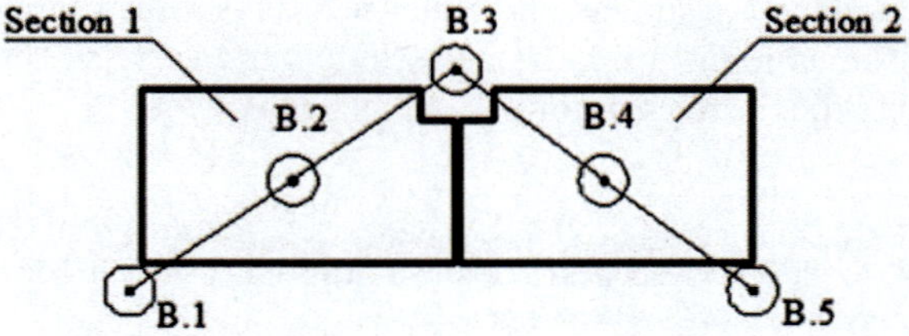

modulus of deformation E is indicated at the top of the EGE-2 sands compacted by tamping down crushed stone by a roller and forming a 200 mm crushed stone bed).

Reinforced concrete foundation slabs with a thickness of 1000 mm are used as the foundations of the residential buildings. The bottom of the foundation slabs is taken to be at the absolute elevation of 102.8 m. The engineering geological section with conventionally arranged foundation slabs is shown in Fig. 3.

The soil base of the foundations is Engineering Geological Element (EGE) number 2, EGE-2, fine-grained sands, varying from moist to saturated, of medium density, with the modulus of deformation of $E = 28.0$ MPa in a water-saturated condition (according to the report on geotechnical investigations). However, based on the additional stamp tests at the top of the EGE-2 sands, which were compacted by tamping down crushed stone by a road roller to form a 200 mm thick crushed stone bed, the total modulus of deformation is equal to $E = 9.0$ MPa, which was introduced in the calculations. Figure 4 shows a photo of the base of the foundations being prepared.

The maximum average pressure under the foundation slab only from the dead weight of the erected structures without the introduction of multiplying factors does not exceed the design resistance of the soil base of $p = 230.02$ kPa $< R = 388$ kPa; therefore, the theoretical calculations of the settlement of the building were performed using a model of a linear elastic layer of finite width with a constraint against the compressible mass and horizontal displacements in plan [3, 8].

According to the calculations [5], the compressible mass of the base is taken to be as follows: Hc = 9.5 m at the average pressure of $p = 65.21$ kPa under the bottom of the foundation slab, which corresponds to the dead weight of the structures with three floors erected; Hc = 10.5 m at $p = 98.55$ kPa (with six floors erected); Hc = 13.5 m at $p = 153.21$ kPa (with eleven floors erected); and Hc = 18.0 m at $p =$ 230.02 kPa (with seventeen floors erected).

Table 1 Designation and description of the soils

EGE number	Designation and description of the soils
EGE-1	Bulk soils: loams, sands with the inclusion of construction debris, dumped in a dry way, of heterogeneous composition, uneven density and compressibility, with the remains of buried foundations, densely packed and not packed, with the thickness of the layer of up to 2.6 m
EGE-2	Yellow-gray and gray, fine and fine-grained sands, varying from moist to saturated, of medium density (friable on top), homogeneous, with interlayers of loams and sandy loams, with the thickness of the layer of up to 5.6 m
EGE-3	Greenish-gray loams, rigidly plastic, soft-plastic in a water-saturated condition, with lenses of dusty sands in the bottom, with the thickness of the layer of up to 2.9 m
EGE-5a	Gray, greenish-gray and bluish-gray fine-grained water-saturated sands, with the thickness of the layer of up to 5.0 m
EGE-8	Bluish-gray clays, varying from semi-hard to rigid-plastic consistency, with interlayers of sandstones, with the thickness of the layer of up to 11.8 m

Table 2 Physical and mechanical properties of the soils

Soil parameter	Legend	Unit of measurement	Layer				
			1: Bulk soil	2: Fine-grained sands varying from moist to saturated	3: Loams, rigidly plastic, soft-plastic in a water-saturated condition	5a: Fine-grained water-saturated dense non-homogeneous sands	8: Clays varying from semi-hard to rigid-plastic consistency
Specific gravity of soil	γ	kN/m^3	–	18.88	19.40	20.23	16.74
Specific gravity of water-saturated soil	γ_{sat}	kN/m^3	–	19.45	19.60	19.22	16.88
Specific gravity of soil fragments	γ_s	kN/m^3	–	26.09	26.68	20.15	26.74
Specific weight of soil suspended in water	γ_{sb}	kN/m^3	–	9.97	9.99	10.65	7.47
Natural moisture content	W	Unit	–	0.19	0.22	0.19	0.38
Porosity coefficient	e	Unit	–	0.63	0.68	0.54	1.06
Specific soil adhesion	c	kPa	–	2	23	4	37
Angle of internal friction	φ	°	–	33	18	36	17
Modulus of deformation in the water-saturated condition	E	MPa	–	28 (9[a])	15	38	13

(continued)

Table 2 (continued)

Soil parameter	Legend	Unit of measurement	Layer				
			1: Bulk soil	2: Fine-grained sands varying from moist to saturated	3: Loams, rigidly plastic, soft-plastic in a water-saturated condition	5a: Fine-grained water-saturated dense non-homogeneous sands	8: Clays varying from semi-hard to rigid-plastic consistency
Fluidity index of water-saturated soil	I_L	Unit	–	–	0.60	–	0.24
Plasticity number	I_p	Unit	–	–	0.11	–	0.28

[a] The modulus of deformation E is determined based on the stamp tests of the top of the EGE-2 sands, which were compacted by tamping down crushed stone by a roller and forming a 200 mm thick crushed stone bed

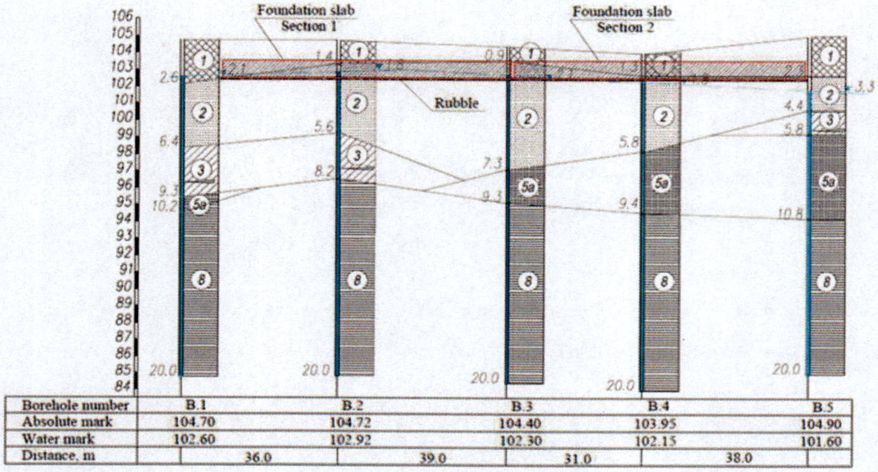

Fig. 3 Engineering geological Sections 1–5

Fig. 4 Base of the foundations being prepared

3 Simulating the Base–Foundation–Building System and Calculation Results

The base–foundation–building system was simulated and calculated in the SOFiSTiK software package. The soil base model was taken as a linear elastic layer of finite width. The depth (thickness) of the layer was limited by the size of the compressible mass for each stage of loading the soil base from the erected building volume. The boundary conditions in plan for the soil mass were taken so that the effect of the rigid bracing constraints of the mass was minimum and equal to 120 × 80 m.

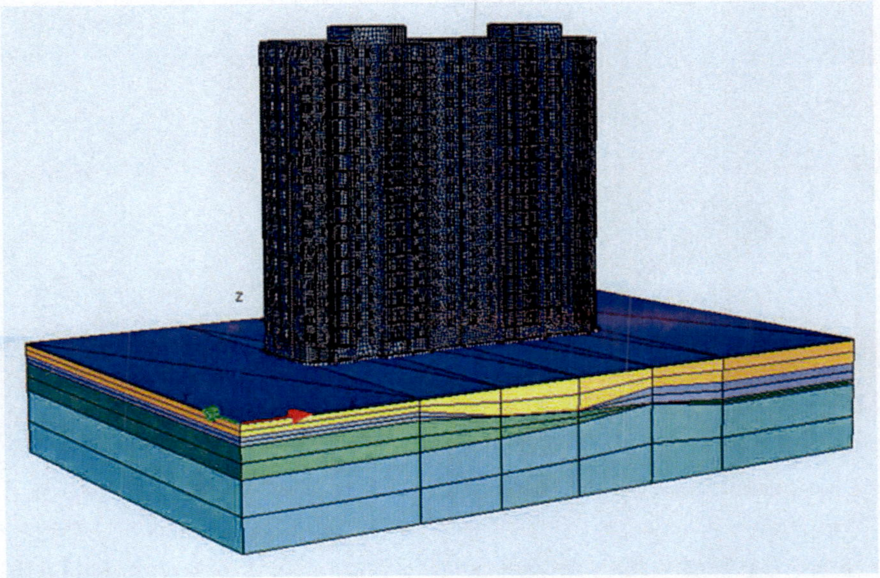

Fig. 5 Residential building model in the base–foundation–building system in the SOFiSTiK software

The foundation slab, floor slabs, and walls were simulated using plate shell elements.

Based on the selected parameters of the model and the initial data, the interaction between the buildings and the above soil base model was simulated. The numerical calculation was performed using the finite element method, which was based on the solutions of the theory of elasticity.

The base–foundation–building calculation model is schematically shown in Fig. 5.

The deformed schematic of the foundations of a two-section building with the soil base being simultaneously loaded by two sections (instant elastic solution) with the total compressible mass of Hc = 18.0 m at $p = 230.02$ kPa (with seventeen floors erected) is shown in Fig. 6.

4 Survey Results

The settlements of the building were surveyed during construction using high-precision Class 2 leveling methods. Survey marks M.1, M.2, M.3, M.4, M.5, M.6, M.7, and M.8, which were laid at the level of the basement floor of both sections of the building, were determined. During construction, six stages of surveys were provided.

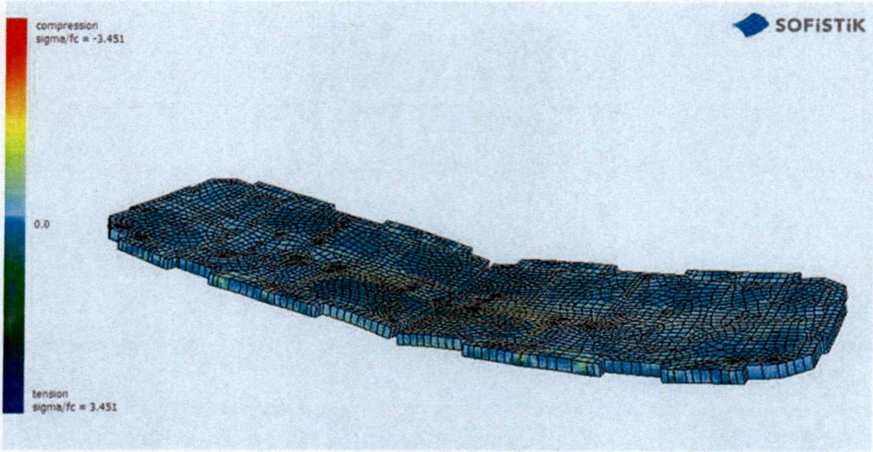

Fig. 6 Schematic showing the deformation of the foundation slabs of the residential buildings

- Stage 1: February 2021. Arranging the basement floor of Section 2 and the 3rd floor of Section 1.
- Stage 2: March 2021. Building the 3rd floor of Section 2 and the 6th floor of Section 1.
- Stage 3: June 2021. Building the 8th floor of Section 2 and the 11th floor of Section 1.
- Stage 4: September 2021. Building the 14th floor of Section 2 and the 17th floor of Section 1.
- Stage 5: November 2021. Building the 17th floor of Section 2 and the 17th floor of Section 1.
- Stage 6: February 2022. Surveying at constant load with 17 floors erected of both Sections 1 and 2.

The H-05 level (see Fig. 5) and the PH-05 invar leveling rod were used for leveling. Leveling was done in both ways forward and backward by matching. At each station, the excesses of ab, af, and their difference $h = \text{ab} - \text{af}$, which did not exceed 0.7 mm, were calculated based on the readings of the main and additional scales. For the forward and backward distances in each section (between the adjacent benchmarks in the distance), the sums of the excesses of the forward $[h]$f and backward $[h]$b distances were calculated. The differences fx did not exceed $\pm 2.1\sqrt{L}$ mm (where L is the length of the distance in km) with the number of stations less than 15. For the accuracy of the calculations, the Microsoft Excel software was used, and the comparison of the results was performed in the ARMGEO package. The responsible executor of geodetic works was Ph.D., Assoc. Prof. Taras Nalyvaiko.

Figures 7 and 8 show the dependency graphs of the values of the actual and calculated settlements and tilts of two sections of the building on the controlled actual loads during construction (only the dead weight of the erected structures without multiplying factors was considered).

Figure 9 shows a view of the expansion joint (at the junction of two sections) at the final stage of construction, which indirectly indicates no local damage to the brick on the upper floors of the buildings near the expansion joint due to the "mutual loading" of the buildings.

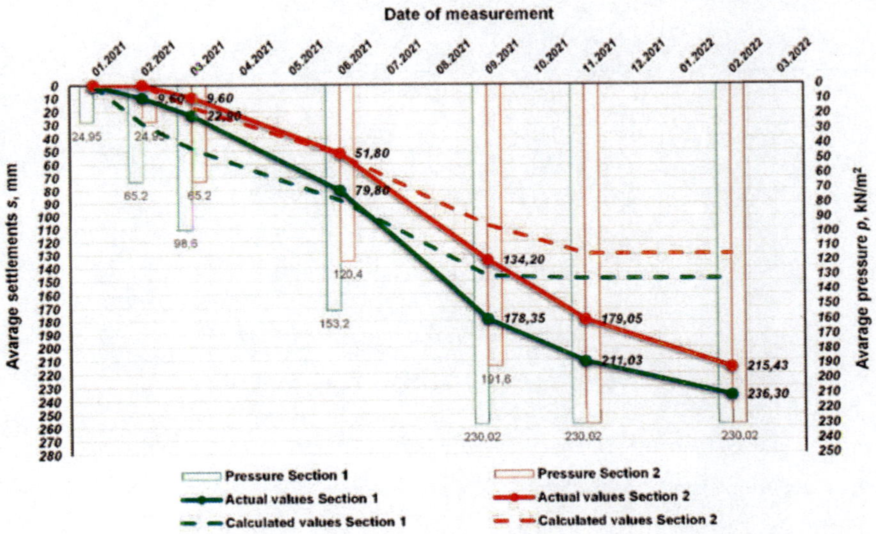

Fig. 7 Average settlements of Sections 1 and 2 due to the average pressure under the bottom of the slabs

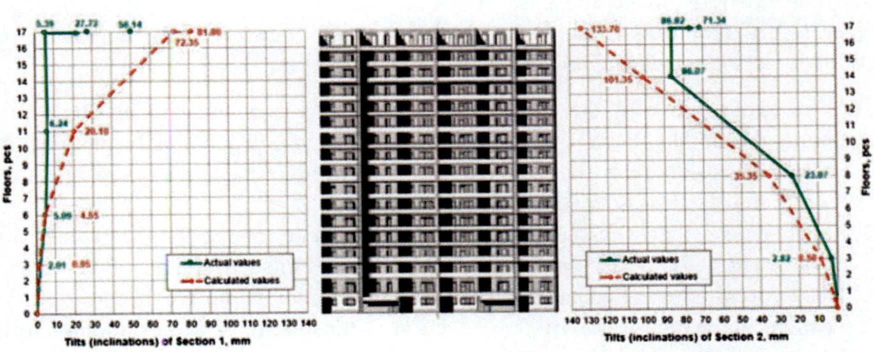

Fig. 8 Tilts (inclinations) of Sections 1 and 2 in height along the building from the erected floors

Fig. 9 Expansion joint at the final stage of construction

5 Conclusions

The deformations of the soil base of two adjacent multi-story buildings on large-sized slab foundations were observed. The analysis of research results allows the following conclusions and recommendations to be made:

1. The applicability of the model of the soil base as a linear elastic layer of finite width has been confirmed for forecasting adequate deformations of sectional buildings on large-sized raft foundations, which are based on weak water-saturated soils on top.
2. In this case, the excesses of the actual settlements of the buildings over the calculated ones are due specifically to the process of seepage consolidation and the process of mechanical suffusion of sand particles into crushed stone bed, which can be considered by the reduction of the moduli of deformation of soils in this soil base model.

3. The development of the tilts of the two sections of the building toward each other confirms the classical theoretical solutions of the mechanics of a solid environment, i.e., the superimposition of stresses (deformations) under mutual effect. However, Fig. 8 clearly shows that in the process of construction of two sections with a difference of several floors (four floors, in this case), the main effect is manifested specifically at the final stage of the construction of two sections, and it is Section 2, which was built last, that has an effect. Therefore, the analysis of the actual tilts near the foundations of sectional buildings indicates the impossibility of adequately considering the mutual effect in case of buildings with different numbers of floors using a model of a continuous medium with the common compressible mass, which requires improving the simulation of the soil base in the base–foundation–building system.

References

1. Bowles JE (2019) Foundation analysis and design, 6th edn. McGraw Hill, New York
2. Jain OP, Chakraborty US (2018) Design of reinforced concrete foundations, 3rd edn. Nem Chand & Bros, New Delhi
3. Luchkovsky IY (2000) Vzaimodeystviye konstruktsiy s osnovaniyem [The interaction of structures with the base]. O.M. Beketov National University of Urban Economy (ITE Journal), Kharkiv (in Russian)
4. Nouri A, Sabouri F, Shahin H (2021) Comparative study of different soil models on the behavior of pile-supported raft foundations. Int J Geomech 21(1)
5. Samorodov AV (2017) Proektirovanie effektivnyh kombinirovannyh svajnyh i plitnyh fundamentov mnogoetazhnyh zdanij: Monografiya [Design of high-performance combined pile-raft foundations for multi-story buildings: monograph]. Typography Madrid, Kharkiv (in Russian)
6. Taly NKS (2019) Behaviour of combined raft and pile foundation on weak soil. In: Proceedings of the Indian geotechnical conference
7. Tan Y, Ng CWW, Lee FH (2021) Comparative study of different soil models for predicting the settlement of raft foundation. Geotech Geol Eng 39(2)
8. Ter-Martirosyan ZG (2009) Mekhanika gruntov [Soil mechanics]. ASV, Moscow (in Russian)
9. Ukrainian National Building Code DBN B.2.1-10:2018 (2018) Osnovy i fundamenty budivel ta sporud. Osnovni polozhennya [Bases and foundations of buildings and structures. Main provisions] (in Ukrainian)
10. Zhang C, Guo X, Xie W, Cui G (2019) Numerical and experimental study on the bearing capacity of composite foundation with a pile-raft system in weak ground. Soil Dyn Earthq Eng 126

Experimental and Numerical Analysis for Enhancing Bearing Capacity of Shallow Foundation Through Geotextile Implementation

Sayan Saha and Sujit Kumar Pal

1 Introduction

The placement of foundations on loose soil can result in structural damage, including settlement, cracks, and collapse. However, the use of deep foundations is often discouraged due to the complexities involved in their installation process and the high construction costs associated with them. In such scenarios, geotextile-reinforced foundations present a cost-effective solution for addressing low-fringe soil conditions. This technique involves incorporating reinforcing materials with high compressive and tensile strengths, such as metal bars, strips, nets, or mats, into the soil. Extensive global research has been undertaken to ascertain the significance of geotextiles in enhancing the bearing capacity of foundations. In our research study, we analyze the bearing capacity and settlement of shallow footings, with the help of experimental and numerical approach with PLAXIS 2D software as this software has been selected due to its capability to conduct finite element analysis (FEA) and tackle complex geotechnical issues. The software delivers reliable and accurate results, allowing for the prediction of soil–structure interaction behavior. Through bearing capacity analysis, the maximum load that a foundation can withstand without failure or excessive settlement is determined, while settlement analysis evaluates soil deformation and foundation displacement. By combining experimental and numerical evaluation with the aid of PLAXIS software, a comprehensive approach is achieved for studying and improving the performance of shallow foundation.

S. Saha (✉) · S. K. Pal
National Institute of Technology Agartala, Tripura 799046, India
e-mail: sayan4131@gmail.com

S. Kolathayar et al. (eds.), *Best Practices in Geotechnical and Pavement Engineering*, Lecture Notes in Civil Engineering 449, https://doi.org/10.1007/978-981-99-8505-0_5

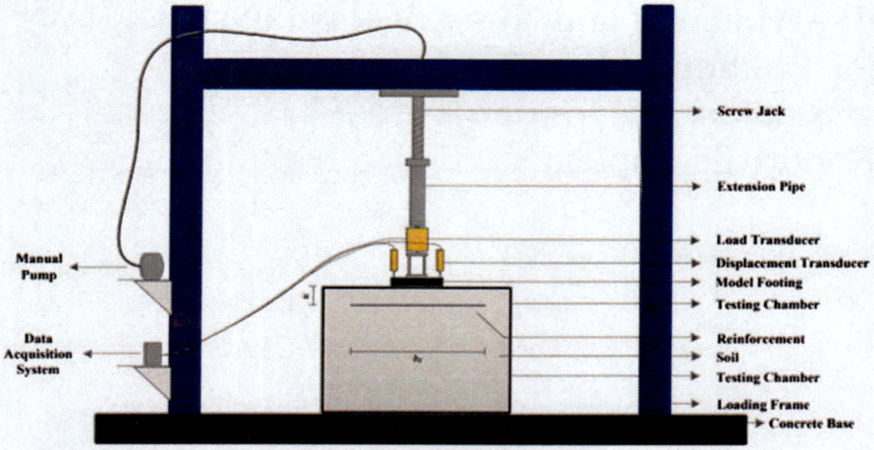

Fig. 1 Schematic view of experimental model arrangement

2 Experimental Study

2.1 Model Test Setup

A 5.0 t hydraulic screw jack was employed to apply a vertical compressive load at a rate of 1.0 mm/min. To measure the vertical load, a standard compression-type load transducer with a capacity of 5.0 kN and an accuracy of 0.01 kN was mounted on a steel plate. Additionally, two displacement transducers with a sensitivity of 0.01 mm were used to measure vertical displacements. For visual reference, please refer to Fig. 1, which depicts the experimental setup employed in this study.

2.2 Test Box and Model Footing

The testing box was constructed using a robust design, consisting of a 24 mm thick transparent Perspex sheet and 10 mm thick steel plates. It had a square cross-section measuring 370 mm × 370 mm, with a height of 600 mm. According to Verghese Chummar [1], in non-cohesive soil, the failure zone extends to 5 times the width of the sidewalls and 3 times the width underneath the model footing. Therefore, model footings measuring 70 mm × 70 mm × 16 mm were employed for this tank size to ensure unobstructed formation of failure zones.

2.3 *Test Medium*

Sand. This study utilized locally available river sand as the sole soil type for testing purposes. The sand was categorized as 'SP' (signifying poorly graded sand) using the Indian Standard Soil Classification System, based on the particle size distribution presented in Fig. 2 and data extracted from Table 1.

Soil Reinforcement. The details of the reinforcement are given in Table 2.

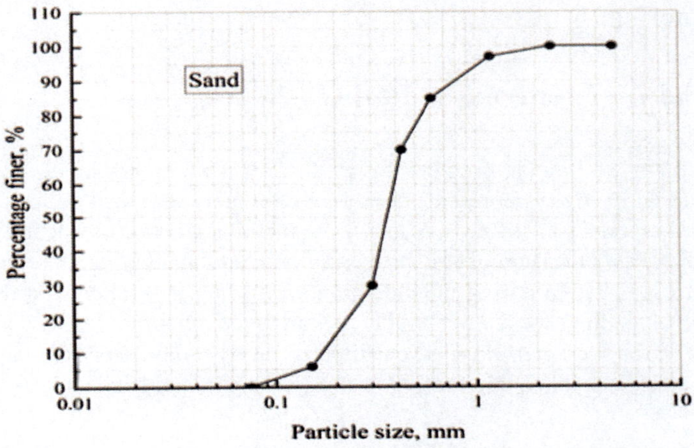

Fig. 2 Particle size distribution of soil

Table 1 Properties of soil

Description	Value
Specific gravity, *G*	2.67
D_{10} (mm)	0.19
D_{30} (mm)	0.26
D_{60} (mm)	0.40
D_{50} (mm)	0.36
Coefficient of curvature (C_c)	0.87
Coefficient of uniformity (C_u)	2.08
Type of soil	SP
Maximum dry unit weight (kN/m^3)	16.4
Minimum dry unit weight (kN/m^3)	14.5
Φ (°)	38.0
e_{max}	0.806
e_{min}	0.597

Table 2 Reinforcement specifications

Description	Value
Chemical composition	Polypropylene
Structure	Nonwoven
Mass per unit area (GSM)	300
Thickness (mm)	2.0
Tensile strength (kN)	2.50
Elongation at break (%)	8.0

3 Numerical Study

3.1 Models of Materials in PLAXIS 2D

The Hardening Soil model used for this PLAXIS 2D which has input stiffness's of $E_{\text{oed}}{}^{\text{ref}}$, and $E_{\text{ur}}{}^{\text{ref}}$. $E_{50}{}^{\text{ref}}$ represents the initial or secant modulus of elasticity and is often used to model soil stiffness. $E_{\text{oed}}{}^{\text{ref}}$ is the modulus of elasticity at smaller strain levels and is also a measure of soil stiffness. $E_{\text{ur}}{}^{\text{ref}}$, on the other hand, represents the ratio of stress to strain during unloading and reloading and models' soil behavior during cyclic loading, while $E_{\text{ur}}{}^{\text{ref}}$ and $E_{\text{oed}}{}^{\text{ref}}$ are typically around $3E_{50}{}^{\text{ref}}$ and $E_{50}{}^{\text{ref}}$, respectively, for average soil types, customized ratios can be used for very soft and very stiff soils, and for square footing, Linear Elastic model is considered.

3.2 PLAXIS 2D Modeling

To analyze a soil element with drainage condition and without considering pore water pressure, 15-node triangular elements were used in the FE calculations. To replicate prototype-scale behavior, the dimensions of the square footing were enlarged by a factor of 15, and the necessary requirements for both reinforced and unreinforced sand were incorporated. Boundary conditions were replicated in the FE model, and load on the footing's plate element was imposed using the point load feature. Details of the soil stratigraphy are presented in PLAXIS model (Fig. 3). The data in Table 3 were obtained by incorporating the EA and EI values of the footing, drawing inspiration from the prior studies conducted by Bindiya et al. [2] and Nasr et al. [3], and utilizing the EA of the geotextile.

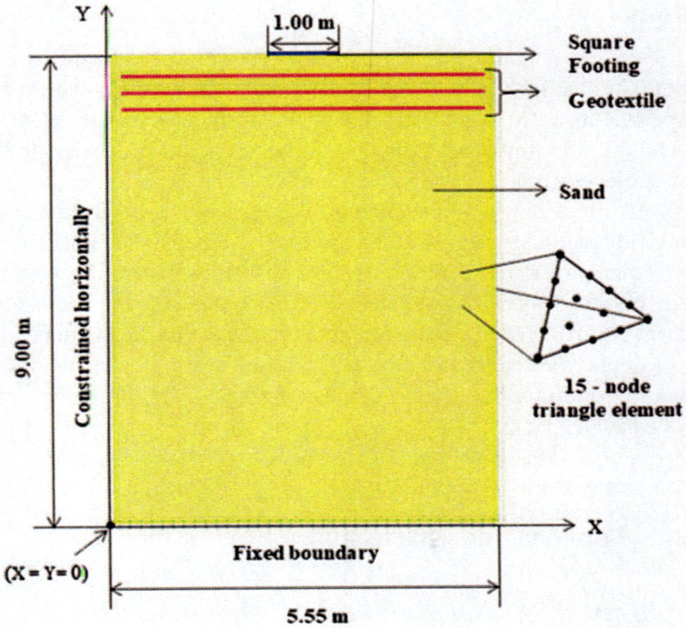

Fig. 3 Model setup in finite element mesh

Table 3 Parameters of materials used for numerical analysis

Parameters	Sand	Footing	Geotextile
Material model	Hardening-soil model	Linear elastic	Elastic
Type of material behavior	Drained	–	–
Primary loading stiffness, $E_{50}{}^{ref}$ (kPa)	30,000	–	–
$E_{oed}{}^{ref}$ (kPa)	30,000	–	–
$E_{ur}{}^{ref}$ (kPa)	90,000	–	–
Poisson's ratio, ν	0.30	0.20	–
Cohesion, C (kPa)	0	–	–
Friction angle, Ø (°)	38.0	–	–
Dilatancy angle, ψ (°)	8.0	–	–
Axial stiffness, EA (kN/m)	–	50.4×10^6	32.0
Flexural stiffness, EI (kPa/m)	–	241,920	–
γ_{unsat} (kN/m^3)	16.4	–	–
γ_{sat} (kN/m^3)	17.6	–	–

4 Methodology

To determine the load-settlement characteristics of a prototype square footing, four sets of experiments were conducted which then replicate in numerical condition on reinforced and unreinforced granular soil with a relative density of $D_r = 50\%$ which was achieved through compaction method. The experiments and numerical conditions were divided into four segments. These segments involved in a static load test on a model footing. Segment A and Segment B evaluated the impact of spacing of first reinforcement layer and spacing between consecutive layers of reinforcement on footing size, respectively. These segments were conducted using a consideration of reinforcement width ($b_g/B = 5$) to identify the optimal value. Segment C investigated the effect of geotextile width variation (b_g) on soil bearing capacity, and Segment D examined the optimum number of reinforcing layers (N) to increase maximum amount of bearing capacity.

5 Results Analysis and Discussion

5.1 *Test Segment A: Determination of Optimum Reinforcing Depth* 'u'

Various studies have proposed different values for u/B, with Chung and Cascante [4] proposing 0.3–0.5, and Yetimoglu et al. [5] proposing 0.25. To ensure the reliability of these findings, u/B was modified in a series of trials. In Fig. 4, we analyze depth values at multiples of 0.2, 0.4, and 0.6 times the width of the footing to ascertain the most suitable depth. The results indicate that for this type of geotextile, $q_{\text{Reinforced}}$ is highest for $u/B = 0.4$ in both modes of analysis for the spacing of first layer of reinforcement. In Fig. 5, there is also a clear indication of a similar increase.

5.2 *Test Segment B: Determination of Optimum Spacing Between Consecutive Reinforcing Layers* 'h'

For sequential layers, variations of 0.2, 0.3, and 0.4 in terms of width of footing were adopted. Omar et al. [6] investigated that the value of square footing using planner geosynthetic-reinforced sand was 0.33*B*. Yetimoglu et al. [5] determined that the ideal value should be between 0.2*B* and 0.4*B*. Adams and Collin [7] reported values ranging from 0.25*B* to 1.25*B* for varied square foundation layouts. In Fig. 6, it is evident that the bearing pressure undergoes fluctuations across different settlement ratios and embedded depths. Notably, the most favorable outcome is observed when h/B equals 0.3, a result consistent with both experimental data and numerical analysis.

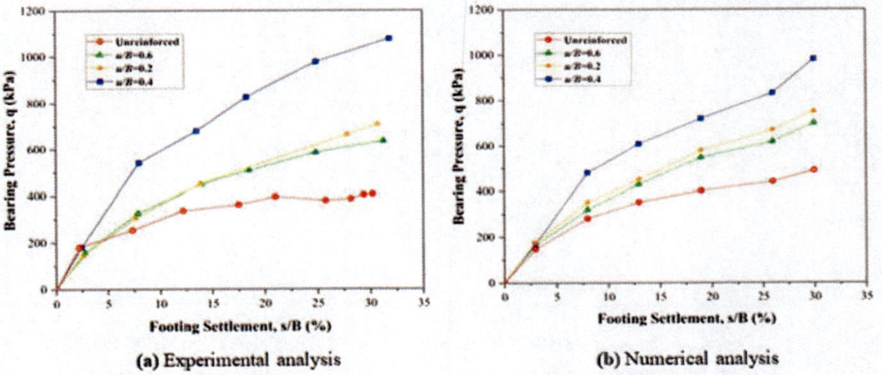

Fig. 4 Bearing pressure versus settlement ratio used for $b_g/B = 5$ and $N = 1$ (Segment A)

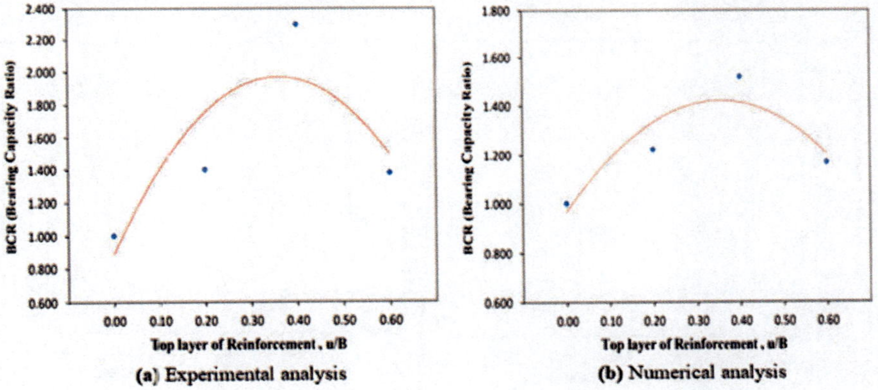

Fig. 5 Bearing capacity ratio versus top layer of reinforcement used for $b_g/B = 5$ and $N = 1$ at s/B (%) = 20 (Segment A)

This alignment can also be observed in the corresponding Fig. 7, where the bearing capacity ratio versus the multilayer reinforcement graph is depicted.

5.3 *Test Segment C: Determination of Width of Reinforcements 'b_g'*

Figure 8 indicates that increasing the width of reinforcing layers (b_g) also increases the bearing capacity of the soil. However, the highest bearing capacity was observed when the soil was in the 5*B* condition, as shown in both analyses according to the

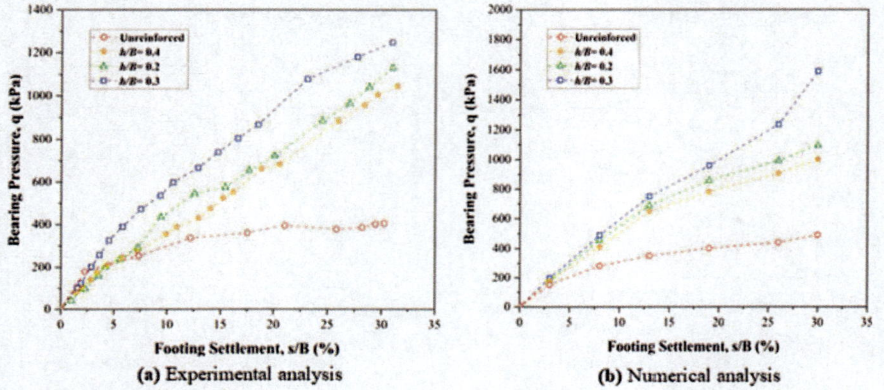

Fig. 6 Bearing pressure versus settlement ratio used for $u/B = 0.4$, $b_g/B = 5$ and $N = 2$ (Segment B)

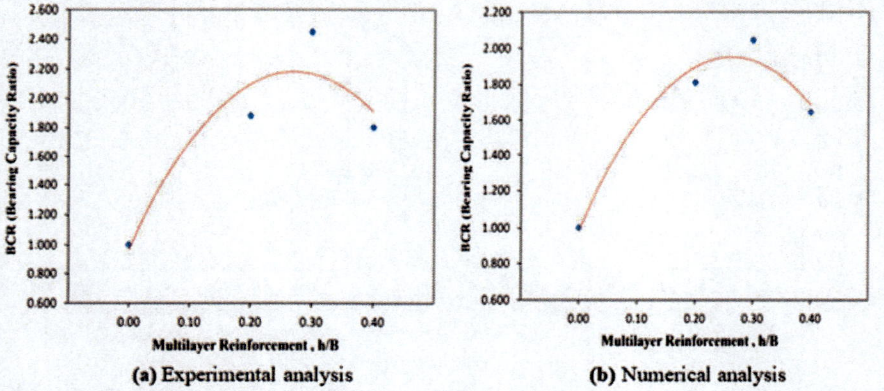

Fig. 7 Bearing capacity ratio versus multilayer reinforcement used for $b_g/B = 5$, $u/B = 0.4$, $N = 2$ at s/B (%) $= 20$ (Segment B)

test result. This is attributed to the wider span of reinforcement, which expands the confined area of the soil, ultimately enhancing the bearing capacity of the soil bed.

5.4 *Test Segment D: Determination of No. of Reinforcements Layers 'N'*

Figure 9 depicts the outcomes for different numbers of reinforcing layers, denoted by N. The spacing between first layer, consecutive layers and width of the reinforcement are optimized based on the above test findings. The figures demonstrate that the

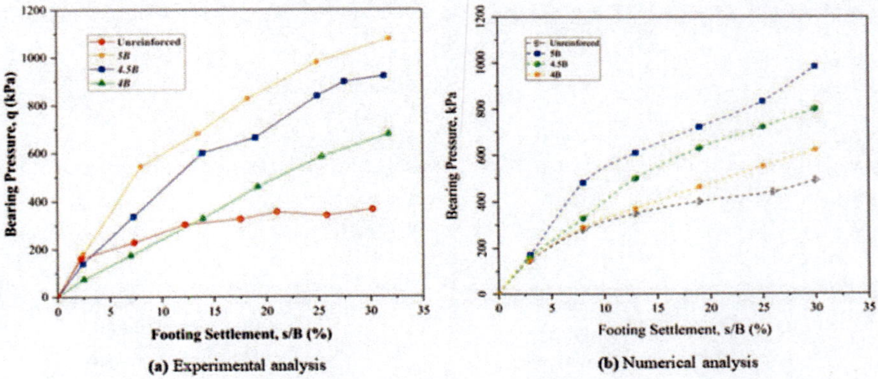

Fig. 8 Bearing pressure versus settlement ratio used for $u/B = 0.4$, $h/B = 0.3$, and $N = 1$ (Segment C)

value of N increases in relation to the bearing capacity ratio (BCR). However, as N surpasses 3, its impact diminishes significantly; a trend that is clearly evident in Fig. 10. This is because the depth of impact under the loaded footing is limited, beyond which additional reinforcement does not significantly enhance the bearing capacity.

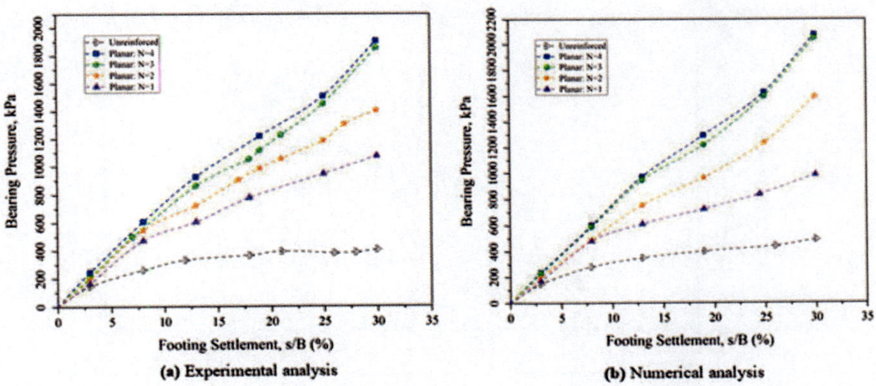

Fig. 9 Bearing pressure versus settlement ratio used for u/B = 0.4, h/B = 0.3, and b_g/B = 5 (Segment D)

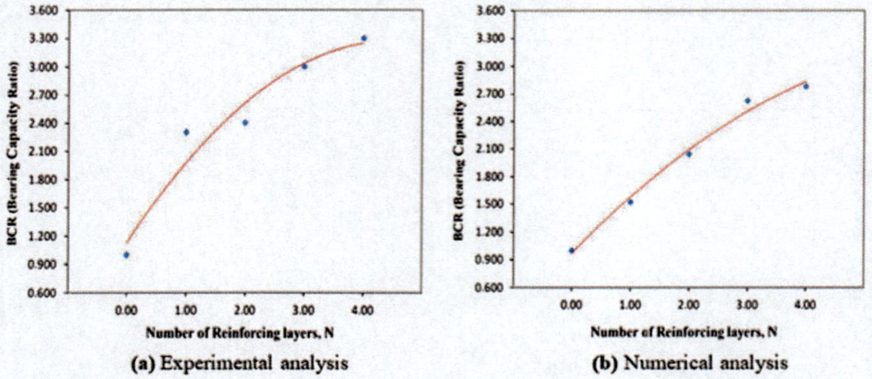

Fig. 10 Bearing capacity ratio versus number of reinforcing layers used for $u/B = 0.4$, $h/B = 0.3$, $b_g/B = 5$ at s/B (%) = 20 (Segment D)

6 Validation

Figure 11 depicts a strong concurrence between the experimental and numerical outcomes, signifying the precision and dependability of the numerical analysis in forecasting the system's response to lateral loads.

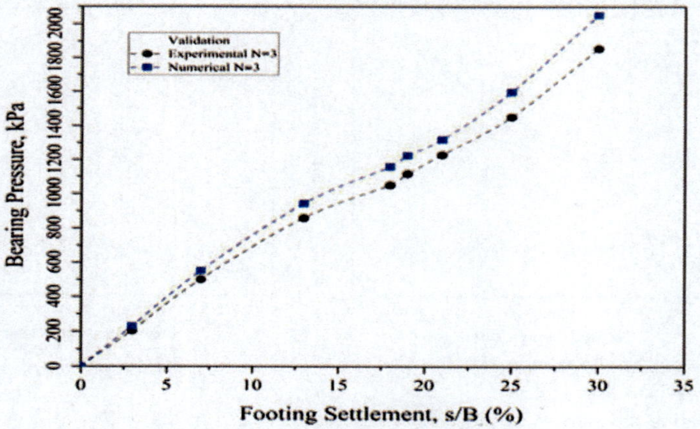

Fig. 11 Pressure-settlement curve for experimental test data and developed numerical model

7 Conclusion

Based on this present study of both experimental and numerical investigations, the following conclusions are drawn:

- Effective depth of anchorage-reinforced zone (d_r) with inclusion of geotextile in granular soil bed has found to be 1*B*. After that, further reinforcement is unlikely to make much of a difference.
- The performance of a reinforced system is significantly influenced by the spacing of the uppermost layer, and the present study determined that the optimal value of spacing for geotextile-reinforced sand bed is $u/B = 0.4$ for the uppermost layer.
- Within the effective reinforced zone consecutive spacing between the layers of reinforcement ascertained to be 0.3 times the footing width.
- This study concludes that optimum reinforcement width is 5*B* for the present arrangement.
- The study recommends a maximum of three layers of reinforcement for a square footing arrangement on a sandy soil bed to achieve optimal BCR values.
- The deviation between the numerical outcomes and the experimental findings was found to be in the range of 9.0–11.0% in pressure-settlement curve.
- The findings indicate that the introduction of geotextile within a sandy soil bed yields a substantial enhancement in soil bearing capacity, with values ranging from 1.7 to 2.3 times higher when compared to the unreinforced condition.

References

1. Verghese Chummar A (1972) Bearing capacity theory from experimental results. J Soil Mech Found Div 98:1311–1324. https://doi.org/10.1061/JSFEAQ.0001816
2. Bindiya K, Gangadhara S, Muddaraju HC, Tejaswini BR (2014) Numerical study of behavior of square footing on geogrid-reinforced flyash beds under static loading. Int J Res Eng Technol 03(18):82–85
3. Nasr AMA, Azzam WR (2016) Behaviour of eccentrically loaded strip footings resting on sand. Int J Phys Model Geotech 1–18. https://doi.org/10.1680/jphmg.16.00008
4. Chung W, Cascante G (2007) Experimental and numerical study of soil-reinforcement effects on the low-strain stiffness and bearing capacity of shallow foundations. Geotech Geol Eng 25:265–281. https://doi.org/10.1007/s10706-006-9109-0
5. Yetimoglu T, Wu JT, Saglamer A (1993) Bearing capacity of rectangular footings on geogrid reinforced sand. J Geotech Eng 120:2083–2099
6. Omar MT, Das BM, Puri VK, Yen SC (1993) Ultimate bearing capacity of shallow foundations on sand with geogrid reinforcement
7. Adams MT, Collin JG (1997) Large model spread footing load tests on geosynthetic reinforced soil foundations. J Geotech Geoenviron Eng 123:66–72. https://doi.org/10.1061/(ASCE)1090-0241(1997)123:1(66)

Retrofit of Foundations Supporting Structures That Have Undergone Settlement Located in Seismically Active Zones

Y. K. Guruprasad

1 Introduction

Reinforced concrete structures [1] are designed to safely resist external loads due to live loads [2, 3], loads induced due to fire [4–6], blast loads [7, 8], wind loads [9–11] and earthquake loads [12–14]. The deformations that take place in the superstructure would not cause complete instability of the entire structure when translations and rotations take place in structural elements (beams, columns and slabs) due to externally applied loads. The superstructure housing structural elements that have undergone damage, due to deformations can be repaired when damage is within limits, adopting various repair strategies. When foundation settlement takes place, the structure's stability becomes a matter of grave concern, as these foundation settlements can lead towards collapse of the structure. The settlement of foundations takes place predominantly due to error in design of the foundation by not considering the type and dimensions of the foundation required for the loads acting, especially gravity loads, earthquake loads and wind loads. In certain cases, the settlement is taken into consideration into the design phase when the structure is founded on soils that have lower bearing capacity and tend to deform when subjected to loading. The repair of foundations that have undergone settlement due to different reasons is comparatively difficult when compared to the repair of structural elements present in the superstructure. In spite of the difficulty in repair of foundations, engineers have applied repair and retrofit strategies to restore foundations that have undergone settlement in the recent past. In the present work, a practical case of retrofit of a slab-type reinforced concrete raft foundation supporting a mercantile building and founded on soil (medium sand) having a soil subgrade reaction—10,000 kN/m^2 that has undergone settlement and deformation at few locations due to erroneous design

Y. K. Guruprasad (✉)
Associate Professor, Department of Civil Engineering, BMS College of Engineering, Bangalore 560019, India
e-mail: guruprasad.civil.iisc@gmail.com

S. Kolathayar et al. (eds.), *Best Practices in Geotechnical and Pavement Engineering*, Lecture Notes in Civil Engineering 449, https://doi.org/10.1007/978-981-99-8505-0_7

and previously experienced an earthquake has been taken up as a retrofit case study. The structure was constructed 12 years ago. The settlement that has taken place at various locations in the raft is assessed, and retrofitting of the existing raft foundation is carried out by introducing beam slab type of retrofit, where new RC beams and RC jacketing of the raft slab were carried out with additional new reinforcement by strengthening the existing raft slab. The structural effectiveness of the retrofit was assessed based on the identification of optimized thickness of RC jacket slab and optimum depth of newly introduced RC beams, by checking for any possible further settlements considering gravity and earthquake loads in post-retrofit conditions.

2 Analysis of Raft Foundation

Analysis of the existing raft foundation along with the superstructure is modelled and the analysis of the raft slab retrofitted adopting RC jacket (new RC slab) and new RC raft beams, both provided above the existing raft slab is carried out in ETABS.

2.1 Details of Structural Model of the Structure and Raft Slab Considered for Analysis Before and After Retrofitting

2.1.1 Details of the Structure and Existing Raft Slab Without Retrofit

The structure is a mercantile building constructed 12 years ago, measuring 30 m × 30 m in plan having ten storeys, with each storey height equal to 3 m, with five bays along *X*- and *Y*-directions in plan with each bay having a span of 6 m, as shown in Fig. 1a. The live load considered is 4.5 kN/m^2 as per IS 875 (Part 2) for mercantile building. Seismic analysis adopting response spectrum is carried out along *X*- and *Y*-axes in plan based on IS 1893 part 1 2016. The following data is based on the information received from the existing structure and adopted for carrying out the analysis: soil corresponding to type-II, importance factor equal to 1.5, zone: IV. M30 grade of concrete has been used for columns, beams, slabs, raft and Fe415 steel has been adopted for all structural members in the existing structure that was constructed earlier. The load combinations for earthquake are considered as per IS 1893 (2016) part 1 and for determination of the settlement of the raft foundation service load combinations are considered in the analysis. The geometrical/cross-sectional dimensions of structural elements in the structure are: column dimensions: (300 × 450) mm, beam dimensions: (300 × 600) mm (along both *X*- and *Y*-directions in plan), thickness of roof slab: 175 mm, the **existing raft foundation** is a **slab type** (modelled as thick plate) having **slab thickness** of **325 mm,** with plan dimensions of 34 m × 34 m, founded below the ground level at 2.7 m depth. **Medium sand** is the **soil existing below the raft** is modelled as area springs having a spring constant—soil

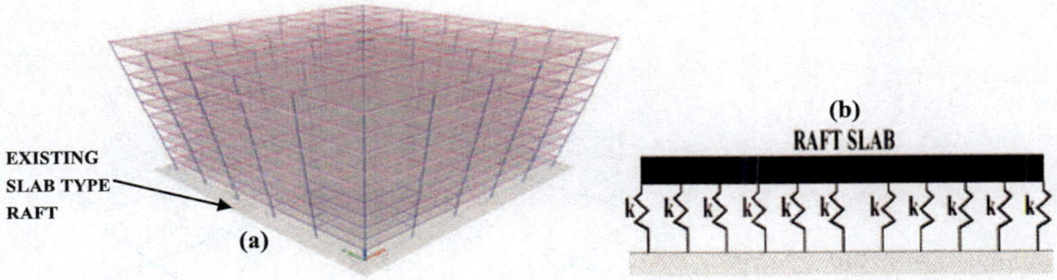

Fig. 1 **a** Structure with un-retrofitted raft slab, **b** soil below the raft is modelled as area springs having soil subgrade reaction $k = 10{,}000$ kN/m^2/m

subgrade reaction equal to $k = 10{,}000$ kN/m^2/m that corresponds to the soil subgrade reaction soil as shown in Fig. 1b. The SBC of soil is 250 kN/m^2. Reinforcement in the existing raft slab is 20Ø@275 mmc/c on the top and bottom layers.

2.1.2 Details of the Retrofitted Raft Slab Retrofitted Present in the Structure

Figure 2a shows the structure with the retrofitted raft slab. Thickness of the RC slab or RC jacket having M35 grade concrete is included individually in the analysis (modelled as thick plate) above the existing raft slab are: t = 125, 275, 425 and 625 mm. Minimum thickness of the RC jacket adopted is based on guidelines for seismic retrofitting of existing structures [15, 16]. Concrete cover of the existing raft slab is removed to weld the reinforcing steel in the new RC jacket slab with the reinforcement present in the existing raft slab. Depth of ribs of new RC raft beam of M35 grade concrete, having a width of 300 mm has been provided above the new RC jacket slab are: D_{retro} = 150, 300, 450, 600 mm. The new RC jacket slab is placed above the existing raft slab. The concrete cover of the existing raft slab is removed to weld the reinforcements present in the new RC jacket slab and new RC raft beam ribs with the reinforcement of the existing raft slab. Figure 2b, c shows the typical details of raft slab retrofitted with RC jacket slab having a certain thickness t and new RC raft beam rib with depth D_{retro} provided above the new RC slab jacket by welding the old reinforcement in the existing raft slab to the new reinforcement in the RC jacket and in the new RC raft beam ribs. Reinforcement provided in the new RC slab is 20Ø@150 mmc/c at the top and bottom faces. Reinforcement provided in the new RC raft beam is 5#-20Ø at top and bottom faces with shear reinforcement of 2L-10Ø@200 mmc/c.

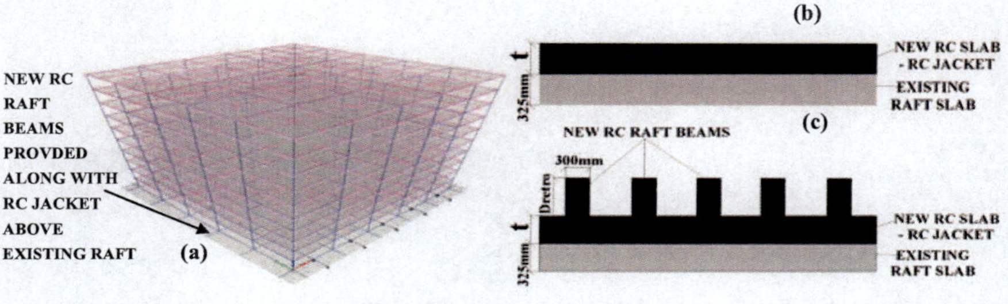

Fig. 2 **a** Structure with retrofitted raft slab—RC slab—RC jacket and new RC raft beams. **b** Typical details of raft slab retrofitted with only RC jacket having a certain thickness *t*. **c** Typical details of raft slab retrofitted with RC jacket having a certain thickness *t* along with new RC raft beam rib with depth D_{retro} provided above the new RC jacket slab

3 Results and Discussions

Maximum settlements along with soil pressures developed in the raft footing that is retrofitted by addition of only an RC jacket slab having different thicknesses provided above the existing raft slab are shown in Table 1. The soil below the raft being medium sand [17], having a dry unit weight of 18 kN/m^3, angle of friction of 37° and specific gravity of 2.66. The SBC and soil subgrade reaction considered in the analysis correspond to medium sand. The reinforcement of the RC jacket is welded to the reinforcement of existing raft slab after removal of existing concrete cover. Figure 3a, b shows the maximum settlement in the raft footing without retrofit and retrofitted with 425 mm thick RC jacket slab provided above the existing RC raft slab. Figure 3c, d shows the maximum soil pressure in the raft footing without retrofit and retrofitted with 425 mm thick RC jacket slab provided above the existing RC raft slab. The maximum settlements and soil pressures developed in the raft footing that is retrofitted by addition of an RC jacket slab having different thicknesses and addition of RC raft beams, with 300 mm width and having different depths of the beam rib provided from the top surface of the RC jacket is tabulated in Table 2.

The reinforcement especially stirrups and main reinforcement of the RC raft beams as well as the reinforcing steel present in RC jacket are welded to steel reinforcement of existing raft slab after removal of existing concrete cover. Figure 4a shows the retrofitted raft slab RC beams layout in the retrofitted raft footing: beams spaced at 3 m along *y*-axis and 6 m centre to centre parallel to *x*-axis, Fig. 4b shows the settlement in retrofitted raft with 425 mm thick RC slab—RC jacket along with RC beam 300 mm × 300 mm both provided above the existing RC raft slab. Figure 4c shows the maximum soil pressure in the raft footing retrofitted with 425 mm thick RC slab—RC jacket along with RC beam 300 mm × 300 mm both provided above the existing RC raft slab.

Observing the results, retrofit of the raft slab adopting only the RC jacket slab that is 425 mm thick provided over the existing raft, the settlement in the raft slab was found to reduce up to 52 mm. Though the settlement in this retrofitted raft slab

Table 1 Maximum settlements and soil pressures developed in the raft footing that is retrofitted by addition of only an RC jacket slab having different thicknesses

Retrofit RC slab—RC jacket thickness (mm)	Maximum settlement in vertical direction at central portion under a column (mm)	Maximum soil pressure (kN/m^2)
Without retrofit		
325 mm	86	858
With retrofit—RC slab—RC jacket		
125 mm (450–325)	52	618
275 mm (600–325)	56	543
425 mm (750–325)	52	538
625 mm (950–325)	50	530

reduced the soil pressure is found to be high greater (in Table 1) than the SBC of the soil, due to the same there is high probability of the columns to punch through the retrofitted raft slab. Therefore, the raft slabs retrofitted considering only addition of new RC slab as RC jacket are not suitable to be considered as the final retrofitting technique.

Therefore, it is observed from the results that, beam slab type of retrofit of the existing raft slab was effective. A RC jacket raft slab thickness of 425 mm and a rib height of 300 mm (width 300 mm) from the top of the existing raft slab were found to be the optimum retrofit of the existing raft as it was observed from the results that the settlement [18] in the raft slab was found to be less than the maximum settlement of 75 mm [19], and the soil pressure in the retrofitted raft slab was less than (in Table 2) the SBC (250 kN/m^2) of the soil at the central portion under a column and in the periphery (edges).

Figure 5a, b shows the bending moment and Fig. 6a, b, shows the shear force in the un-retrofitted RC raft slab and in the retrofitted RC raft with RC jacket (425 mm thick) with RC beam rib 300 mm deep. It is observed from Figs. 5a, b and 6a, b, bending moment and shear force in the raft slab, was found to have reduced after application of retrofit by inclusion of beam ribs 300 mm × 300 mm and RC jacket 425 mm thick. The retrofit that comprises of beam ribs 300 mm × 300 mm and 425 mm thick RC jacket tend to carry the excess bending moment and shear.

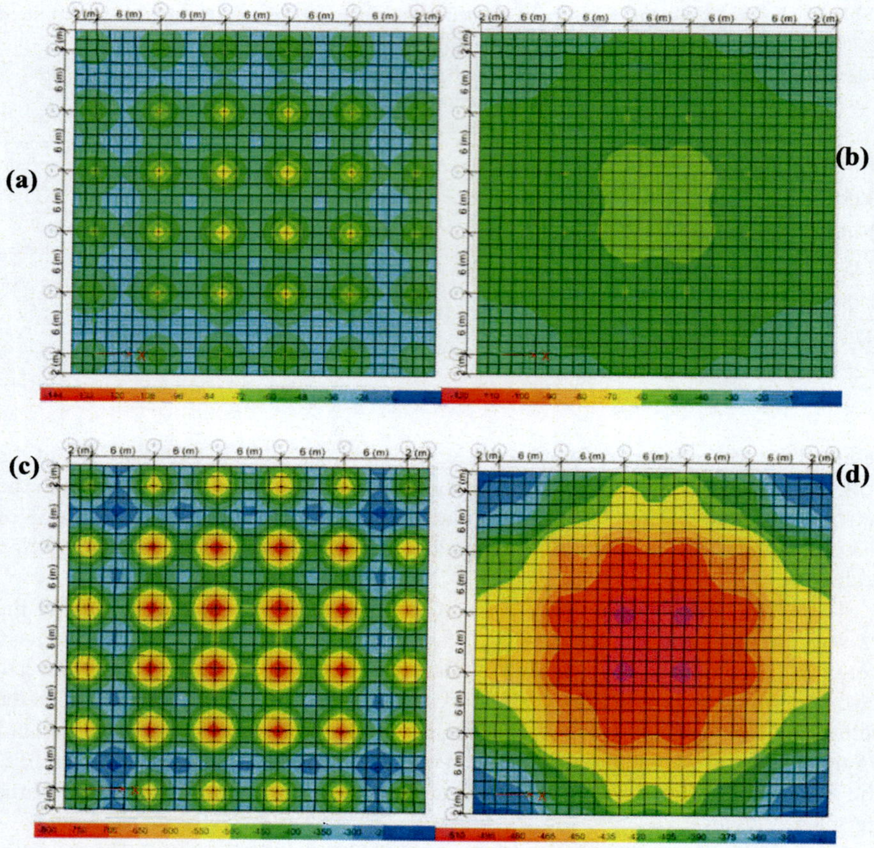

Fig. 3 Maximum settlement in the raft footing: **a** without retrofit, **b** retrofitted with 425 mm thick RC jacket slab provided above the existing RC raft slab. Maximum soil pressure in the raft footing: **c** without retrofit, **d** retrofitted with 425 mm thick RC jacket slab provided above the existing RC raft slab

Table 2 Maximum settlements and soil pressures developed in the raft footing retrofitted by addition of an RC jacket slab having different thicknesses and addition of RC raft beams, with 300 mm width and having different depths of the beam rib

Retrofit RC jacket slab thickness (mm) + RC beam (300 × D_{retro}) mm		Maximum settlement in vertical direction (mm)	Maximum soil pressure (kN/m^2)
Without retrofit: retrofit RC slab—RC jacket			
325 mm existing raft slab		89	858
With retrofit: retrofit RC jacket slab thickness (mm) + RC beam (300 × D_{retro}) mm			
Slab: 125 mm (450–325)	D_{retro} = 150 mm	29 at centre, 43 at the edge	415 (at edge) 247 (under column)
	D_{retro} = 300 mm	24 at centre, 37 at the edge	403 (at edge) 235 (under column)
	D_{retro} = 450 mm	20 at centre, 34 at the edge	370 (at edge) 224 (under column)
Slab: 275 mm (600–325)	D_{retro} = 150 mm	16 at centre, 35 at the edge	330 (at edge) 115 (under column)
	D_{retro} = 300 mm	13 at centre, 31 at the edge	306 (at edge) 109 (under column)
	D_{retro} = 450 mm	11 at centre, 28 at the edge	298 (at edge) 106 (under column)
Slab: 425 mm (750–325)	D_{retro} = 150 mm	10 at centre, 30 at the edge	289 (at edge) 65 (under column)
	D_{retro} = 300 mm	6.3 at centre, 26 at the edge	240 (at edge) 62 (under column)
	D_{retro} = 450 mm	6.1 at centre, 21.5 at the edge	234 (at edge) 59 (under column)
Slab: 625 mm (950–325)	D_{retro} = 150 mm	5 at centre, 21 at the edge	205 (at edge) 33 (under column)
	D_{retro} = 300 mm	4.2 at centre, 17.9 at the edge	192 (at edge) 31 (under column)
	D_{retro} = 450 mm	3.9 at centre, 16.7 at the edge	124 (at edge) 29 (under column)

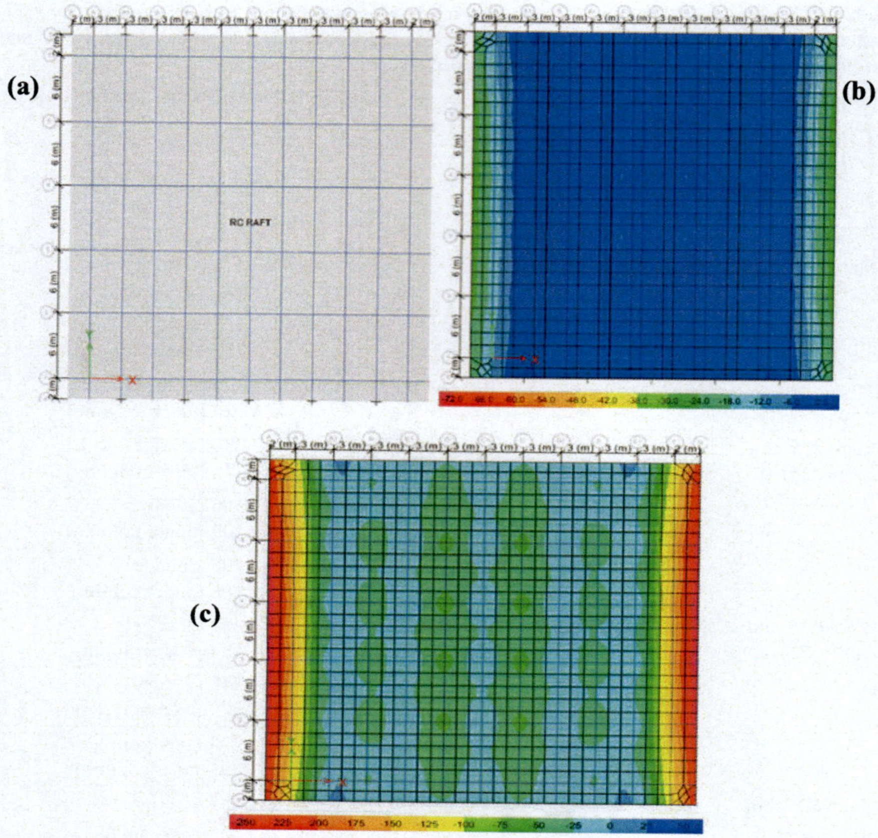

Fig. 4 **a** Retrofitted RC raft slab beam layout. **b** Settlement in raft slab retrofitted with 425 mm thick RC slab along with RC beam 300 mm × 300 mm both provided above the existing raft slab. **c** Maximum soil pressure in the raft footing retrofitted with 425 mm thick RC jacket slab along with RC beam 300 mm × 300 mm both provided above the existing RC raft slab

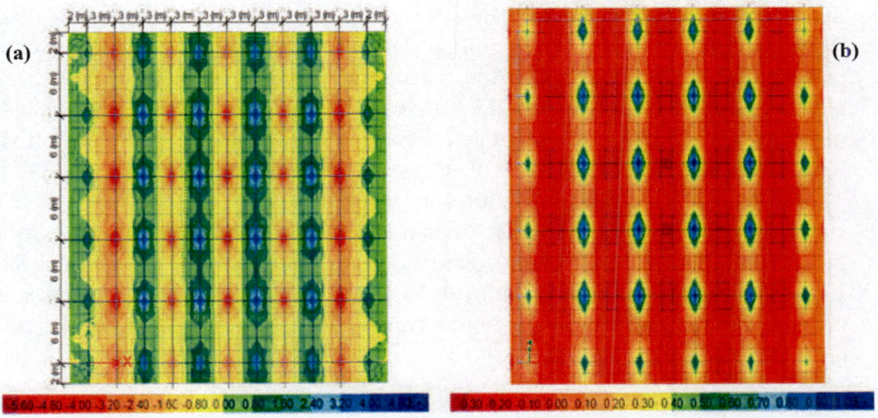

Fig. 5 **a** Bending moment in the un-retrofitted RC raft slab. **b** Bending moment in the retrofitted RC raft with RC jacket (425 mm thick) along with RC beam rib 300 mm × 300 mm

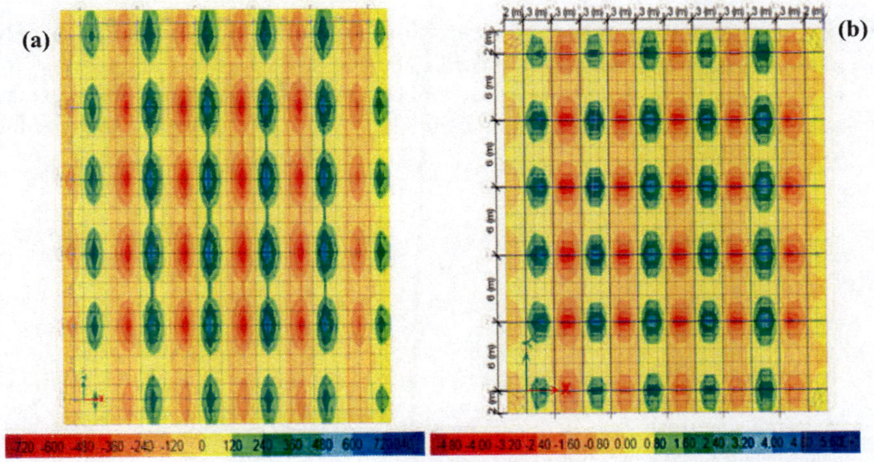

Fig. 6 **a** Shear force in the un-retrofitted RC raft slab. **b** Shear force in the retrofitted RC raft with RC jacket (425 mm thick) along with RC beam rib 300 mm × 300 mm

4 Conclusion

- A maximum settlement of 89 mm and maximum soil pressure 858 MPa in the raft footing having a thickness of 325 mm without retrofitting have been observed from the results.
- When the raft foundation was retrofitted with addition of an RC slab—RC jacket slab provided above the existing raft slab, the settlement was observed to reduce. For an RC jacket thickness of 425 mm provided above the existing raft foundation

that is 325 mm thick, the settlement observed was 52 mm that was less than the maximum settlement (75 mm [19]). Though the settlement reduced, when a check on the soil pressure was made, a maximum soil pressure of 536 kN/m^2 was developed which is high that would cause the SBC to be exceeded. Therefore, just provision of RC slab—RC jacket as a retrofit measure would not be an effective retrofit measure.

- Therefore, when the raft foundation was retrofitted by addition of an RC jacket slab thickness of 425 mm and RC beam having a rib height of 300 mm both provided above the existing raft foundation having a thickness of 325 mm, the settlement reduced to a value equal to 6.1 mm. In this case, a maximum soil pressure of 240 kN/m^2 was developed which was less than the SBC of the soil (250 kN/m^2).

Therefore, adopting a retrofit measure by providing a thickness of 425 mm to the new RC jacket slab and RC beam having rib height of 300 mm and rib width of 300 mm above the existing raft slab is identified as the optimized dimensions of the new RC jacket and RC beam leading to an effective retrofit solution of the raft foundation.

- It is observed from this work that, while retrofitting a raft foundation only increasing the slab thickness by providing the RC jacket slab may not be sufficient and in such cases introduction of new raft beams along with the new RC slab may be essential to restore the raft foundation that has already undergone distress due to settlement.

References

1. IS 456 (2000) Plain & reinforced concrete code of practice. BIS (Reaff2005)
2. IS 875 (Part 1) (1989) Code of practice for design loads (other than earthquake) for buildings and structures. Part 1, dead loads. Bureau of Indian Standards (BIS)
3. IS 875 (Part 2) (1989) Code of practice for design loads (other than earthquake) for buildings and structures. Part 2 live-loads. BIS
4. Murtiadi S, Akmaluddin A (2023) Performance of concrete building structure exposed to localized fire. Civ Eng Archit 11(1):114–122
5. Qin D, Gao PK, Aslam F, Sufian M, Alabduljabbar H (2022) A comprehensive review on fire damage assessment of reinforced concrete structures. Case Stud Constr Mater 16:e00843
6. Guruprasad YK (2021) Effect of size and shape of concrete column elements exposed to high temperature. In: Proceedings of SECON2020. Lecture notes in civil engineering, vol 97. Springer, pp 929–937
7. Ross CA, Nash PT, Griner GR (1979) Failure of underground concrete structures subjected to blast loadings. Shock Vibr Inf Cent Shock Vibr Bull Pt 3:1–9
8. Bharath TK, Guruprasad YK (2021) Response of multi-storeyed structures subjected to blast loading. In: Recent advances in structural engineering. Lecture notes in civil engineering, vol 135. Springer, pp 213–222
9. IS 875 (Part 3) (1989) Code of practice for design loads (other than earthquake) for buildings and structures. BIS

10. Rama Raju K, Shereef MI, Iyer NR, Gopalakrishnan S (2013) Analysis of tall building subjected to wind and seismic loads. In: National conference on emerging technologies in civil engineering (ETCE'13), pp 103–111
11. Shukle AA, Guruprasad YK (2021) A comparative study of wind effects on high-rise structures having different shapes in plan. In: Recent advances in structural engineering. Lecture notes in civil engineering, vol 135. Springer, pp 243–252
12. Meghashree L, Guruprasad YK (2023) Enhancement of lateral stability of high-rise structures adopting effective lateral load resisting systems. In: Recent advances in materials, mechanics and structures. Lecture notes in civil engineering, vol 269. Springer, pp 473–485
13. IS 1893 (Part 1) (2016) Criteria for earthquake resistant design of structures. BIS
14. Derecho AT, Kianoush MR (2001) Seismic design of reinforced concrete structures. In: Naeim F (ed) The seismic design handbook. Springer, Boston, MA. https://doi.org/10.1007/978-1-4615-1693-4_10
15. Retrofitting of Existing Buildings (2008) CAMTECH/2008/C/RETROFITTING/1.0. Government of India
16. IS 15988 (2013) Seismic evaluation and strengthening of existing reinforced concrete buildings—guidelines. Bureau of Indian Standards
17. Al-Neami MA (2018) Investigation of sampling error on soil testing results. Int J Civ Eng Technol 9(13):579–589
18. Bilagi N, Guruprasad YK (2023) Retrofitting of RC footings subjected to settlement due to earthquake. In: Recent advances in materials, mechanics and structures. Lecture notes in civil engineering, vol 269. Springer, pp 487–499
19. IS 1904 (2006) Code of practice for design and construction of foundations in soils: general requirements. Reaffirmed

Sustainable Ground Improvement Techniques

Assessment of Stability of Embankments on Soft Ground Using Matsuo Chart and SLOPE/W

Manda Sushma, J. Y. V. Shiva Bhushan, and Madhira R. Madhav

1 Introduction

Several coastal regions and river valleys across the world have thick layers of soft ground. These soils exhibit poor geotechnical properties such as high compressibility, excessive moisture content, low permeability and weak undrained strength. Rapid increase in infrastructure development in countries like INDIA demands the construction of embankments on such soft terrain for expressways, railroads, and runways for airports. Due to the increase in vehicular usage in urban areas and avoid traffic in congestions, embankment construction is time bound. These demands result in increased rate of construction of embankment on soft ground. Hence, assessing the embankment stability during and post-construction is necessary.

Slope stability of an embankment is affected by soil's shear strength, unit weight, and geometry of embankment like height of fill and steepness of the slope. The rate of plastic flow and settlement of subsoil due to shear deformation and consolidation respectively, influence the stability of embankments on soft ground. When embankment construction is carried at rapid pace, the rate of lateral deformation may exceed the consolidation settlement, which leads to embankment failure. Matsuo and Kawamura [1] proposed construction control chart by plotting settlement, s as ordinate and ratio of lateral displacement, δ to settlement as abscissa based on field measurements. The degree of safety of an embankment during the construction itself can be estimated from the chart presented in Fig. 1. Matsuo charts are suitable to homogeneous and isotropic behaviour of soils, ignore progressive failure and are limited for complex embankment slope geometries. Research is necessary to assess relevance and accuracy of the Matsuo chart in assessing the embankment stability during construction under various geometrical and subsurface conditions itself.

M. Sushma · J. Y. V. Shiva Bhushan (✉) · M. R. Madhav
Department of Civil Engineering, VNR Vignana Jyothi Institute of Engineering and Technology, Hyderabad 500090, India
e-mail: shivabhushan_jyv@vnrvjiet.in

S. Kolathayar et al. (eds.), *Best Practices in Geotechnical and Pavement Engineering*, Lecture Notes in Civil Engineering 449, https://doi.org/10.1007/978-981-99-8505-0_8

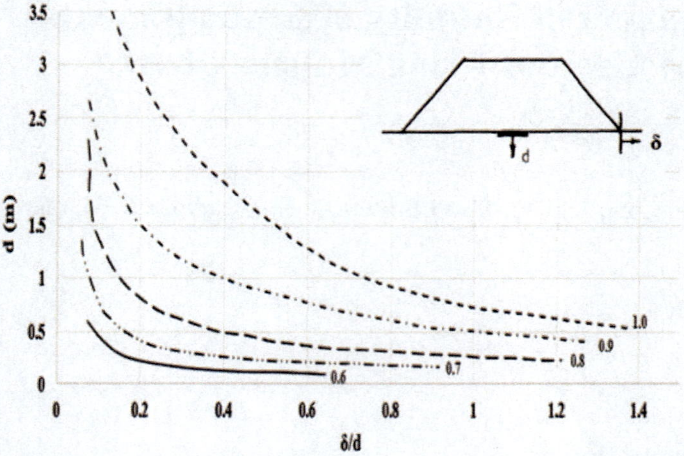

Fig. 1 Settlement versus ratio of lateral deformation to settlement (after Matsuo and Kawamura [1])

The embankment stability can be assessed using field-monitored data and numerical techniques like limit equilibrium methods (LEM). Several researchers have employed LEM for slope stability analysis [2–4] while recently many have used finite element method (FEM) [5–7] and finite element analysis (FEA) [8, 9] for the assessment of embankment slopes. Additionally, some researchers have utilized the strength reduction method [3, 10] for embankment stability analysis. Abhinay Kumar et al. [11, 12] analysed the lateral displacements and settlements of soft ground under embankment loading using PLAXIS 2D.

Matsuo charts based on field displacements and finite element software—GeoStudio are utilized in the present study. FEM can be used to determine the embankment's deformation characteristics as well as its factor of safety. Many different methods of analysis and failure surfaces are postulated in the limit equilibrium method (LEM), so that the factor of safety (FOS) can be determined. The main objective of the current study is to assess the stability of an embankment for its construction control using Matsuo chart and validated by LEM using GeoStudio software.

2 Case Studies

Case A: An instrumented embankment is built on soft Malaysian marine clay [13]. The subsurface profile consists of 2 m thick crust on the top of 16.5 m of soft silty clay. Soil beneath is comprised of very soft to soft silty clay. The thickness of embankment fill varies from 2 to 5 m. Lateral displacements increased gradually with fill thickness, and sharply once its height reached 5 m. The lateral displacements and vertical deformations at various fill thickness of 2.0, 3.0, 4.0, and 5.0 m were analysed. Failure was observed at fill thickness of 5.5 m. Table 1 presents the properties of embankment soil for all the case studies considered.

Case B: A trail embankment is constructed with side slopes of 1 V:1.5 H and a height of 5.4 m [8]. The subsoil of thickness of 22 m is considered for the analysis. The weathered crust is 2.0 m thick overlying very soft silty clay of thickness of 2–7 m followed by 7–12 m of silt, and 12–18 m of clay and sandy clay. The undrained strength at different depths of the soil is obtained using Vane shear test. The surface settlement is observed at a fill thickness of 2.5 and 5.4 m.

Case C: Test embankments constructed on soft soil deposit located in Lian Yun-Gang, China [5] are considered in this study. The length and base width of the embankments are 45 and 42 m, respectively. A berm of 8 m width is provided on both sides above a 0.5 m thick sand mat. The embankment slopes are 1 V:1.75 H. The critical height of the embankment on natural subsoil was 4.04 m. The lateral displacements and settlements were observed at fill thicknesses of 2.6, 3.7 and 4.35 m.

Case D: Three distinct sublayers exist within the topmost soil strata at Bangkok site [14]. The top 2 m thick sublayer of the crust was pseudo-consolidated reddish-brown clay. Below this layer lies a soft greyish clay sublayer, extending to a depth of approximately 8 m, with a moisture content ranging from 60 to 80% overlying 3.0 m thick medium-stiff clay sublayer consisting of silt and fine sand.

The level of groundwater at the test site changes from 0.50 to 1.50 m during the rainy and dry seasons, respectively. Embankments have side slope of 1.5 V to 1.0 H. A canal of 2.0 m width and 1.8 m deep was dug along the toe of the embankment, to secure the desired location of the failure and to lower the necessary fill height. 1.0 m-thick berm supports the embankment.

Case E: An embankment is constructed over soft ground for Shi-Man Expressway in China [6]. The embankment has 24 m wide crest and 6 m height. The in situ soil is composed of two soft soil layers that were overlain by a fairly stiff layer. The soft clay is covered with 0.5 m thick sand bed.

In all the case studies, the rate of construction of test embankment is known and the field displacements (both horizontal at toe and vertical at the centre) during its construction have been measured and reported.

Table 1 Soil parameters of the embankment

Case studies	Unit weight (kN/m^3)	Cohesion (kPa)	Angle of shearing resistance (°)
Case A	20.5	19	26
Case B	20.5	19	26
Case C	19	5	25
Case D	18.5	15	30
Case E	20.4	32	28.5

3 Methodology

Limit equilibrium study is carried out using SLOPE/W software. Heterogeneous soil conditions and all possible slip surfaces can be modelled with SLOPE/W. SLOPE/W employs a wide range of surface geometries and variable pore-water pressure conditions. Slope stability can be assessed using deterministic approaches or by incorporating input variables with uncertainty. Ordinary method of slices, Bishop [15] and Janbu's [16] methods are been deployed to determine the FOS values. The entry and exit method is utilized to determine the factor of safety.

The Fellenius method of slices involves dividing the soil mass above the failure surface into a number of slices, without considering interslice forces. Bishop's method presumes that shear forces in the upward and directions are equal. Janbu's method satisfies both force and moment equilibrium conditions, whereas Bishop's method satisfies only moment equilibrium. Janbu's approach is also be applicable to non-circular failure surfaces.

Matsuo charts are drawn based on lateral deformations at the toes and vertical displacements beneath the centre of embankments for various heights of the embankment during the construction phase, for all five case studies (A–E). Stability analysis is carried out using GeoStudio—SLOPE/W software in the second stage and both the results are compared.

4 Results and Discussion

Figures 2 and 3 present the variations of lateral displacement and settlement at toe and centre of the embankment respectively with fill thickness for the five cases analysed.

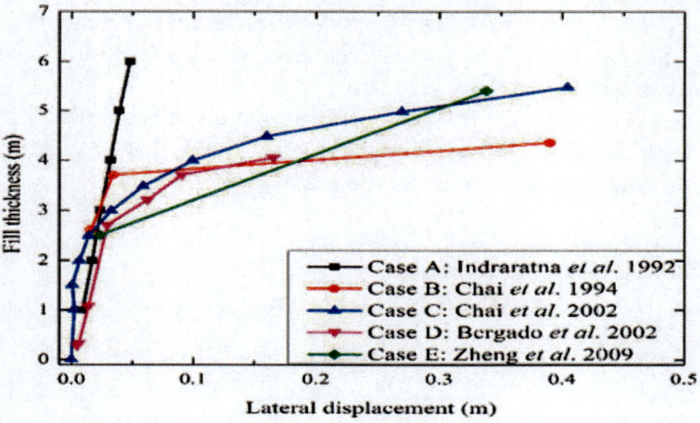

Fig. 2 Lateral displacement versus embankment fill height

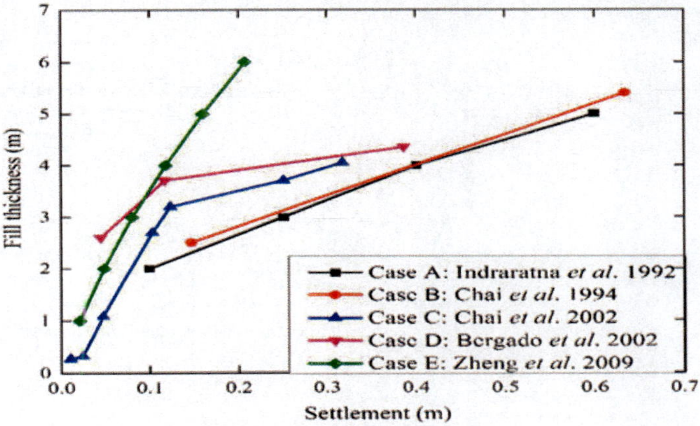

Fig. 3 Settlement versus embankment fill height

Lateral displacement and settlement increase linearly with embankment height reaching to about 0.05 m and 0.6 m respectively for 6.0 m height at site A. In case of site B, where the fill thickness varied between 2.5 and 5.4 m, the horizontal and vertical displacement were analysed. It was observed that vertical displacement exhibited a linear increase for all the embankment heights, whereas horizontal displacement initially increased linearly up to a height of 3.8 m. Beyond this point, lateral deformation of soil at toe sharply escalated to 0.4 m, suggesting a potential failure of embankment at a fill height of 4.35 m.

At site C, settlements increased linearly till a height of 3.2 m and then exhibited a sharp increase beyond that. However, lateral displacement showed relatively smaller increments initially but sharp increase after reaching a height of 3 m. For site D, both vertical and lateral displacements increased gradually initially and significantly beyond an embankment height of 3.7 m. Settlements increased gradually and lateral displacement rapidly with varying fill thickness, in case of site E. Horizontal of 0.34 m is higher than the vertical settlement of 0.18 m for fill thickness of 5.4 m.

The variations of settlement with ratio of lateral displacement to settlement are plotted for different cases A–E (Fig. 4). Matsuo chart indicates factor of safety (reciprocal of the values given in the chart) of 1.2, 1.17, 1.14, 1.31 and 1.63 respectively for cases A, B, C, D and E. It is inferred that sites D and E are most stable, while sites A, B and C, tend to fail should construction progress further.

The results obtained from SLOPE/W software for various sites A–E are discussed. Embankment constructed for a height of 6.0 m over a 0.5 m thick sand cushion, is modelled with half the top crest width of 24 m (Fig. 5). Factors of safety with Janbu, Bishop and Ordinary Method of slices are 1.81, 1.87 and 1.74, respectively, and larger than 1.63 from Matsuo's chart (Fig. 5). Similar analyses was conducting using SLOPE/W for the remaining sites (A, B, C and D).

Factors of safety for different cases and based on different methods of analysed are compared in Fig. 6. Ordinary method tends to provide FOS values closer to those

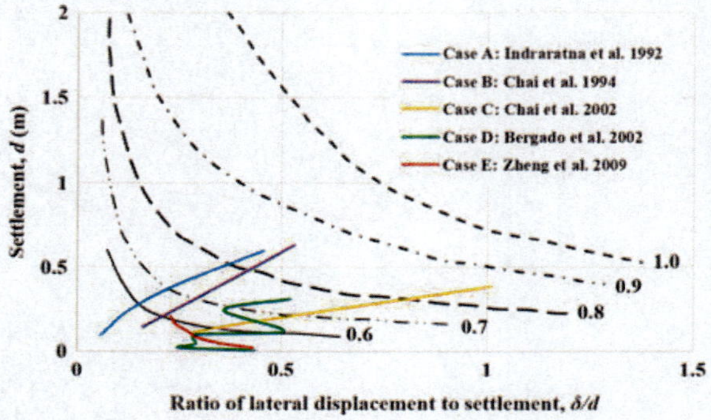

Fig. 4 Matsuo plots obtained for cases A–E

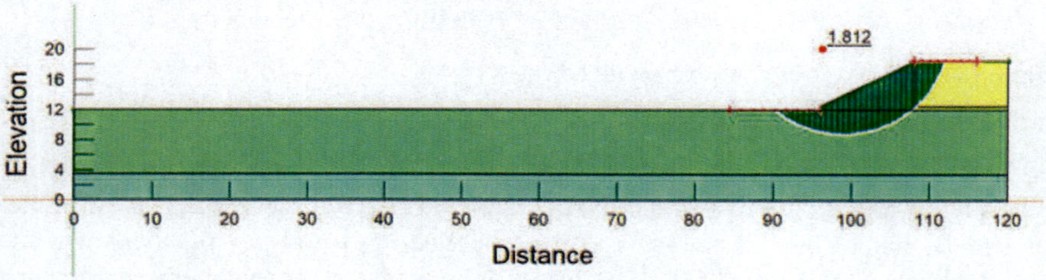

Fig. 5 Critical slip surface for Janbu simplified method for site E

from the Matsuo plot for all sites except Case A, for which Janbu method yields FOS values closer to the Matsuo plot.

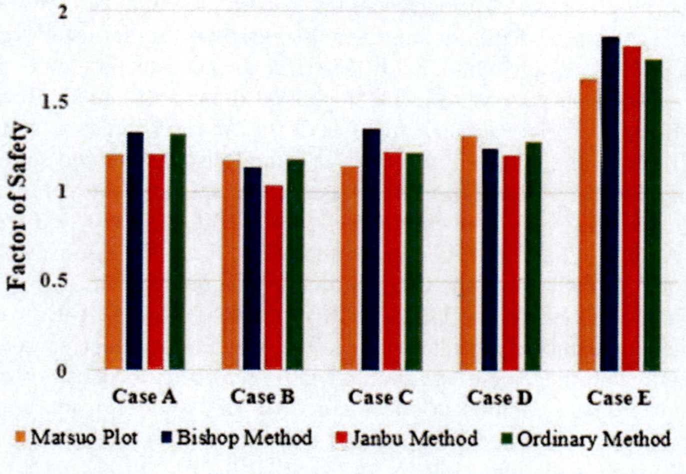

Fig. 6 Comparison of factor of safety values for different sites using different methods

5 Conclusions

Stability of five embankments is assessed using in situ data (settlement and lateral displacement data during the construction) using Matsuo plots. Their stability is also estimated by limit equilibrium methods using SLOPE/W software. Matsuo's chart is a good construction control as the factors of safety observed from it are either comparable or in some cases slightly lower than the limit equilibrium approaches. It is not often possible to identify approaching failure but the one presented comes close to it.

Further research can involve the application of advanced numerical techniques, to evaluate the effectiveness and reliability of construction control charts in slope stability analysis.

References

1. Matsuo M, Kawamura K (1977) Diagram for construction control of embankment on soft ground. Soils Found 17:37–52. https://doi.org/10.3208/sandf1972.17.3_37
2. Hammouri NA, Husein Malkawi AI, Yamin MMA (2008) Stability analysis of slopes using the finite element method and limiting equilibrium approach. Bull Eng Geol Environ 67:471–478. https://doi.org/10.1007/s10064-008-0156-z
3. Maula BH, Zhang L (2011) Assessment of embankment factor safety using two commercially available programs in slope stability analysis. Procedia Eng 14:559–566. https://doi.org/10.1016/j.proeng.2011.07.070
4. Tantri P, Lastiasih Y (2015) Slope stability evaluation using limit equilibrium method (LEM) and finite element method (FEM) for Indonesia soft soil. In: The 3rd Bali international seminar on science and technology, Bali, Indonesia, pp 1–8
5. Chai JC, Miura N, Shen SL (2002) Performance of embankments with and without reinforcement on soft subsoil. Can Geotech J 39:838–848. https://doi.org/10.1139/t02-033
6. Zheng J, Chen B, Lu YE, Abusharar SW, Yin J (2009) The performance of an embankment on soft ground reinforced with geosynthetics and pile walls. Geosynth Int 16:173–182. https://doi.org/10.1680/gein.2009.16.1.1
7. Jiang X, Chen X, Fu Y, Gu H, Hu J, Qiu X, Qiu Y (2021) Analysis of settlement behaviour of soft ground under wide embankment. Balt J Road Bridge Eng 16:153–175. https://doi.org/10.7250/bjrbe.2021-16.543
8. Chai JC, Sakajo S, Miura N (1994) Stability analysis of embankment on soft ground (a case study). Soils Found 34:107–114. https://doi.org/10.3208/sandf1972.34.2_107
9. Hird CC, Pyrah LC, Russell D, Cinicioglu F (1995) Modelling the effect of vertical drains in two-dimensional element analysis. Can Geotech J 807:795–807. https://doi.org/10.1139/t95-077
10. Lim K, Li AJ, Schmid A, Lyamin AV (2017) Slope-stability assessments using finite-element limit-analysis methods. Int J Geomech 17:1–8. https://doi.org/10.1061/(asce)gm.1943-5622.0000715
11. Abhinay Kumar C, Shiva Bhushan JYV (2022) Deformation analysis of soft ground treated with PVDs—a numerical study. IOP Conf Ser Earth Environ Sci 982. https://doi.org/10.1088/1755-1315/982/1/012056
12. Abhinay Kumar C, Shiva Bhushan JYV, Madhira M (2023) Lateral displacements of soft ground treated with PVDs under embankment loading. In: Muthukkumaran K, Ayothiraman R, Kolathayar S (eds) Soil dynamics, earthquake and computational geotechnical engineering.

IGC 2021. Lecture notes in civil engineering, vol 300. Springer, Singapore. https://doi.org/10.1007/978-981-19-6998-0_12
13. Indraratna BB (1992) Performance of test embankment constructed to failure on soft marine clay. J Geotech Eng 118:12–33. https://doi.org/10.1061/(ASCE)0733-9410(1992)118:1(12)
14. Bergado DT, Lorenzo GA, Long PV (2002) Limit equilibrium method back analyses of geotextile-reinforced embankments on soft Bangkok clay—a case study. Geosynth Int 9:217–245. https://doi.org/10.1680/gein.9.0216
15. Bishop AW (1955) The use of the slip circle in the stability analysis of slopes. Géotechnique 5:7–17. https://doi.org/10.1680/geot.1955.5.1.7
16. Janbu N (1973) Slope stability computations. In: Hirschfeld RC, Poulos SJ (eds) Embankment dam engineering. Wiley, New York, pp 47–86

Exploring the Correlation Between Compressive Strength and Hydraulic Conductivity in Soft Soil with Metakaolin as a Stabilizing Agent: An Experimental Study

Himanshu Jangde, Farhan Khan, and Abdul Ghaffar

1 Introduction

The qualities and textures of various kinds of soil have significant effects on construction. Poor mechanical characteristics and a lack of weight-bearing capacity are major issues in the construction industry due to soft ground [1, 2]. The qualities and textures of various kinds of soil have significant effects on construction. Poor mechanical characteristics and a lack of weight-bearing capacity are major issues in the construction industry due to soft ground [3–9]. Kaolin transforms into metakaolin when heated. This material may be utilized to enhance the productive capabilities of soil [10–12]. Metakaolin and calcium hydroxide react to form a glue that accomplishes this [13–16]. Finding a correlation between hydraulic conductivity and soil hardness is the primary focus of this investigation. This is crucial for regulating the amount of water that seeps underground [17–20]. Understanding how metakaolin impacts hydraulic conductivity is crucial for maintaining earth's stability. The goal of the research is to discover how metakaolin may be utilized to stabilize soft soils in order to propose new eco-friendly construction techniques. By making it simpler to compact sandy soils and construct sturdy buildings, the new results will advance the field of geotechnical engineering.

H. Jangde · F. Khan (✉)
Department of Civil Engineering, Rungta College of Engineering and Technology, Bhilai, Chhattisgarh, India
e-mail: Farhan1@rungta.ac.in

A. Ghaffar
S.B. Jain Institute of Technology, Management and Research, Nagpur, India

S. Kolathayar et al. (eds.), *Best Practices in Geotechnical and Pavement Engineering*, Lecture Notes in Civil Engineering 449, https://doi.org/10.1007/978-981-99-8505-0_9

2 Research Structure

The assigned research paper will include both a comprehensive literature assessment of metakaolin's use as a soil stabilizer and an analysis of the correlation between hydraulic conductivity and compression strength. The sampling and testing procedures that formed the backbone of this investigation will be discussed in length. The study's findings will be shown by analyzing the correlation between hydraulic conductivity and soil strength after treatment with metakaolin. The implications for geotechnical engineering will be analyzed in depth, and potential research gaps will be identified. The study's workflow is mapped out (Fig. 1).

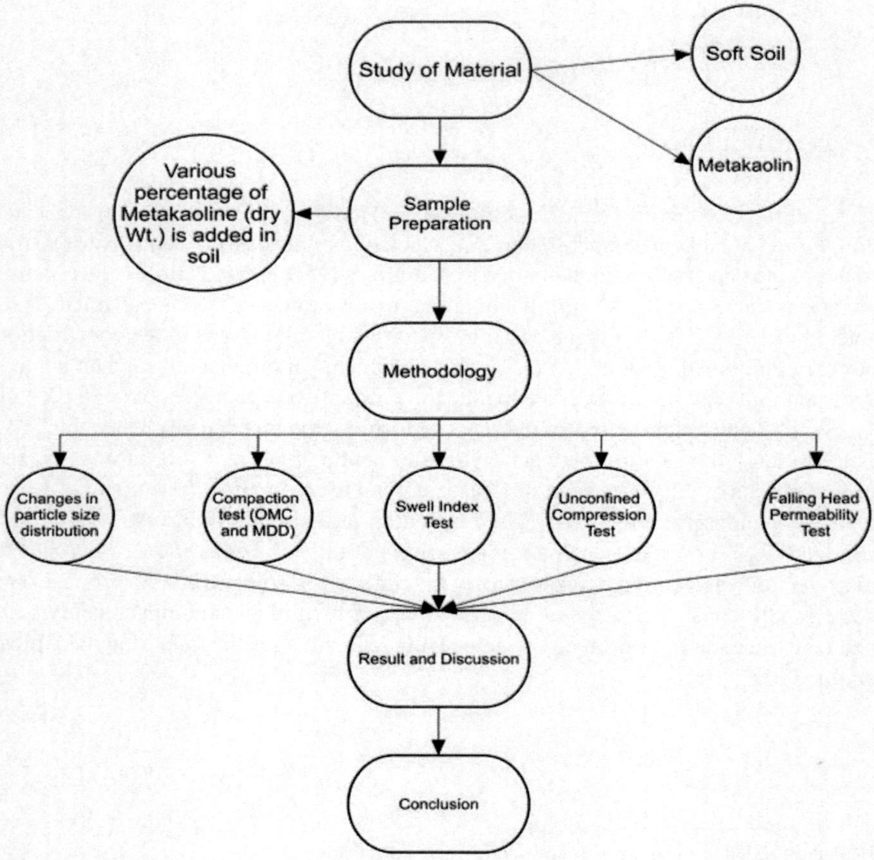

Fig. 1 Flow of the research

3 Material and Preparation of Sample

3.1 *Material*

Soil. This kind of soil is often found in areas that are characterized by a high level of moisture, such as marshes and river deltas. It is also resistant to the passage of liquids like water and can bend in a variety of orientations. Certain materials, for example, have a lower degree of stability because they may shift shape, generate sedimentation, or prevent water from moving through them. The term "soft soil" refers to ground that is easily compacted, is weak, and swells and shrinks depending on the amount of precipitation that falls on it. In order to conduct this research, soil samples were taken from a field located in Bhilai, Chhattisgarh, which is next to Rungta College. In the following paragraphs, we will talk about the real components that make up the soil (Table 1).

Metakaolin. The kaolin clay is burnt at relatively high temperature in order to form reactive pozzolan metakaolin to eliminate the crystal water that is contained within the kaolin clay. It is a highly reactive chemical that has an amorphous structure and is abundant in silica and alumina. When water molecules are removed, a reactive material is produced that has a temperature range of between 600 and 850 °C. By heating kaolin, one may transform it into metakaolin, and the corresponding chemical formula is as follows:

$$\underset{\text{(Kaolin)}}{Al_2Si_2O_5(OH_4)} \xrightarrow{700^\circ} \underset{\text{(Metakaolin)}}{Al_2O_3 \cdot 2SiO_2} + 2H_2O$$

Table 1 Property (physical) of soil (Bhilai)

Property	Index value	Specification
Liquid limit	41	IS 2720:1985 (Part-5)
Plastic limit	27	IS 2720:1985 (Part-5)
Plasticity index	14	–
Maximum dry density	18 kN/m^3	IS 2720:1983 (Part-8)
Optimum moisture content	13.6%	
Passing no. 200	94%	–
Percent of sand	6%	IS 2720:1985 (Part-4)
Percent of clay	59	
Percent of silt	41	
Unconfined compressive strength	0.48 MPa	IS 2720:1991 (Part-10)
Specific gravity	2.67	IS 2720 (Part III) 1980

Source Author

Table 2 Properties (physical) of metakaolin

Properties	Metakaolin
Moisture (%)	0.5
Oil absorption (%)	60 ± 2
Average (d50)	1 ± 1
pH of 10% solution	6 ± 1
Brightness	90 ± 1
< 10 μm (%)	95 ± 2
< 2 μm (%)	80 ± 1
Specific gravity	2 ± 0.1
Bulk density (g/l)	320 ± 20

Source From Manufacturer

Table 3 Chemical properties of metakaolin

Chemical properties	Percentage
Al_2O_3	36.3
Fe_2O_3	4.21
CaO	< 0.10
SiO_2	52.8
K_2O	1.41
MgO	0.81
Loss on ignition, LOI (%)	3.53

Source From Manufacturer

After the hydroxyl groups in kaolin clay have been removed, what is left behind is a powder that is white in color, odorless, and has a large surface area. As a highly dynamic pozzolan substance, it boosts the compression strength, durability, workability, and setting time of concrete. Additionally, it makes concrete easier to work with. The metakaolin that was used in this study came from AJ Corporation in Mumbai, a business that is well-known for manufacturing and selling metakaolin. This corporation was the supplier of the metakaolin. The list of some of the atomic, molecular, and physical features of metakaolin (Tables 2 and 3).

3.2 *Sample Preparation*

When casting soil samples, a cylindrical form with height of 100 and 44 mm in width is used. A sodium hydroxide solution with a concentration of 10 M is made. Dry soil and several different concentrations of metakaolin (MK) are completely mixed together before the consistency can be determined. Rehydrating the dried material with an alkali solution that is 38% strength results in the production of the sample.

Table 4 Outcomes of compaction tests [21] conducted on soft soil that has been treated with varying percentages of metakaolin

Moisture content	Dry densities (kN/m^3) at different percentages of metakaolin			
	0% metakaolin	5% metakaolin	10% metakaolin	15% metakaolin
9	15.39	15.91	15.94	16.8
10	15.71	16.26	16.81	17.46
11	15.41	16.01	16.53	17.17
12	15.12	15.76	16.37	16.78
13	14.82	15.51	16.12	16.29
14	14.53	15.25	15.85	16.15

Source From Author

It is necessary to activate metakaolin in order to bind the grime together. Increasing the proportion of metakaolin contributed to the polymerization process, which in turn helped to reinforce the globe. Accurate testing requires meticulous specimen preparation on the part of the patient.

4 Findings and Analysis

4.1 Maximum Dry Density and Optimum Moisture Content (Compaction Test)

Findings from Table 4, as has been demonstrated, increasing the amount of metakaolin that is added to dried soil results in an increase in that soil's density. This is because metakaolin eliminates space between dirt particles, which causes the dirt to become more compact and bound together. The highest dry density may be accomplished with a moisture content of 10%, and this holds true regardless of the quantity of metakaolin that is used. Because the soil achieves its possible maximum dry density at this amount of moisture, we refer to this as the OMC.

Compaction curves are valuable for analyzing the behavior of soil during compaction experiments. A *compaction curve* illustrated in Fig. 2 depicts how the dry density and moisture content change over compaction time.

4.2 Swelling Potential

From Table 5, the swell index is decreased, and the soil is strengthened by the addition of metakaolin (MK). Cementitious chemicals are produced when it reacts with

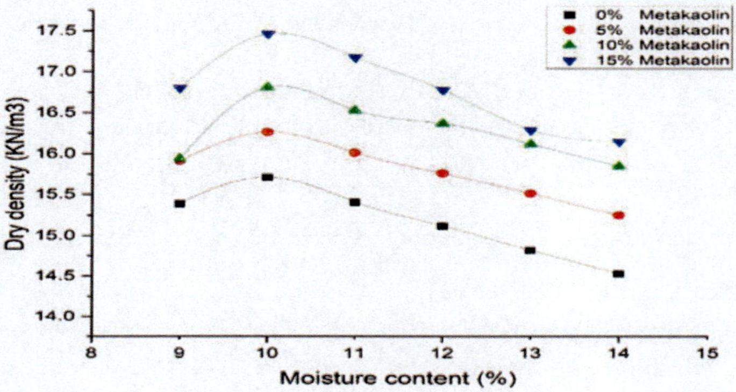

Fig. 2 Compaction test findings for various soil samples with varying proportions of metakaolin. *Source* From Author

Table 5 Variations in the soil's swell index when utilizing metakaolin as a stabilizing agent

Metakaolin percentage (%)	Plasticity index (%)	Liquid limit (%)	Plastic limit (%)	Swell index
0	19	49	28	41
5	18	47	29	33
10	13	37	24	23
15	12	34	22	22

Source From Author

calcium hydroxide in the soil, decreasing the soil's pliability. Soil's swelling potential (46.34%) and flexibility index (36.84%) decrease when metakaolin is introduced at a 15% concentration. By binding free calcium and forming solid mineral phases, metakaolin makes porous soil more adaptable and resistant to stress. Due to its pozzolanic nature, metakaolin is used to create cementitious compounds that harden soil (as Table 5 shows). Metakaolin improves soil stability and safety in geotechnical applications like foundation design.

4.3 Unconfined Compression Test

In a stress test performed without confinement, metakaolin causes the earth to have more resistance. Because it contains a high percentage of metakaolin, it is more long-lasting. Those containing neither metakaolin nor any needed 88 kPa of pressure when subjected to 20% strain, but those containing 15% metakaolin required 146 kPa of pressure. The power has been increased by a factor of 0.75 (Fig. 3 and Table 6). The addition of calcium hydroxide to metakaolin results in the formation of a gel that is constituted of calcium silicate hydrate. According to the findings of this study,

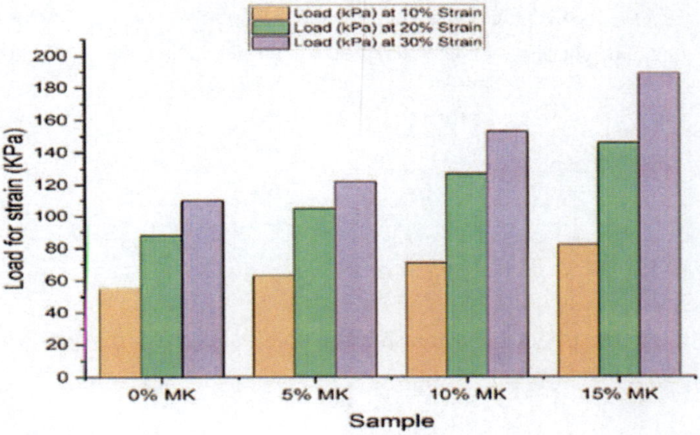

Fig. 3 Experimental findings from unconfined compression tests conducted on soft soil samples containing different percentages of metakaolin, while subjecting them to various strain conditions. *Source* From Author

Table 6 Results of unconfined compressive strength (UCS) data obtained under varying strain conditions

Sample	Metakaolin percentage (%)	Hydraulic conductivity (cm/s)	Initial water level (cm)	Length (cm)	Final water level (cm)	Time (s)
1	0	2.29×10^{-3}	45	12.2	25	523
2	5	1.42×10^{-3}	45	12.2	25	845
3	10	1.06×10^{-3}	45	12.2	25	1132
4	15	7.30×10^{-4}	45	12.2	25	1648

Source From Author

metakaolin may be used to improve the quality of soil without the need to add a significant amount of extra ingredients.

4.4 *Permeability Test (Falling Head)*

Table 7 shows the water-conducting capabilities of soft earth are enhanced when metakaolin is included. The sample having 15% metakaolin had the least hydraulic conductivity of all the samples. When compared to soil that had not been treated, the hydraulic conductivity was reduced by 68.12%. Metakaolin and pozzolanic soil work together to lower the permeability of the soil and its ability to transport water via the forming of a calcium silicate hydrate (C-S-H) gel. Metakaolin has the ability to make soil more impermeable, thick, and impervious when it is coupled with water.

Table 7 Changes in the hydraulic conductivity of soft soil as the percentage of metakaolin varies

Sample	Initial height (mm)	Load (kPa) at 30% strain	Load (kPa) at 20% strain	Load (kPa) at 10% strain
0% MK	20	110	88	55
5% MK	20	122	105	63
10% MK	20	153	127	71
15% MK	20	189	146	82

Source From Author

Metakaolin has the effect of decreasing the soil's ability to transport water when applied to loose, low-density soil.

4.5 The Connection Between Hydraulic Conductivity and Compressive Strength of Soft Soil When Metakaolin Is Used as a Stabilizing Agent

The relationship between the unconfined compression test and the falling head permeability test results for soft soil is visually represented through a scatter plot in Fig. 4. The graph depicting a scatter plot shows that there is a connection between the metakaolin treatment, the unconfined compression test (UCS), and the permeability test (falling head). According to the findings, adding metakaolin to loose soil boosts both the soil strength and its capacity to allow water to pass through it. Metakaolin has the capacity to efficiently increase soil strength while simultaneously lowering the water permeability of the soil. When seen by the curve, there is a discernible trend toward the material being more rigid and less permeable when the amount of metakaolin is increased.

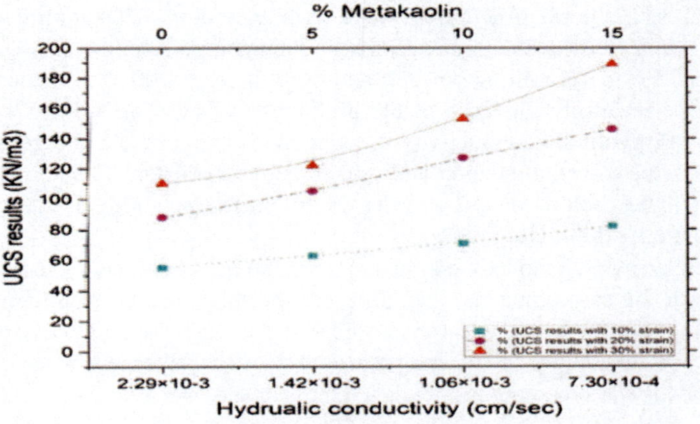

Fig. 4 Correlation between unconfined compressive strength (UCS) data and the hydraulic conductivity of soft soil. *Source* From Author

5 Conclusion

Based on the experimental study on the "interrelationship between hydraulic conductivity and compressive strength of soft soil using metakaolin as a stabilizer." The incorporation of metakaolin causes significant alterations to take place in both the mechanical and hydrodynamic characteristics of soft earth. These statistics were derived from the association that exists in the hydraulic conductivity of a soil and its resistance to being compressed. According to the results, using metakaolin to make soil more compact and less permeable to water may be an effective method for enhancing the quality of soil that is otherwise deficient. The following presents some of the results.

1. The addition of metakaolin to soil with low compaction strength boosts the soil's overall strength. The addition of 15% metakaolin causes an increase of 11.14 percentage points in the maximum dry density. When metakaolin pieces are added to soil and the pozzolanic reaction is seen, the soil may be rendered more stable, less likely to settle, and more able to support increased loads of weight. Further testing is necessary in order to determine how well it will function throughout the course of its lifetime.
2. After being combined with MK, the loose soil had a reduction of 36.84% in its flexibility index and a reduction of 46.34% in its spread index. The pozzolanic interaction that occurs between metakaolin and natural lime is what makes soil less flexible.
3. When added to loose soil, metakaolin produces a 75% increase in the soil's compression strength under any and all conditions of pressure. The interaction of pozzolanic material with earth lime results in the formation of a gel known as calcium silicate hydrate, abbreviated as C-S-H. This gel helps to enhance the fit

as well as the pores. According to the tests, including metakaolin in a material results in a considerable improvement in the material's compression strength.

4. When 15% metakaolin is put into soft soil, there is a 68.12% reduction in the soil's permeability. Because of the production of calcium silicate hydrate (C-S-H) gel, hydraulic conductivity is reduced. This is because the gel makes the pores in the soil smaller and the dirt itself is more resistant. When metakaolin is mixed to the mixture of soft soil, the water-conducting ability of the soft earth is substantially diminished.
5. The experiment demonstrated that the permeability of soft soil could be reduced while at the same time its strength could be increased by stabilizing the soil with metakaolin. More research is needed to ascertain the length of time the soil maintains its stability and to identify any changes that take place when metakaolin is added in varied amounts.

References

1. Subramani T, Ramesh KS (2015) Experimental study on partial replacement of cement with fly ash and complete replacement of sand with M sand. Int J Appl Innov Eng Manage (IJAIEM) 4(5):313–322
2. Wang S et al (2021) Silty clay stabilization using metakaolin-based geopolymer binder. Front Phys 9. https://doi.org/10.3389/fphy.2021.769786
3. Al-Alaily HS, Hassan AAA (2016) Time-dependence of chloride diffusion for concrete containing metakaolin. J Build Eng 7:159–169. https://doi.org/10.1016/j.jobe.2016.06.003
4. Saha A, Pan S, Pan S (2014) Strength development characteristics of high strength concrete incorporating an Indian fly ash. Int J Sci Technol 2(5)
5. Barbhuiya S, Chow P, Memon S (2015) Microstructure, hydration and nanomechanical properties of concrete containing metakaolin. Constr Build Mater 95:696–702. https://doi.org/10.1016/j.conbuildmat.2015.07.101
6. Chavali BT, Karunakar P (2016) Effect of varying quantities of metakaolin and fly ash on strength characteristics of concrete. Int J Technol Res Eng 4(2):282–288
7. Teja Kiran CH, Prudvi K (2016) Strengthening of concrete by partial replacement of cement with flyash and metakaoline mix. Int J Adv Technol Innov Res 8(3):0459–0463
8. John N (2013) Strength properties of metakaolin admixed concrete. Int J Sci Res Publ 3(6)
9. Malagavelli V, Angadi S, Prasad JSR, Joshi S (2018) Influence of metakaolin in concrete as partial replacement of cement. Int J Civ Eng Technol (IJCIET) 9(7):105–111
10. Narmatha M, Felix Kala T (2017) Metakaolin and flyash with partial replacement of cement using in HPC. Int J Civ Struct Eng Res 4(2):101–106
11. Ahmed MD, Hamza NA (2015) Effect of metakaolin on the geotechnical properties of expansive soil. J Eng 21
12. Mermerdaş K et al (2012) Strength development of concretes incorporated with metakaolin and different types of calcined kaolins. Constr Build Mater 37:766–774. https://doi.org/10.1016/j.conbuildmat.2012.07.077
13. Mirmoghtadaei R et al (2015) The impact of surface preparation on the bond strength of repaired concrete by metakaolin containing concrete. Constr Build Mater 80:76–83. https://doi.org/10.1016/j.conbuildmat.2015.01.018
14. Mohanta NR, Samantaray S (2019) Study of combined effect of metakaolin and steel fiber on mechanical properties of concrete. Pertanika J Sci Technol 26(3):1381–1396

15. Kirange P, Misal AG, Jadhav SP, Metakari SA, Nilankar OD (2019) Effect of metakaolin and flyash in strength of concrete. IJIERT 65–66
16. Issac S, Paul A (2018) A literature review on the effect of metakaolin and fly ash on strength characteristics of concrete. IJARIIE 4(2):2395–4396
17. Rashad AM (2013) Metakaolin as cementitious material: history, scours, production and composition—a comprehensive overview. Constr Build Mater 41:303–318. https://doi.org/10.1016/j.conbuildmat.2012.12.001
18. Sakr M, Azzam W, Meguid M, Ghoneim H (2021) An experimental study on the effect of micrometakaolin on the strength and swelling characteristics of expansive soils. https://doi.org/10.21203/rs.3.rs-544440/v1
19. Abdulkareem SO, Abbas JM (2021) Effect of adding metakaolin based geopolymer to improve soft clay under different conditions. In: Second international conference on geotechnical engineering-Iraq. https://doi.org/10.1088/1755-1315/856/1/012011
20. Shi X et al (2022) Experimental study on the mechanical properties and microstructure of metakaolin-based geopolymer modified clay. Molecules 27(15). https://doi.org/10.3390/molecules27154805
21. Available from: https://www.civilengineeringforum.me/standard-proctor-compaction-test/

Laboratory Investigation of Soil Stabilization Using Terrazyme and Cement

K. Sneha Sagari, V. Arun, V. Srinivasa, K. Bindiya, G. S. Mallikarjuna, and G. H. Anusha Patil

1 Introduction

Road stabilization is the approach of improving the capability to the native soil against highly traffic condition and to reduce road damages at a different climatic condition. The technique used entails adding admixtures, compacting the soil and disinfecting it. Industrial wastes, cement, and fly ash are examples of additives [1]. A technique called soil stabilization is used to change various soil characteristics and enhance the performance of the soil for engineering purposes. Strongest and durability are advantages of cement base paved surfaces [2]. It is the greatest material for soil stabilization and is easily available. The soil and hydrated cement product combined to create the cement-stabilized base [3]. The strength of stabilized soil depends on how much cement is injected. The quantity of cement needed in soil determine based on the type of soil. Portland cement is widely used as a stabilizing material of soil because of its easy manageable and quality control properties. Cement stabilization is more suitable for clay soil.

Cement is commonly used in stabilization. The inclusion of a little cement quantity, up to 2%, modifies soil characteristics, but the addition of a significant amount of cement causes some modifications in the soil properties [4]. Except for organic soil material and clay with extreme plasticity, cement has the ability to stabilize

K. Sneha Sagari (✉) · V. Arun · V. Srinivasa · K. Bindiya · G. S. Mallikarjuna · G. H. Anusha Patil
Department of Civil Engineering, JNNCE, Shivamogga, India
e-mail: snehasagari142@gmail.com

V. Arun
e-mail: arun.v@jnnce.ac.in

V. Srinivasa
e-mail: srinivasajetty.v@gmail.com

K. Bindiya
e-mail: bindiyakn@gmail.com

S. Kolathayar et al. (eds.), *Best Practices in Geotechnical and Pavement Engineering*, Lecture Notes in Civil Engineering 449, https://doi.org/10.1007/978-981-99-8505-0_10

any sort of soil. The soil's OMC and maximum dry unit weight are changed by combining cement. The compression tensile and fracture strength of the soil–cement mixture increased as cement was added to all soil types [5]. We produce an excellent bond between soil particles and cement particles due to the cement distribution and soil's grain sizes. Cement is widely utilized in pavement construction, with inter-layer stabilization with cement being among the most common applications [5]. Soil stabilization with cement remains a well-known approach for improving pavement interlayers by the addition of cement, which is based on the formation of primary and secondary products.

1.1 Objectives of the Study

To study the behaviour of soils by using terrazyme-5x, 11x and cement on soil cement blocks by unconfined compression strength test.

- To determine the optimum dosage of cement with Terrazyme 5x, and 11x stabilizer on problematic soil.

1.2 Materials

Sandy clay soil: Clay soil was found at JNNCE campus Shivamogga. In the current study, the experimental examination aims at enhancing the engineering characteristics of natural sandy clay soil [1]. The soil samples were taken at a depth of 1–1.5 m from ground level. Table 1 gives soil properties.

Terrazyme: When terrazyme is sprayed to soil, it facilitates the decomposition of organic compounds and improves the soil particles' ability to moist and connect [6].

Table 1 Properties of soil used in the study

S. No.	Properties	Results	Remarks
1	Gravel	0.96%	IS:2720 (Part-IV) 1985
2	Sand	78.2%	
3	Silt and clay	20.84%	
4	Liquid limit	37.62%	IS:2720 (Part-V) 1985
5	Plastic limit	4.9%	
6	Plasticity index	32.73%	
7	Unified soil classification	SC	
8	Max dry density	1.840 g/cc	IS:2720 (Part-VII) 1980
9	OMC	13%	IS:2720 (Part-VII) 1983

Table 2 Properties of terrazyme [6]

S. No.	Property	Description
1	Density	Terrazyme
2	Appearance/dour	Dark brown liquid/non-obnoxious
3	Hazardous components	None
4	Boiling point	100 °C
5	Rate of evaporation	Equivalent to water
6	Saturation in water	Complete
7	Melting point	Liquid
8	Reactivity data	Stable
9	Storage	Below 74 °C
10	Shelf life	3 years proper storage conditions
11	Not flammable	Will not burn

Table 3 Properties of cement

S. No.	Property	Description
1	Grade	53
2	Type	OPC
3	Packing size	50 kg
4	Specific gravity	3.12
5	Magnesia	1.07%
6	Loss of ignition	1.72%

It makes it possible for soil components to be more easily wet and compacted. Additionally, the strengthened chemical bonds aid in bonding the soil atoms together to form a far durable a structure impervious to wear, deterioration, and water infiltration and properties of terrazyme are shown in Table 2.

Cement: In the present study, cement is used to change or improve the soil, or it can be made into a cemented mass that is more durable and strong. Portland cement is a type of hydraulic cement created by crushing a mixture of clay and limestone heated in a kiln, and the properties of cement are tabulated in Table 3.

2 Methodology

In the current study, the collected materials are cement, soil, terrazyme 5x and 11x are introduced to basic tests, which help analyse the engineering properties like California bearing ratio (CBR) and unconfined compressive strength (UCS) of soil–cement stabilization with enzymes 5x, 11x respectively by using these materials [3]. The methodology adopted in the present work is shown in Fig. 1.

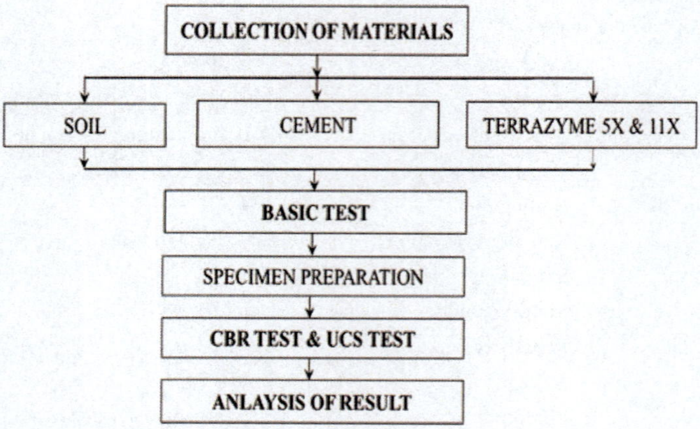

Fig. 1 Flowchart represents the methodology

For determining the terrazyme dosage by conducting CBR and UCS test

In the current work, standard test apparatus and a protocol for the California bearing ratio test are used in accordance with IS 2720 (part 16) 1979 "Laboratory Determination of CBR". In order to test for high compaction, both OMC and wet conditions are used [4]. The CBR test for two types of Terrazyme-5x is done at 200 ml for 2.5, 2.6, 2.8 and 3.0 m^3, and Terrazyme-11x is done at 200 ml for 5.2, 5.6, 6.0 and 6.6 m^3 for the soaking period of 4 days.

- The optimum dosage of Terrazyme-5x and Terrazyme-11x is found to be 200 ml for 2.8 m^3 and 6.6 m^3 respectively for both CBR and UCS test.

Specimens' preparation

Different proportions of soil, terrazyme, and cement mix samples are prepared at the corresponding MDD and OMC values [7]. The method of preparation used to obtain specimens of the standard 15 × 15 × 15 cm dimension [5, 7]. Two types of terrazyme such as 5x and 11x is to be 200 ml for 2.8 m^3 and 6.6 m^3 are respectively with the cement should be added a percentage of 3, 3.5, 4, 4.5 and 5% by the dry weight of soils are combined with water, which are then quickly compacted in moulds shown in Figs. 2 and 3.

Method of curing

After the preparation of samples, the demoulded cubes are covered by plastic and kept in curing with the help of gunny bags, the curing period adopted is 7, 14 and 28 days. Following the desired curing period, the prepared moulds are tested as shown in Figs. 4 and 5, specimens are examined for unconfined compressive strength (UCS) [7].

Fig. 2 Soil with cement

Fig. 3 Cube casting

Fig. 4 Demoulded cubes

Fig. 5 Curing of moulds

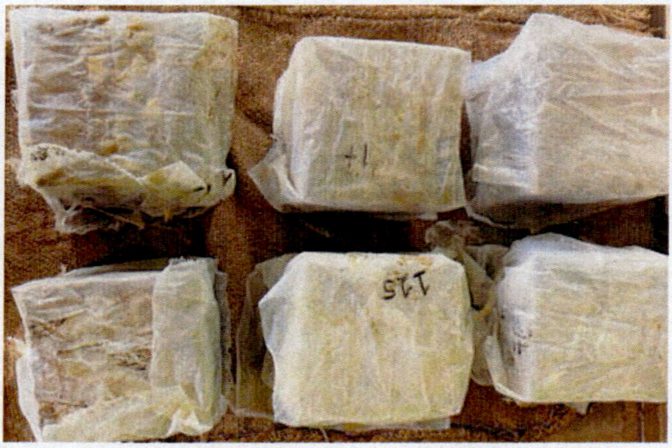

Method of testing

After the specified period of curing, the samples are tested as per procedure shown in Fig. 6. Unconfined compression test is one of the least expensive and fastest ways to gauge the compressive strength of soil [4].

$$F = \frac{P}{A} \text{ in N/mm}^2 \quad (1)$$

where F = Compressive strength of specimen (MPa)
P = Maximum load applied on the specimen (N)
A = Cross-sectional area of the specimen (mm^2).

3 Results and Discussion

California Bearing Ratio

In the current investigation, the standard test apparatus and procedure for the CBR test [IS 2720 (part 16) 1979] were applied. The test was run under wet conditions for high compaction as well as OMC conditions. The results of the soaked CBR tests for the stabilized soil samples are given in Tables 5 and 6. The CBR tests are performed using cement 3, 3.5, 4, 4.5, 5% of 200 ml for 2.8 m^3 and 3, 3.5, 4, 4.5, 5% of 200 ml for 6.6 m^3 of soil over a 4-day soaking period. Table 4 represents the CBR calculation load along with penetration.

$$\text{C.B.R.} = (\text{PT/PS}) \times 100 \text{ in } \% \quad (2)$$

where PT = Corrected test load sustained by the specimen at 2.5 mm or 5 mm penetration from the load penetration curve.

Fig. 6 Cube testing after curing periods

PS = Standard load sustained by the specimen at the corresponding penetration taken from Table 4.

From Table 5 and Fig. 7 (graph), it is observed that when terrazyme 5x is mixed with 4.5% of cement and soil have an increased CBR value of 79.02%.

Table 4 CBR calculation load along with penetration

Penetration of plunger (mm)	Standard load (kg)
2.5	1370
5.0	2055

Table 5 CBR test result for terrazyme 5x with different percentages of cement

S. No.	Cement dosage	CBR % value
1	Untreated soil	10
2	3.0% cement for soil	77.90
3	3.5% cement for soil	73.96
4	4.0% cement for soil	78.90
5	4.5% cement for soil	79.02
6	5.0% cement for soil	78.73

Table 6 CBR test result for terrazyme 11x with different percentages of cement

S. No.	Cement dosage	CBR value (%)
1	Untreated soil	10
2	3.0% cement for soil	30.12
3	3.5% cement for soil	28.51
4	4.0% cement for soil	29.42
5	4.5% cement for soil	29.75
6	5.0% cement for soil	30.25

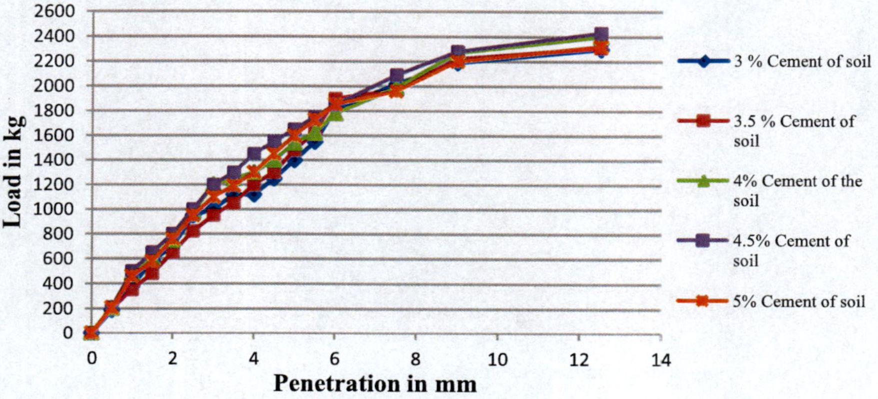

Fig. 7 Comparison of CBR results for different percentages of cement with terrazyme 5x

From Table 6 and Fig. 8, it is observed that when terrazyme 11x is mixed with 5.0% of cement and soil have an increased CBR value of 30.21%.

Unconfined Compressive Strength

In the current experiment, the standard test apparatus and UCS test technique (IS 2720 (Part 10): 1991) were applied. The UCS tests are performed using cement is 3, 3.5, 4, 4.5, 5% of 200 ml for 2.8 m^3 and 3, 3.5, 4.5, 5% of 200 ml for 6.6 m^3 of soil for the curing period of 7, 14 and 28 days are tabulated in Tables 7 and 8 represent in N/mm^2 [8].

Figure 9 represents the compressive strength of the soil–cement stabilized blocks made by using the different percentages of cement 3–5% by the dry weight of the soil. At the end of 7, 14, 28 days of curing of these the optimum dosage of 4.5% along with cement gave maximum strength for terrazyme 5x chemical when it is compared with different dosage of cement.

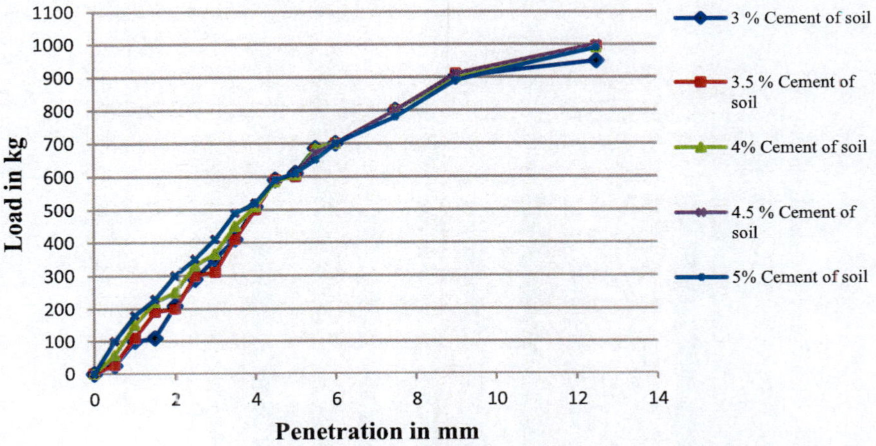

Fig. 8 Comparison of CBR results for different percentages of cement with terrazyme 11x

Table 7 UCS test result for different dosage of cement with terrazyme 5x

S. No.	% of cement dosage	UCS value (N/mm^2)		
		7 days	14 days	28 days
1	3.0% cement for soil	1.93	2.31	2.11
2	3.5% cement for soil	1.74	2.16	2.36
3	4.0% cement for soil	1.86	2.24	2.90
4	4.5% cement for soil	2.20	3.11	3.63
5	5.0% cement for soil	1.89	2.87	2.93

Table 8 UCS test result for different dosage of cement with terrazyme 11x

S. No.	% of cement dosage	UCS value (N/mm^2)		
		7 days	14 days	28 days
1	3.0% cement for soil	1.29	1.53	1.80
2	3.5% cement for soil	1.81	1.92	2.08
3	4.0% cement for soil	1.44	1.94	2.06
4	4.5% cement for soil	1.54	1.98	2.10
5	5.0% cement for soil	2.47	2.76	2.98

Figure 10 represents the compressive strength of the soil–cement stabilized blocks made by using the different percentages of cement 3–5% by the dry weight of the soil. At the end of 7, 14, 28 days of curing of these the optimum dosage of 5% along with cement gave maximum strength for terrazyme 11x chemical when it is compared with different dosage of cement.

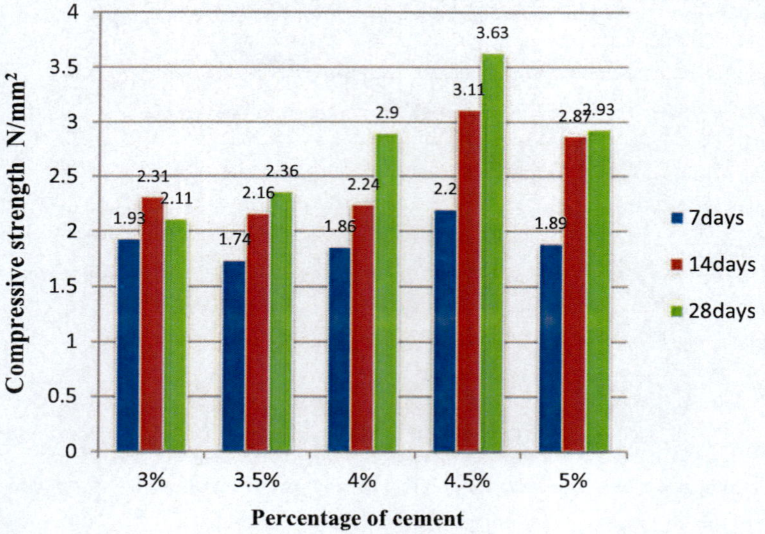

Fig. 9 Bar chart represents the difference of UCS for different percentages of cement (Terrazyme 5X)

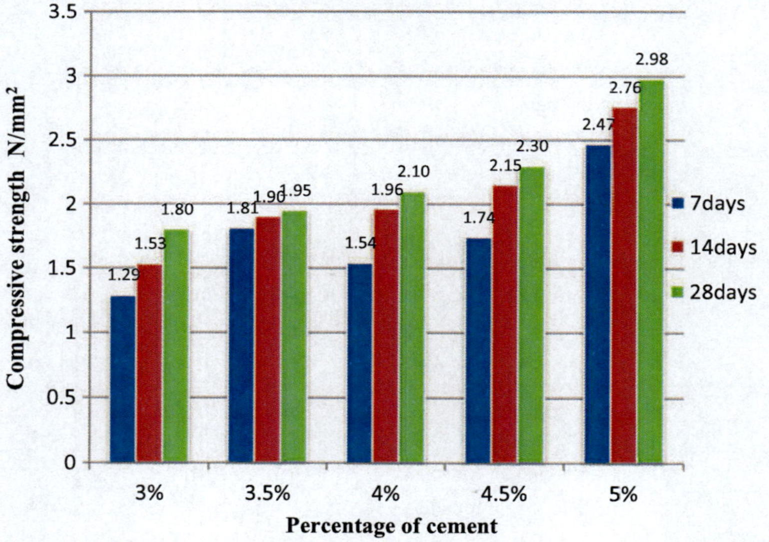

Fig. 10 Bar chart represents the difference of UCS for different percentages of cement (Terrazyme 11X)

4 Conclusions

The goal of this research is to enhance the geotechnical qualities of problematic soil. The soil is mixed with water, two types of terrazyme chemical such as 5x, 11x and cement, and curing is done for 7, 14, and 28 days to conduct the strength tests. The following conclusions are drawn from this experimental study.

- When soil is mixed with terrazyme 5x and then the optimum dosage of 4.5% along with cement, the CBR value is increased to 79.02%, and UCS value is 3.63 MPa for 28 days.
- When soil is mixed with terrazyme 11x and then the optimum dosage of 5% along with cement, the CBR value is increased to 30.21%, and UCS value is 2.98 MPa for 28 days.
- Even small additions of terrazyme and Portland cement can greatly increase the soil's tensile strength.
- The unconfined compressive strength of soil shows good results with increasing in curing periods when treated with terrazyme 5x, 11x and cement.

Scope

The following elements might be researched further:

- Other bio-enzymes and different dosages of cement and their effect on soil can be studied.
- To study the terrazyme behaviour on various types of soil.
- The terrazyme-treated soil can be used for other pavement layers that can be also studied.

References

1. Ramaji AE (2012) A review on the soil stabilization using low-cost methods. J Appl Sci Res 8(4):193–2196. https://www.researchgate.net/publication/232322828
2. Afrin H (2017) A review on different types soil stabilization techniques. Int J Transp Eng Technol 3(2):19–24. https://doi.org/10.11648/j.ijtet.20170302.12
3. Zumrawi MME (2015) Stabilization of pavement subgrade by using fly ash activated by cement. Am J Civ Eng Archit 3:218–22. https://doi.org/10.12691/ajcea-3-6-5
4. Pourakbar S, Asadi A, Huat BBK, Fasihnikoutalab MH (2015) Stabilization of clayey soil using ultrafine palm oil fuel ash (POFA) and cement. Transp Geotech 24–35. https://doi.org/10.1016/j.trgeo.2015.01.002
5. Pandey A, Rabbani A (2017) Soil stabilization using cement. Int J Civ Eng Technol (IJCIET) 8:316–322. http://www.iaeme.com/IJCIET/index.asp
6. Solihu H (2020) Cement soil stabilization as an improvement technique for rail track subgrade, and highway subbase and base courses: a review. J Civ Environ Eng 10:3. https://doi.org/10.37421/jcce.2020.10.344
7. Ghadir P, Ranjbar N (2018) Clayey soil stabilization using geopolymer and Portland cement. Constr Build Mater 361–371. https://doi.org/10.1016/j.conbuildmat.2018.07.207

8. IS 2720-PART 16. Method of test for soil: determination of California bearing ratio value. Bureau of Indian Standards, New Delhi
9. Roy A (2014) Soil stabilization using rice husk ash and cement. Int J Civ Eng Res 5:49–54
10. Shooshpasha I, Shirvani RA (2015) Effect of cement stabilization on geotechnical properties of sandy soils. Geomech Eng 8:17–31. https://doi.org/10.12989/gae.2015.8.1.017
11. Adeyanju EA, Okeke CA (2019) Clay soil stabilization using cement kiln dust. IOP Conf Ser Mater Sci Eng. https://doi.org/10.1088/1757-899X/640/1/012080
12. ASTM D 2487-17. Standard practice for classification of soils for engineering purposes (unified soil classification system). ASTM International, West Conshohocken

A Framework for Geotechnical Engineering Practice and Education

Altaf Usmani, Atul Nanda, and S.K.Jain

1 Introduction

In the present scenario of rapid increase in growth of infrastructure sector and subsequent demand of skilled professionals, it is relevant to analyze what geotechnical knowledge is required in practice for a geotechnical engineer and what is the best way to educate and train engineers so that they can successfully endure in this field. This becomes even more urgent in view of the requirements of the need to develop massive infrastructure on one hand and also the need to address the climate crisis. Unfortunately, there are no clear-cut answers to this question, as it is not clear what constitutes geotechnical knowledge explicitly. Without a clear understanding of this, it will not be possible to establish the best way to acquire, store and transmit the geotechnical knowledge required for practice. Further with major developments in technology expected in the near future along with the requirements of sustainability, a clear understanding of the geotechnical knowledge required in both education and practice is even more relevant.

In recent years there have been considerable interests in the education and training of geotechnical engineers. A detailed discussion on the education and training of geotechnical engineers is available in journal article [1]. There has also been considerable interest in future developments in geotechnical engineering which have been covered in detail by [2]. Further a detailed review of sustainability in geotechnical engineering is presented by [3]. Engineering is a profession and its practice requires specific knowledge which is mainly acquired on the job. However, for successful practice a good and relevant education is required which also constitutes knowhow

A. Usmani (✉) · A. Nanda
EIL, New Delhi, India
e-mail: altaf.usmani@eil.co.in

S.K.Jain
J.P. University, Uttar Pradesh, Anupshahr, India

S. Kolathayar et al. (eds.), *Best Practices in Geotechnical and Pavement Engineering*, Lecture Notes in Civil Engineering 449, https://doi.org/10.1007/978-981-99-8505-0_11

of advancing technological changes and requirements of sustainability. But unfortunately, at present a separation exist between approaches adopted by practitioners and academicians.

This work presents a collaborative approach of nurturing geotechnical engineers to bridge the gap between the practice and academician. The suggested outline is a culmination of experience of several decades in practice and education and identifies the key aspects of geotechnical knowledge required for grounding of future geotechnical engineers. In this paper a different approach is deliberated towards both education and training of geotechnical engineers for their multidisciplinary development. Starting from main activities of geotechnical engineers; i.e., from the exploration phase, analysis and design, management aspects and finally construction, some key sources of geotechnical knowledge and their application in engineering practice is addressed. These include the engineering sciences, models, codes of practice, uncertainty, heuristics and engineering judgement. Salient features of these are discussed with a focus on engineering practice. Further it also deliberates some broad trends which are expected to impact the profession in the coming decades such as development and incorporation of new technology, execution of complex projects requiring multi-disciplinary teams and sustainable design. Study concludes with some key requirements in the education and training of geotechnical engineers including topics in geotechnical masters' programs and innovative teaching methods.

2 Main Activities of Geotechnical Engineers

The main activities of geotechnical engineers can be split into four broad categories.

2.1 Exploration, Testing and Interpretation

It may be noted that the ground is not a manufactured product and is the result of historical geological processes which are only very approximately known. Exploration is a tool for those who really know how to use it. The objectives of the exploration plan along with the expected outcomes should be clearly defined and understood before executing any investigating plan. It requires an understanding of geology along with investigation procedures and in many cases the engineers need to work closely with geologists to focus those areas which can be potentially problematic in future. The traditions and practices of engineers and geologists are very different and can in many cases lead to problems due to a mutual lack of understanding.

2.2 Analysis and Design

Geotechnical analysis and design are based on both strength of materials approach from the engineering sciences of soil and rock mechanics along with empirical approaches. Further various rules of thumb (heuristics) and codes of practice are used along with experience from previously similar successful approaches. In addition, the design is adapted to suit the available construction equipment's and technologies. In most cases a simplified ground engineering model derived from the exploration phase along with various ground parameters from empirical correlations are used to develop a model which is used in the analysis and design. This involves considerable simplifications as most geotechnical models are a very approximate interpretation of reality. Thus, the initial uncertainties of the ground engineering model are further compounded in view of considerable uncertainty in the material properties and loadings. These epistemic uncertainties are indirectly linked to project time cycle and cost estimates. These along with the inherent uncertainty result in changes during construction, which can have contractual implications. In geotechnical design, it is important to understand that not only magnitude of uncertainties changes during the project, but there may also be different objectives to satisfy at different times, which require an adaptive and flexible approach.

2.3 Management

This includes planning, contracting, costing, engineering and project management and these activities are common to all civil engineering works. However, because of the variability and uncertainty in all geotechnical works, specific changes are required in contract and management of these works. For example, today geotechnical risk management is used in many large geotechnical projects [4]. These activities along with legal and ethical issues will not be discussed further, but is should be noted that most engineers spend a major portion of their time in these activities. Another very important factor for successful execution of projects lies in the need to interact with specialists from related field such as geology, geophysics, hydrogeology, structures, etc. as projects become increasing complex requiring multi-disciplinary teams working closely with other.

2.4 Construction and Monitoring

Engineers have to supervise and manage the construction, so as to ensure that the design is implemented as intended. This requires knowledge of construction equipment's and technologies. They also have to ensure that all necessary testing and quality control requirements are met. Geotechnical projects frequently encounter unexpected geological conditions or unexpected behaviour of the system under execution, which may require design modifications with consequent time and cost implications. This is common in geotechnical works due to inadequate investigations, poor interpretation of data, lack of experience, poor judgement and the inherent uncertainty always present in these works. Aside from specific construction requirements and quality control, many geotechnical works require monitoring to validate and update the design during construction. The application of the observational method is one such example [5]. Geotechnical monitoring has recently seen as an important part of any underground project and engineers need to understand its significance as a tool of design validation during construction.

3 Multi-disciplinary Approach in Geotechnical Engineering: Underground Storage Caverns

Owing to increasing complexity of problems, when dealing with Mother Nature, the present conceptions in geotechnical engineering problems requires more accurate analyses, simulating filed conditions and construction sequences as closely as possible. Further different interfaces involved in any geotechnical engineering project; structure, soil, rock, liners etc. along with availability of construction equipment's, timelines etc. calls for multidisciplinary involvement of specialists to work out most optimum solutions. Thus, a multidisciplinary support of specialists is necessary and these interference between fields of activity have to be known by the young geotechnical engineers as a tool for better understanding of the problem. This section briefly presents an example from a case study of unlined underground rock caverns constructed in southern coast of India for storage of crude oil using the principle of hydraulic containment. The projects included construction of several large unlined rock caverns of 30 m $\times$ 20 m cross section and a kilometer length along with associated access and water curtain tunnels and shafts [6]. Successful execution of such mega engineering project depends on many activities that need to be considered over the project time period. These can be broadly summarized in a typical flowchart form as shown for a typical underground cavern project in Fig. 1. This flow chart highlights different stages of project along with main associated activities. Flowchart presentation list out main aspects that need to be considered by an engineer for proper management and successful execution of the project, as it progresses from its conceptualization to completion stage. Each stage of the project requires multidisciplinary teams working together. Investigations include geophysical, geological, hydro-geological, geotechnical and rock testing and studies. Based on the results of the investigations a geological, hydro-geological and geotechnical model is prepared. Based on these models a basic design and layout is prepared in collaboration with process and construction engineers. The final design execution involves not only the requirement of geo-engineers but also structural, mechanical, process, planning and other engineers working seamlessly together for completion of project. During the

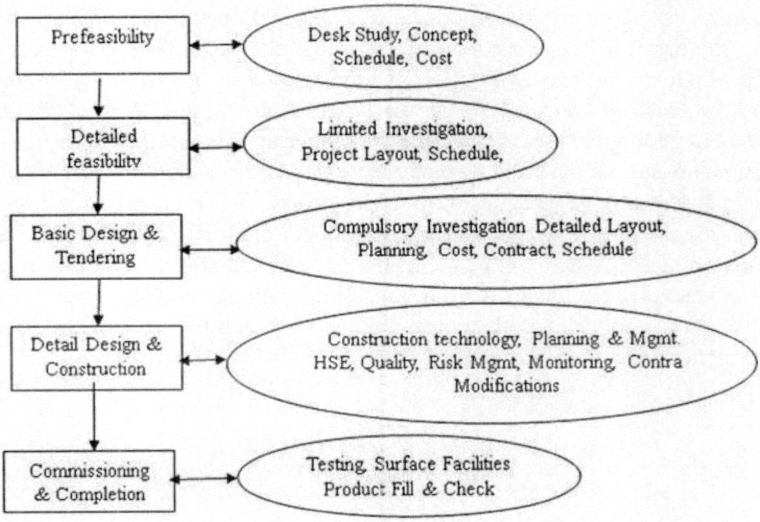

Fig. 1 Flow chart of a typical underground cavern project

construction stage the observational approach is adopted to update the engineering design in tune with construction requirements along with available resources and time schedules. This involves close interaction between the design and construction teams for successful execution of project.

4 Key Aspects of Geotechnical Knowledge

In this section some key sources of geotechnical knowledge and their application in engineering practice is discussed. Some of these are common to all civil engineering, while some are unique to geotechnical engineering. Each of the items discussed below constitute a vast subject on their own, hence only salient features will be discussed here.

4.1 Engineering Sciences

As every practicing engineer knows, engineering is not applied science. In fact, science plays a small part in the practice of engineering. While there is no clear definition of engineering, broadly engineering is about building things. This involves development of the concept of the required system, analysis and design, detailing and specification, management, planning, costing, contracting, construction and finally operation. The engineering sciences of soil and rock mechanics along with engineering geology form the basis of our engineering knowledge [7, 8]. However, unlike science which is concerned with understanding nature, the geotechnical engineering sciences are mainly concerned with the analysis and design and construction of structures in geological media. While the engineering sciences give us a framework to understand and analyze the ground, these constitute a fraction of geotechnical knowledge. Knowledge of codes of practice, heuristics and engineering judgement along with adequate experience in investigations, interpretation, analysis, design, construction and management of geotechnical works is also required. Researchers [9] advocated that challenge for the education of geotechnical engineering is to not only teach a good theoretical base, but also to teach the students to apply risk management during the entire design process.

4.2 Models

Typically, geotechnical engineering works require two types of models. The first is the ground engineering model which is derived from the investigations and testing. This typically includes the topography, strata, the range of the main ground parameters from field and laboratory testing, ground water and other relevant parameters. The second model is the geotechnical or analysis model. It is typically a simplified version of the first model and in addition to the observed and measured parameters includes parameters obtained from correlations. Average values of key analysis parameters such as strength and stiffness are estimated for each layer or zone. In some cases, advanced constitutive (ground behaviour) models may be used and these may require additional testing to obtain the parameters.

4.3 Codes of Practice

Today various codes of practice exist for practically every aspect of geotechnical engineering. Bureau of Indian Standards (BIS) has codes of practice covering subsurface investigation, testing, materials, design and construction for a wide range of geotechnical works. These codes are based on the engineering sciences and on the knowledge obtained over the years in the various geotechnical works. In additions to the codes of practice, there exist several guidelines for various types of geotechnical works published by engineering societies, academic and research institutions and governmental and private parties. Organization's such as Indian Road Congress (IRC) has standards and guidelines for highways; Indian Geotechnical Society (IGS) publishes documents on soil mechanics while International Society of Rock Mechanics (ISRM) has practice guidelines for testing of rocks, While codes of practice incorporate a substantial knowledge of the profession in a form easily usable by engineers, these on their own are not sufficient. Due to constant new developments, most codes tend to lag behind the latest technologies available. Further in some cases the codes can be unnecessary restrictive and engineers need a thorough understanding of the subject so that they can navigate between the various requirements.

4.4 Heuristics and Judgement

A heuristic is anything that provides a plausible aid or direction in a solution of a problem but is in the final analysis unjustified, incapable of justification and potentially fallible [10]. These include rules of thumb, factors of safety and performance parameters, design basis or procedures, etc. which have been successfully used in the past. The design basis or procedures could be thought of as Meta rules of thumb which have been successfully used in the past. A substantial part of the engineering knowledge is in the form of heuristics as engineering is essentially an empirical body of knowledge. Most of these heuristics are learnt on the job [11]. In addition, engineers use judgement at every stage of the work from concept to commissioning to achieve the requirements. There is no clear definition of engineering judgement, but every experienced engineer is familiar with it and uses judgement in his work. Engineering judgement is not arbitrary or procedural but is a capacity to recognize and quickly and competently address complex situations [12]. It may be noted here that engineering education provides rules and young engineers use and follow these rules. However, with experience, engineers develop the capability to quickly home in on the key issues and address the problems and provide workable solutions. Engineering judgement is developed on the job and over a period of several years. Researchers agree that tactic knowledge residing in expert engineer head is a very 'messy' and esoteric concept. Therefore, capturing it is a task fraught with difficulties. But if captured and put into explicit form, tacit knowledge is a driving force behind any sort of innovation, be it new technology, new process or a and new technique. The role of heuristics and judgement is even more important in geotechnical engineering, as most codes and guidelines give considerable flexibility and freedom in selection of design parameters and design methods.

4.5 Uncertainty

While uncertainty is prevalent in all engineering, it is the degree and the extent in geotechnical engineering that is unique. In general, it is understood that uncertainty permeates and defines geotechnical engineering and as can be seen from the above discussion, almost every geotechnical activity involves considerable uncertainty and handling this is one of the major concerns of geotechnical engineers [13]. The uncertainties can be separated into two broad categories: Aleatory—related to chance and epistemic—related to our knowledge. The first typically involves the randomness such as in loading and material properties and the second involves items such as our lack of knowledge of the ground due to limited investigations, material behaviour, models and analysis. Epistemic uncertainly can be reduced by gaining additional information about the geological conditions of a site and can help in achieving a more controlled project execution.

5 Future Trends

While it is difficult to predict the future, by looking at the past some general observations can be presented. Looking at the past half century, it can be said that geotechnical engineering profession has been impacted by technologies developed outside it, the most obvious being computing and new materials. The profession has also risen to the challenges thrust on it, an example being the geo-environmental problems. Similar trends in the future are expected with growing demand. While it is not possible to predict specific items, the following are expected; Complex projects with very tight technical and non-technical requirements, large scale problems due to global environmental factors and new technologies. The first will pose major challenges to the profession and will require the adaption of existing approaches along with the development of new technical and management methodologies. Execution of complex projects will require multi-disciplinary teams working closely together to successfully complete the works meeting the technical, financial and time criteria. Future infrastructure will need to simultaneously address issues of environment, economics, equity and engineering. The requirements of sustainable design and engineering will be an essential part of every project. Underground space utilization is an emerging area having vast potential in different fields including creating storages for hydrocarbons, waste and energy management and is expected to be developed as a major engineering domain in near future. The new technologies will mostly arise outside the profession and will have to be adapted to suit the problems at hand. These could include new materials, ground improvement, investigation technology, modelling and computing, sensors, large data and AI etc. Further, as technology continuously evolves, engineers will need to update their knowledge throughout their professional career. It is clear that the conventional approach of learning theory in school and practice at job will no longer work. Young engineers today are expected to perform on the job at the earliest and experienced engineers are expected to continuously update their knowledge quotient.

6 Education and Training

Based on the above it can be observed that there are wide variety of sources of geotechnical knowledge and considerable experience and judgement is required in its application. In consideration of above discussion, there is a critical need to educate young engineers and train existing engineers so that they can rise to the challenges which they will face in the coming years. In our view the academic institutions do a good job in teaching the engineering sciences and analysis techniques. However, it could significantly improve the education of young engineers by incorporating some of the items discussed here in the geotechnical engineering program. The details are left to the academic institutions, however in our view a better understanding of the empirical nature of engineering, interpretation of exploration data, assumptions and basis of modelling, codes and guidelines, probability, construction technology and inclusion of field visits and case studies in the geotechnical engineering program would definitely benefit the young engineers. Greater academia-industry interaction is also necessary to expose the students to geotechnical practice. The minimum educational requirement for geotechnical engineer is a two years master degree. A four-year undergraduate program provides at best a very limited introduction to geotechnical engineering which is inadequate for engineering practice. The two-year program should include at least six courses per semester for three semesters to cover specialized areas of geotechnical engineering domain. However, the exact number and organization of the courses should be best judged by academic institutions. The course work in the master's program should endeavor to introduce the students to all the major areas of geotechnical engineering and specialization should be avoided at this stage. Finally, the last semester should focus on a project of practical significance. Ideally this should cover interpretation of data, analysis, design and constructability aspects. Pure research projects should be avoided as the objective of master's program should be to produce engineers who can comfortably enter engineering practice. For courses involving design and investigations, incorporation of the case study methods [14] and project-based learning [15] will be better suited to teach the items discussed in the previous sections. Case histories have always been used by both practicing engineers and academics in geotechnical engineering, however a systematic incorporation of these in the course work would be beneficial to the students along with incorporation of mini projects in design courses, where teams of students work together to solve problems or create designs. These approaches will foster teamwork, leadership and encourage independent and creative thinking in the students. The use of case studies should also be encouraged along with set of learning outcomes to make students understand theory in its applied context, thus motivating students learning skills [16]. There is also a role for the engineering societies in the training of engineers. These societies do a good job by organizing conferences in their areas of specialization. These help engineers to keep up to date with the developments in their fields. More workshops and training programs will be required and societies such as ISSMGE, IGS, ISRM, ISEG and others increase their coordination and collaboration with each other so that their members can develop multi-disciplinary skills in related areas which will be required to handle the new technologies and new challenges. These platforms also help to shape young minds to challenges of future and nurture them into a more comprehensive geotechnical engineer.

7 Conclusions

In this paper a basic framework was presented for engineering practice, education and training requirements of geotechnical engineers.

Some of the broad conclusions of the study are:

1. Infrastructure work requires multi-disciplinary teams for execution of different projects at site within the available resources and schedules and geotechnical engineers need to be trained to work with cross disciplinary issues.
2. Substantial part of geotechnical engineering knowledge is in the form of heuristics and these are mostly leaned on the job.
3. Future presents both challenges and opportunity, which needs to be looked from the lens of sustainability. Underground storages for sustainable energy management, disposal of hazardous waste, climate change mitigation are some of the emerging areas in geotechnical engineering which will require use of geotechnical engineering beyond its contemporary domain.
4. Geotechnical engineering education should be conscripted on a broad spectrum of case base studies and project-based learning on the paradigm of technological progress. This would significantly improve the education of young engineers and prepare them for the future challenges.
5. Continuous academia-industry interaction involving industry experts in the education of students as well as exposing teachers to practice and training of practicing engineers to update their knowledge in the latest developments is needed of the hour.

Acknowledgements The authors would like to thank Engineers India Limited for their support and encouragement.

References

1. Jaksa MB, Ho KKS, Woodward MA (2009) Management, training and education in geotechnical engineering. In: Proceedings of 17th international conference on soil mechanics and geotechnical engineering, Alexandria, Egypt, 5–9 Oct 2009
2. Simpson B, Tatsuoka F (2008) The next 60 years. Geotechnique 58(5):357–368
3. Basuter D, Misra A, Puppala A (2015) Sustainability and geotechnical engineering: perspectives and review. Can Geotech J 52(1):96–113. https://doi.org/10.1139/cgj-2013-0120
4. International tunnelling Insurance Group (2006) A code of practice for risk management of tunnel works
5. Peck RB (1969) Advantages and limitations of the observational method in applied soil mechanics. Geotechnique 19 (2)
6. Nanda A, Rath R, Usmani A (2015) Underground storage technologies. Engineers India Ltd., New Delhi. ISBN No. 978-93-5254-383-0
7. Terzaghi K (1943) Theoretical soil mechanics. Wiley, New York, USA
8. Goodman RE (1989) Introduction to rock mechanics. Wiley, Chichester

9. Van Tol AF, Korff M, van Staveren MTh (2009). The education of geotechnical engineers should incorporate risk management. In: Proceedings of 17th international conference on soil mechanics and geotechnical engineering, pp 2741–2744
10. Koen BV (2003) Discussion of the method, conducting the engineers approach to problem solving. Oxford University Press, New York
11. Gainsburg J, Rodriguez-Lluesma C, Bailey DE (2010) A knowledge profile of anengineering occupation: temporal patterns in the use of engineering knowledge. Eng Stud 2(3):197–219
12. Bulleit W, Schmidt J, Alvi I, Nelson E, Rodriguex-Niki T (2015) Philosophy of engineering: what it is and why it matters. ASCE, J Profess Issues Eng Educ Practice
13. Christian JT (2004) Geotechnical engineering reliability: how well do we know what we are doing. J Geotech Eng. ASCE, vol 130
14. Davis C, Yadav A (2014) Case studies in engineering. Cambridge handbook of engineering education research
15. Uzaik J (2016) A project-based learning approach in an engineering curriculum. Global J Eng Educ 18(2)
16. Orr TLL, Pantazidou M (2012) Use of case studies in geotechnical courses: learning outcomes and suitable cases. In: Proceedings of international conference on shaking the foundations of geo-engineering education, Galway, Ireland, 4–6 July 2012, pp 105–110

Forsterite-Treated Silt as a Liner Material

K. A. Deepa, S. Gangadhara, and Chandrashekar S. Patil

1 Introduction

Siltation is a significant issue that affects the health of lakes and can cause a range of problems, such as reduced water quality, decreased water storage capacity, and loss of aquatic habitats. Restoring and rejuvenating lakes often involve implementing anti-siltation measures to reduce the amount of sediment and other debris that accumulates in the lake over time. Dredging is a commonly adopted anti-siltation measure that can help to restore and rejuvenate lakes, improving their water quality and ecological health. Disposing of this dredged silt from lake beds has been challenging for local governments and organizations involved in lake restoration projects. Reusing this dredged silt in other applications is one potential solution for this problem. This study explores the suitability of this dredged silt as a landfill liner material.

Large quantities of municipal solid waste (MSW) are generated in cities, which involves a series of steps to separate, treat, and dispose of the residual waste which cannot be processed by any other measures safely and effectively. The residual waste is usually transported to a landfill for final disposal. Landfill liners prevent the leaching of waste into the soil and groundwater. Materials to be used in the construction of landfill liners must have low hydraulic conductivity and adequate strength [3].

Considerable research has been done in the past to produce commercially feasible liner materials. The possible use of modified locally available soil as a liner material is widely explored [7, 18, 21, 23]. Researchers have utilized bentonite in combination with a secondary additive to modify the properties of local soils to facilitate its use as a liner material. Bentonite is commonly added to reduce the hydraulic conductivity of local soil. The addition of bentonite may adversely affect the plasticity and strength of the soil, and hence, stabilization of the soil-bentonite mixture using suitable additives

K. A. Deepa (✉) · S. Gangadhara · C. S. Patil
University Visvesvaraya College of Engineering, Bangalore University, Bengaluru, India
e-mail: karuthedathdeepa@gmail.com

S. Kolathayar et al. (eds.), *Best Practices in Geotechnical and Pavement Engineering*, Lecture Notes in Civil Engineering 449, https://doi.org/10.1007/978-981-99-8505-0_12

like lime, cement, etc., was attempted [15, 17, 20, 25]. However, additives such as lime and cement are frequently criticized for the carbon dioxide emissions associated with their production. Recent researchers have attempted to use biopolymers such as polysaccharides, xanthan gum, and guar gum [2, 16], by-products such as fly ash [1, 10, 24], rice husk ash [19], palm oil fuel ash [4], ground granulated blast furnace slag [5] to modify soil properties indicating the significance of using sustainable and environmentally friendly materials for soil stabilization.

The possibility of utilization of dredged silt from lake beds as a landfill liner material is investigated in this study. Silt samples collected from two different lake beds were used in this study. According to U.S. EPA guidelines, soil liners for an inert landfill must have hydraulic conductivity less than or equal to 1×10^{-7} cm/s and an unconfined compressive strength value greater than 200 kPa [26]. Forsterite or magnesium-rich olivine is used as an additive to modify the properties of these silt samples so that they may be used as a liner material.

Olivine is a sustainable, economical, and naturally occurring nesosilicate mineral available abundantly in the earth's subsurface. Forsterite or magnesium-rich olivine may be effective in stabilizing soil due to high amounts of magnesium oxide. Magnesium oxide has been successfully used in the past by various researchers to improve soil strength. Previous studies have used magnesium oxide for soil stabilization [9, 14, 27]. It has significant environmental benefits when compared to commonly used additives such as lime and cement. Due to its ability to adsorb atmospheric carbon dioxide, olivine is widely used for carbon dioxide sequestration. Additionally, the financial and environmental costs of processing olivine are reported to be far lower than those of cement and lime [22].

Studies conducted on marine clay demonstrated that olivine undergoes hydration reaction, leading to the development of Magnesium Silicate Hydrate (M–S–H) and Magnesium Aluminate Hydrate (M–A–H) gel phases. These gel phases contribute significantly to enhancing engineering properties of the stabilized soil [8]. According to existing literature, olivine rich in magnesium exhibits the potential to improve the hydro-mechanical properties of marine clay. Apart from formation of hydration products, this improvement is attributed to its affinity for carbon dioxide adsorption, resulting in the formation of stable carbonates [6]. Although these studies primarily focused on marine clay samples indicate that forsterite could serve as a sustainable alternative for enhancing the engineering properties of soil, further investigation is required to ascertain the effect of olivine inclusion on the properties of the dredged silt from lake beds. To evaluate the viability of forsterite-treated silt as a liner material, the compaction characteristics, hydraulic conductivity, and unconfined compressive strength of the treated samples were examined.

2 Materials and Methods

2.1 Materials Used

Silt

Silt samples collected from lake beds in two different locations (Lingasaguru, Raichur District, Karnataka, and Ulsoor, Bangalore, Karnataka) were used in this study. The silt samples collected from Lingasaguru, Raichur district, and Ulsoor, Bangalore, are designated as S1 and S2, respectively, in order to facilitate the presentation of results. The specific gravity values for S1 and S2 are 2.51 and 2.25, respectively. Table 1 shows the significant geotechnical properties of silt samples in comparison with standard values required for liner materials for an inert landfill as specified by the U.S EPA [26]. It can be seen that the permeability values for both the silt samples need to be modified in order to satisfy the minimum required criteria for landfill liner material. Although the UCS value of S2 was found to be greater than 200 kPa, the strength of S1 needs to be improved.

Forsterite

Magnesium-rich olivine or forsterite used in the current study was supplied by M/S Navbhan Exporters, Bangalore. Dunite rocks rich in forsterite obtained from mines in Salem, Tamil Nadu, were processed and powdered into finer particles. The particle size of olivine significantly influences the reaction rate and hence the improvement of soil properties [8]. Finer olivine sizes were reported to be more effective, and hence, olivine samples passing 75 μm IS sieve were used for testing.

Olivine used in the current study is non-plastic, containing majorly silt-sized particles. The specific gravity of olivine used in this study was found to be 2.75. Table 2 presents the major oxides in olivine. Standard methods such as titration methods as outlined in the IS code were employed to determine the major oxides present in olivine sample used in this study. It is observed from Table 2 that olivine used in the current study has significant quantities of magnesium oxide (42%). The presence of magnesium oxide is expected to contribute for the improved performance of forsterite-treated silts.

Table 1 Properties of the silt samples in comparison with the required range for a liner material

Property	Required range [26]	S1	S2
% fines	> 20	82.5	77.4
Hydraulic conductivity (cm/s)	$\leq 10^{-7}$	1.5×10^{-6}	2.27×10^{-6}
Unconfined compressive strength (kPa)	> 200	154	280

Table 2 Major oxides in olivine

Major oxides	Concentration (wt %)
SiO_2	37.38
TiO_2	< 0.10
Al_2O_3	0.4
Fe_2O_3	8.87
CaO	1.55
MgO	41.81

2.2 *Methods of Testing*

The effect of olivine inclusion on the compaction characteristics, unconfined compressive strength, and hydraulic conductivity of the treated samples were studied. Oven-dried samples of silt (S1 and S2) and olivine were used for testing. Olivine dosages were chosen, and forsterite-treated silt sample mixtures were prepared [8]. In order to obtain forsterite-treated silt sample mixtures, samples S1 and S2 were mixed with various olivine dosages (0, 10, 20, 30, 40%) by dry weight of the sample required for various tests. A 10% forsterite-treated silt sample mixture consists of 10% olivine and 90% silt by dry weight. The forsterite-treated silt samples are designated in a specific way in this paper in order to facilitate presentation of results. For example, a 20% forsterite-treated S1 silt sample is represented as "S1 + 20% olivine", and 10% forsterite-treated S2 silt sample is represented as "S2 + 10% olivine".

Compaction Test

The compaction characteristics of silt samples (S1 and S2) and forsterite-treated silt samples were determined using the standard proctor test following procedures outlined in IS-2720 (Part-7) 1980 (Reaffirmed in 2021).

Unconfined Compressive Strength Test

Unconfined compressive strength (UCS) tests were conducted on silt samples (S1 and S2) and forsterite-treated silt samples in accordance with IS 2720 (Part 10) 1991 (Reaffirmed in 2020). The laboratory test setup is represented in Fig. 1. Specimens of 76 mm height and 38 mm diameter were prepared at maximum dry density and optimum moisture content. These specimens were tested up to failure or until 20% strain at a constant strain rate of 1.25 mm/min. Tests were repeated on three identical specimens. The stress–strain curve was plotted, and peak stress was obtained. The unconfined compressive strength reported in this paper is the average of three values obtained from identical specimens.

Fig. 1 UCS test setup

One-Dimensional Consolidation Test

One-dimensional consolidation tests were performed on silt samples (S1 and S2) and forsterite-treated silt samples in accordance with IS: 2720 (Part 15) 1986 (Reaffirmed in 2016). The laboratory test setup is represented in Fig. 2. The samples were compacted at their respective OMC and MDD in the consolidation ring up to a height of 18 mm. The system was connected to a water reservoir and allowed to swell freely under an initial seating pressure of 6.25 kPa until no further change in readings or an equilibrium state was attained. Once there were no further changes in the dial gauge readings, the samples were loaded in stages (pressure increment ratio of 1). Each imposed load was maintained until no change in dial gauge reading was recorded. Similarly, the samples were unloaded in stages such that 1/4 of the load was retained at each stage until no change in dial gauge reading was recorded.

Fig. 2 Consolidation test setup

The coefficient of consolidation (C_v) values was calculated for each test using Taylor's square root time method. In this study, the hydraulic conductivity of silt samples as well as the forsterite-treated samples was determined indirectly from the consolidation test results. The coefficient of permeability (k) was calculated using the following equation:

$$k = c_v \times m_v \times \gamma_w, \tag{1}$$

where k is the coefficient of permeability, c_v is the coefficient of consolidation, m_v is the coefficient of volume change, and γ_w is the unit weight of water.

The one-dimensional consolidation tests were conducted as a part of a research program to study the effect of olivine inclusion on the consolidation characteristics of the silt samples S1 and S2. While the falling head permeability test provides a direct measurement of soil permeability, the consolidation test offers additional insights into soil behavior. The consolidation characteristics are useful in predicting the settlement behavior of treated silt samples and hence essential for evaluating the stability of the proposed liner material. This paper focuses only on discussing the changes in permeability values of the silt samples on olivine treatment as obtained from the test results of one-dimensional consolidation tests explained earlier.

3 Results and Discussion

3.1 Effect of Forsterite on Compaction Characteristics

Figures 3 and 4 show the variation of maximum dry density (MDD) and optimum moisture content (OMC) for silt samples treated with various olivine dosages, respectively. It can be observed that an increase in olivine content led to an increase in MDD and a decrease in the OMC values of S1 samples. However, the changes in values are negligible in the case of S2 samples.

The specific gravity of olivine is more than that of untreated silt samples, and hence with the addition of olivine, an increase in MDD values of forsterite-treated samples was observed. The increase in MDD indicates a possible improvement in soil strength; hence, higher densities are desirable. The reduction in OMC with the increase in olivine dosage may be due to the low water affinity of olivine [8]. The specimens prepared at OMC and MDD were further analyzed for strength and permeability characteristics.

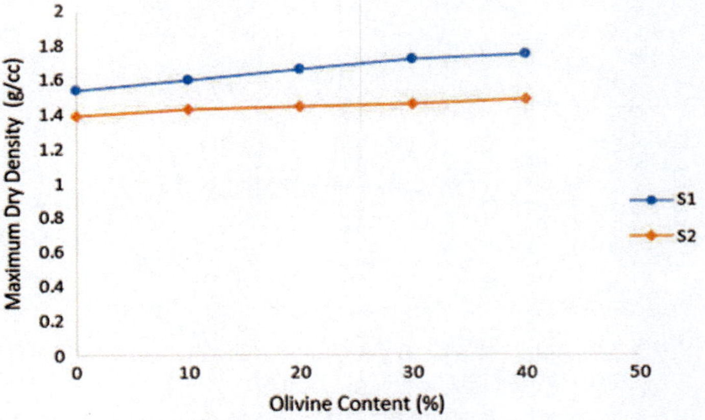

Fig. 3 MDD values of silt samples treated with various olivine dosages

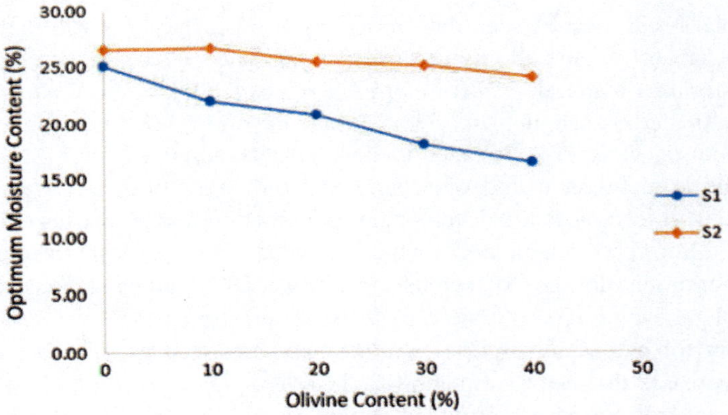

Fig. 4 OMC values of silt samples treated with various olivine dosages

3.2 *Effect of Forsterite on Unconfined Compressive Strength*

Landfill liner specifications require that the liner material should have sufficient strength to sustain the static load exerted by overlying body of waste. Figure 5 shows the effect of olivine inclusions in the unconfined compressive strength (UCS) of silt samples used in the study. The unconfined compressive strength of untreated silt samples, S1 and S2, used in the study was 154 kPa and 280 kPa, respectively.

It can be seen from Fig. 5 that the UCS values of S1 increased marginally on olivine treatment. However, these values do not satisfy the strength criteria required for liner material (> 200 kPa as per the U.S EPA guidelines) [26]. The increment in UCS values of S1 with olivine content was found to be minimal beyond 20% olivine dosage, and hence, 20% olivine dosage was identified as the optimum dosage for

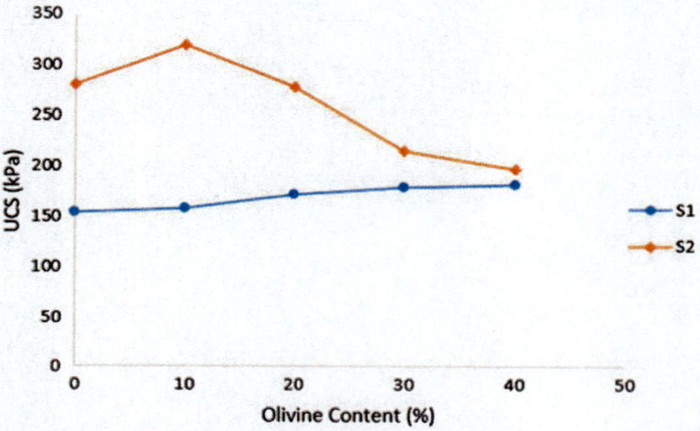

Fig. 5 UCS values of silt samples treated with various olivine dosages

S1 sample. It can also be seen that in the case of S2, the UCS values are greater than 200 kPa for almost all olivine dosages and hence satisfy the strength criteria required for liner material. A maximum UCS value of 319 kPa was obtained for 10% forsterite-treated S2 sample. An olivine dosage beyond 10% resulted in a reduction in UCS values. Thus, 10% was identified as the optimum dosage for S2 sample.

Olivine is reported to undergo chemical interactions with silica and alumina present in soil leading to the development of cementitious gel phases (M-S–H and M-A-H) filling the available pores which contributed to the strength improvement up to an optimum dosage [8]. The decrease in the UCS values of forsterite-treated S2 samples beyond 10% olivine dosage is because only 10% olivine is utilized effectively for hydration reactions and the excess olivine particles are unbonded which decreases the overall strength [8]. The UCS values of forsterite-treated silt samples without curing are provided in this paper. Curing these specimens may enhance the hydration reactions of olivine which may further improve the strength of forsterite-treated samples.

3.3 *Effect of Forsterite on Hydraulic Conductivity*

The hydraulic conductivity or coefficient of permeability (k) values of silt samples treated with their respective optimum dosages of olivine were estimated using the coefficient of consolidation (Cv) values obtained from one-dimensional consolidation tests. It can be seen from Fig. 6 that the coefficient of permeability (k) values for both the silt samples decreased considerably on olivine treatment. S1 samples treated with optimum olivine dosage (20%) gave 'k' values of 4×10^{-8} cm/s, whereas S2 samples treated with optimum olivine dosage (10%) gave 'k' values of 3.6×10^{-8} cm/

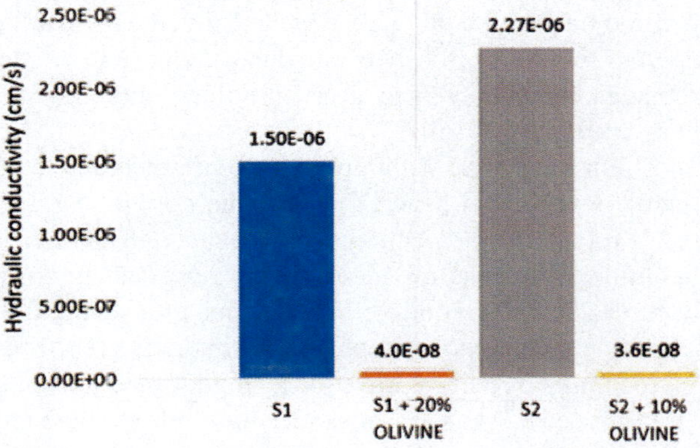

Fig. 6 Variation in hydraulic conductivity values of silt samples treated with optimum olivine dosage

s. Thus, on olivine treatment, both the silts satisfy the permeability criteria required for a landfill liner [26].

The magnesium oxide component in olivine reportedly undergoes hydration reactions with soluble alumina and silica in the soil, resulting in the formation of stable magnesium aluminate hydrate (M-A-H) and magnesium silicate hydrate (M-S-H) gels. These cementing phases bind soil particles, filling the voids, and consequently, may have effectively blocked the flow channels causing a reduction in permeability of the treated silt samples [8]. Magnesium oxide is known to form brucite ($Mg(OH)_2$) in the presence of water. This brucite is reported to react with carbon dioxide present in the atmosphere to form stable carbonates which may also have contributed to the reduction in the hydraulic conductivity of treated soils [6].

4 Conclusions

The following conclusions are drawn on the basis of test results:

1 On treatment with optimum olivine dosage, both the silts satisfy the permeability requirement for a landfill liner ($\leq 1 0^{-7}$ cm/s). Hydraulic conductivity (k) values for both the silt samples decreased considerably on olivine treatment. S1 samples treated with optimum olivine dosage (20%) gave 'k' values of 4×10^{-8} cm/s, whereas S2 samples treated with optimum olivine dosage (10%) gave 'k' values of 3.6×10^{-8} cm/s.

2 The magnesium oxide component in olivine reportedly undergoes hydration reactions with soluble alumina and silica in the soil, resulting in the formation of stable magnesium aluminate hydrate (M-A-H) and magnesium silicate hydrate (M-S-H) gels. These cementing phases bind soil particles, filling the voids, and consequently, may have effectively blocked the flow channels causing a reduction

in permeability of the treated silt samples. Brucite formed as a result of olivine alterations is reported to react with carbon dioxide present in the atmosphere to form stable carbonates which may have also contributed to the reduction in the hydraulic conductivity of treated soils.

3 The increment in UCS values of S1 with olivine content was found to be minimal beyond 20% olivine dosage, and hence, 20% olivine dosage was identified as the optimum dosage for S1 sample. Although UCS value of S1 sample improved marginally on addition of various olivine dosages, these values do not satisfy the strength criteria (> 200 kPa) required for a liner material as per the U.S EPA guidelines. Curing these specimens may enhance the hydration reactions between olivine and soil which may further improve the strength of treated soils.

4 Almost all forsterite-treated S2 samples satisfy the minimum strength requirement as per the U.S EPA guidelines. A maximum UCS value of 319 kPa was obtained for 10% forsterite-treated S2 sample. An olivine dosage beyond 10% resulted in a decrease in UCS values. Thus, 10% was identified as the optimum dosage for S2 sample. The cementation phases formed as a result of hydration reaction are responsible for the improvement in UCS values of forsterite-treated silt samples.

References

1. Baig AA, Sivapullaiah PV (2011) Characterization of lime and gypsum amended class F Fly ashes as liner materials. In Han J, Alzamora DA (eds) Advances in geotechnical engineering geo-frontiers 2011. ASCE 1:1162–1171. https://doi.org/10.1061/41165(397)119
2. Biju MS, Arnepalli DN (2020) Effect of biopolymers on permeability of sand-bentonite mixtures. J Rock Mech Geotechn Eng 12(5):1093–1102. https://doi.org/10.1016/j.jrmge.2020.02.004
3. Bozbey I, Guler E (2006) Laboratory and Field testing for utilization of an excavated soil as landfill liner material. Waste Manage 26(11):1277–1286. https://doi.org/10.1016/j.wasman.2005.10.014
4. Brown OR, Yusof MBBM, Salim, MRB, Ahmed K (2011) Compaction parameters of kaolin clay modified with palm oil fuel ash as landfill liner. In: Proceedings of 2011 IEEE First conference on clean energy and technology (CET 2011). IEEE, Kuala Lumpur, Malasia, pp 119–204. https://doi.org/10.1109/CET.2011.6041463
5. Devarangadi M, Shankar MU (2021) Effect on Engineering Properties of ground granulated blast furnace slag mixed with laterite soil, cement and bentonite mixture as a liner in landfill. J Clean Prod 329:129757. https://doi.org/10.1016/j.jclepro.2021.129757
6. Emmanuel E, Anggraini V (2020) Effects of desiccation-induced cracking and leachate infiltration on the hydraulic conductivity of natural and olivine-treated marine clay. Int J Environ Sci Technol 17:2259–2278. https://doi.org/10.1007/s13762-019-02606-x
7. Emmanuel E, Anggraini V, Gidigasu SSR (2019) A critical reappraisal of residual soils as compacted soil liners. S N Appl Sci 1(5):460. https://doi.org/10.1007/s42452-019-0475-7
8. Emmanual E, Anggraini V, Raghunandan ME, Asadi A, Bouazza A (2022) Improving the engineering properties of a soft marine clay with forsteritic olivine. Eur J Environ Civ Eng 26(2):519–546. https://doi.org/10.1080/19648189.2019.1665593
9. Garcia MA, Chimenos JM, Fernandez AI, Miralles L, Segarra M, Espiell F (2004) Low- grade Magnesium oxide used to stabilize heavy metals in highly contaminated soils. Chemosphere 56(5):481–491. https://doi.org/10.1016/j.chemosphere.2004.04.005

10. Hasan M, Khan MA, Alsabhan AH, Almajid AA, Alam S, Khan MA, Biswas T, Pu J (2022) Geotechnical behavior of fly ash—bentonite used in layers. Appl Sci 12(3):1421. https://doi.org/10.3390/app12031421
11. IS 2720 (Part-7) (1980) Methods of test for soils: determination of water content—dry density relation using light compaction. Indian Standards Institution, New Delhi, India
12. IS 2720 (Part 10) (1991) Methods of test for soils: determination of unconfined compressive strength. Indian Standards Institution, New Delhi, India
13. IS 2720 (Part 15) (1986) Methods of test for soils: Determination of consolidation properties. Indian Standards Institution, New Delhi, India
14. Jegandan S, Liska M, Osmna AAM, Tabba AA (2010) Sustainable binders for soil stabilization. Ground Improv 163(1):53–61. https://doi.org/10.1680/grim.2010.163.1.53
15. Karademir T (2022) Lime Stabilization of clayey landfill base liners: consolidation behavior and hydraulic properties. Environ Res Technol 5(1):1–10. https://doi.org/10.35208/ert.860623
16. Kumar SA, Sujatha ER (2021) An appraisal of the hydro-mechanical behavior of polysaccharides, Xanthan gum, guar gum and β-glucan amended soil. Carbohyd Polym 265:118083. https://doi.org/10.1016/j.carbpol.2021.118083
17. Lakshmikantha H, Sivapullaiah PV (2006) Relative performance of lime stabilized amended clay liners in different pore fluids. Geotech Geol Eng 24(5):1425–1448. https://doi.org/10.1007/s10706-005-0886-7
18. Nath H, Kabir MH, Kafy AA, Rahaman ZA, Rahman MT (2023) Geotechnical proper- ties and applicability of bentonite-modified local soil as landfill and environmental sustain- ability liners. Environ Sustain Indicators 18:100241. https://doi.org/10.1016/j.indic.2023.100241
19. Onyelowe KC, Obianyo II, Onwualu AP, Onyia ME, Moses C (2021) Morphology and mineralogy of rice husk ash treated soil for green and sustainable landfill liner construction. Clean Mater 1:100007. https://doi.org/10.1016/j.clema.2021.100007
20. Quang ND, Chai JC (2015) Permeability of lime and cement-treated clayey soils. Can Geotech J 52(9):1–7. https://doi.org/10.1139/cgj-2014-0134
21. Rashid HMA, Wanigarathna JADK, Kurukulasuriya LC, Priyankara NH, Alagiyawanna AMN, Saito T, Kawamoto K (2017) Characterization of locally available soil as a liner material for solid waste landfills in Sri Lanka. Environ Earth Sci 76:1–13. https://doi.org/10.1007/s12665-017-6717-3
22. Schuiling R, Praagman E (2011) Olivine hills: mineral water against climate change. In: Brunn S (eds) Engineering earth: the impacts of megaengineering projects. Springer, Dordrecht, pp 2201–2206. https://doi.org/10.1007/978-90-481-9920-4_122
23. Sivapullaiah PV, Lakshmikantha H, Kiran M (2003) Geotechnical properties of stabilized Indian red earth. Geotech Geol Eng 21(4):399–413. https://doi.org/10.1023/B:GEGE.0000006051.02215.A6
24. Sivapullaiah PV, Lakshmikantha H (2004) Properties of fly ash as hydraulic barrier. J Soil Sediment Contam 13(5):391–406. https://doi.org/10.1080/10588330490500437
25. Sivapullaiah PV, Lakshmikantha H (2005) Lime Stabilized illite as liner. Ground Improve J 9:39–45. https://doi.org/10.1680/GRIM.2005.9.1.39
26. U.S. EPA (1989) Requirements for hazardous waste landfill design, construction, and closure. In: Seminar Publication. United States Environmental Protection Agency, pp 1–127
27. Yi Y, Liska M, Akinyugha A, Unluer C, Tabbaa AA (2013) Preliminary laboratory-scale model auger installation and testing of carbonated soil-magnesium oxide columns. Geotech Test J 36(3):1–10. https://doi.org/10.1520/GTJ20120052

Evaluating the Influence of Montmorillonite Content on Swelling Behaviour in Relation to Its Plasticity

Peddireddy Sreekanth Reddy, V. Suryaprakash Reddy, Bendadi Hanumantha Rao, and Bijayananda Mohanty

1 Introduction

It is forced to utilize the available land irrespective of its type due to the rapid growth of cities since a decade. Expansive soils are one among them, which pose a severe threat to the infrastructure constructed over it with their swell/shrink behaviour during changes of seasons. In India, approximately 20% of the available land is covered with expansive soils. It reflects that most of the constructible areas may encounter this kind of soil, especially near water bodies. Expansive soils comprise clay content as a major portion. It comprises different minerals, among which montmorillonite is the prime triggering factor behind the plasticity and swelling behaviour. Free swell index (FSI) is an easy method to identify the swelling behaviour of expansive soils [1]. Employing it, Sivapullaiah et al. [2] introduced a modified free swell index for determining swelling behaviour. Further, Prakash and Sridharan [3] investigated the swelling behaviour of expansive soils using FSI. On the other hand, Holtz and Kovacs [4], Mitchell [5], and Skempton [6] predicted swelling behaviour indirectly from plasticity and other parameters of clay soils. Studies also reported that the consolidation apparatus can be employed for the estimation of swell potential in expansive soils [7–10]. However, it is a time-consuming process and not possible to judge the dominant mineral behind the swelling behaviour.

P. S. Reddy (✉)
Acharya Institute of Technology, Soladevanahalli, Bangalore, Karnataka 560107, India
e-mail: sreekanth2648@acharya.ac.in

V. S. Reddy
Institute of Aeronautical Engineering, Dundigal, Hyderabad, Telangana 500043, India

B. H. Rao
Indian Institute of Technology Bhubaneswar, Khordha, Odisha 752050, India

B. Mohanty
National Institute of Technology Mizoram, Aizawl, Mizoram 796012, India

S. Kolathayar et al. (eds.), *Best Practices in Geotechnical and Pavement Engineering*,
Lecture Notes in Civil Engineering 449, https://doi.org/10.1007/978-981-99-8505-0_13

It is evident from the literature that understanding the clay minerals is essential to evaluate the swelling behaviour of expansive soils. In this context, Mitchell and Soga [11] also stated that mineral type plays a vital role in the swelling behaviour and it can be determined by subjecting the soil sample to X-ray diffraction (XRD) analysis. Tahasildar et al. [12] measured the MMC and swelling behaviour from consolidation and reported a linear relationship between these two parameters. In a similar line, Reddy et al. [13] experimented with several expansive soils and measured MMC in clay content and swelling potential from FSI. The results revealed a linear relationship between these parameters. Most of the studies available in the past literature are confined to the establishment of a relationship randomly between MMC (i.e. in clay content/whole soil) and swelling potential (i.e. from FSI/Consolidation). But, very minimal attempts were made to study MMC of expansive soil and swelling behaviour from FSI, in relation to the plastic behaviour. In order to address the same, the current study intends to study the influence of montmorillonite content alone on the swelling behaviour from FSI in relation to the plasticity of expansive soils. In order to avoid problems due to volume change behaviour, it is essential to investigate the mineral behind it. It will give an idea for the selection of materials to arrest or reduce the heaving of soil under different environmental conditions.

2 Materials and Methods

Expansive soils of approximately 36 numbers have been collected from different locations across India (Bhopal, Guntur, Kendrapara, Warangal, Vijayawada, Kakinada, Mysore, Raipur, and Nagpur), which is at a depth of 1 m from the ground level by following the guidelines of ASTM A1452 [14]. The soil samples and their corresponding designations used in the study are listed in Table 1. The topsoil has been discarded to avoid foreign materials in the test samples. Soil samples were dried, pulverized, and downsized to the specific requirement of different tests adopted. Further, the physical properties of all the soil samples were determined, and their corresponding values can be found in Rao et al. [7]. The swelling behaviour of the soil samples was determined by employing the free swell index (FSI) in line with the guidelines of IS 2720 (Part-40) [1], and the values can be found in the work carried by Reddy et al. [7]. The mineralogical contents of expansive soils were determined by subjecting the dry powdered samples to X-ray diffraction followed by quantification of minerals using the software TOPAS 4.2. The detailed procedure related to the quantification and their corresponding values can be found in Rao et al. [9].

Table 1 Region and designations of soil samples used in the current study

Region	Sample ID	Region	Sample ID	Region	Sample ID
Bhopal	B1	Vijayawada	V1	Nagpur	N1
	B2		V5		N2
	B3		V6		N3
	B4		V7		N4
Guntur	G1	Kendrapa	KP1		N5
	G2		KP2		N6
	G4		KP3		N7
	G6		KP4		N8
	G7		KP5	Raipur	R1
Kakinada	K1		KP6		R2
	K2	Warangal	W1		R3
	K3		W2		
Mysore	M1		W3		

3 Results and Discussion

From the obtained test results, a plasticity chart indicating the relationship between liquid limit (LL) and plasticity index (PI) to classify the expansive soils used in the current study as depicted in Fig. 1. From the figure, it can be observed that almost all the soils possess an LL value of > 50% except very few. It indicates that most of the soils possess high compressibility, which is an indirect indication of the presence of swelling minerals, i.e. MMC. Certifying it, many researchers attributed higher plasticity values to the presence of MMC only [4, 9, 12, 13, 15].

Although researchers employed consistency limits as indirect indicators and prediction of swelling behaviour, direct determination offers better accuracy as they are of two different techniques. FSI is an easy and direct technique, which is used to measure the swelling behaviour of expansive soils. The typical photograph of FSI during the test can be seen in Fig. 2. In this context, a relationship is established between MMC and FSI in relation to the plasticity behaviour and depicted in Figs. 3 and 4. From Fig. 3, it can be observed that the swelling increased linearly with MMC up to 85%, irrespective of soil location for CH. The maximum FSI value of 105% is observed with an MMC of 54.58%. In a similar line, Reddy et al. [13] also reported an increment in FSI with the increment of montmorillonite content in clay content. This is due to the fact that the availability of MMC enhances the thickness of the diffuse double layer, in turn, which reflects as swelling. In addition, the pertinent data are collected from the literature and superimposed over the present study. Interestingly, the plotted relationship is in good agreement with the past literature data revealing its authenticity. Supporting this, observations made in the current study are in line with Tahasildar et al. [12] who investigated the influence of MMC on swelling behaviour.

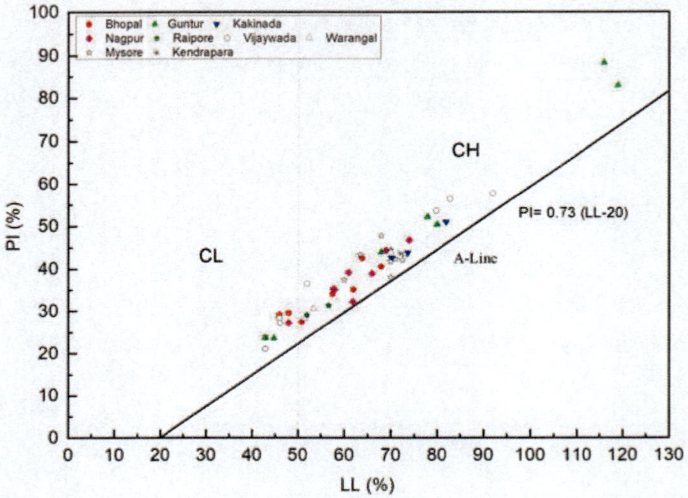

Fig. 1 Plasticity chart classifying the expansive soils used in the study

Fig. 2 Free swell index of different soil samples

In the case of Fig. 4, i.e. CL soils, mimicking the behaviour of CH, the trend increased linearly with MMC. The maximum FSI value of 84% is observed with an MMC of 56.34%. However, significant scatter can be observed. It is due to the marginal variation of MMC with respect to other soil samples. It conveys a fact that

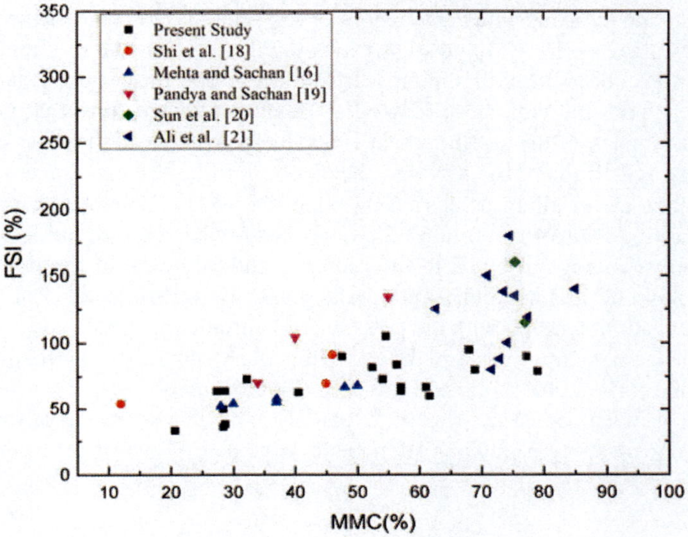

Fig. 3 Variation of FSI with MMC of CH soils

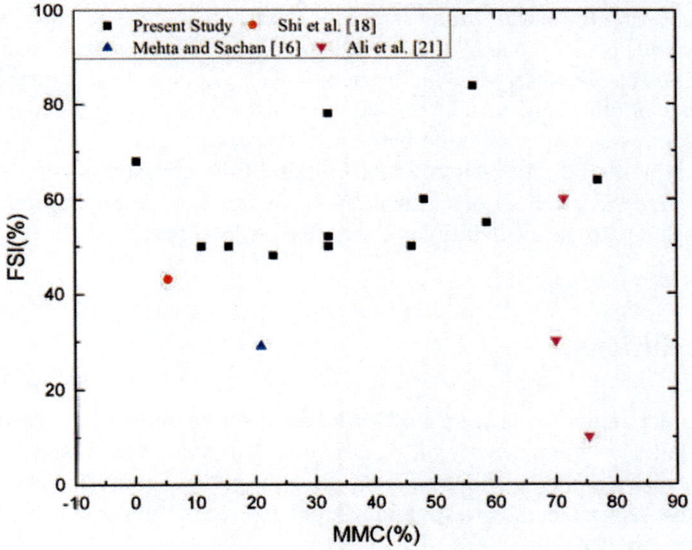

Fig. 4 Variation of FSI with MMC of CL soils used in the study

may not remain the same with the locations of soil samples, and so does the variation, which indicates its priority. Interestingly in both Figs. 3 and 4, two different swelling values can be observed with similar MMC. This is due to the fact that soil does comprise another mineral either may be alone or in a mixed form that can have a significant impact on the swelling behaviour. In this context, Tahasildar et al. [12] and Mehta and Sachan [16] reported that montmorillonite-based mixed minerals can influence the swelling behaviour of expansive soils. The close observation of corresponding MMC of maximum FSI values, later one is greater, but less swelling of 84% only. This is attributed to the presence and influence of the non-swelling mineral kaolinite. It is in confirmation with the studies of Sridharan et al. [17] who reported negative swelling with the presence of kaolinite mineral.

Overall, it can be inferred from the above that the MMC may not remain the same location and have a direct impact on the swelling behaviour of expansive soils. In order to have a safe construction over expansive soils, consideration of mineralogy in addition to swelling behaviour offers better accuracy.

4 Future Scope and Limitations

It is evident from the current study and past literature that the swelling behaviour of expansive soils is controlled by the constituent minerals. Interestingly, two different values of FSI can be observed with the same MMC, which indicates that there is an interference of other minerals, i.e. mixed minerals. Along with individual minerals, soils do comprise montmorillonite-based mixed minerals that can have a significant impact on the swelling behaviour. Identification of these mixed minerals can offer better accuracy in swelling determination. However, their determination is a little tedious task due to the availability of the necessary facilities.

5 Conclusions

In the current study, the influence of MMC on swelling behaviour measured from FSI in relation to the plasticity of expansive soils is evaluated. For achieving the same, expansive soils are collected from different parts of India and are subjected to extensive experimental investigation. From the results obtained, the following conclusions are derived.

(1) The expansive soils used in the current study are falling in the category of CH and CL, which comprise MMC in the range of 0–79%. The MMC contents varied with the location, and so does the swelling behaviour.

(2) The maximum FSI value of 105 and 84% is obtained for MMC of 54.58 and 56.34% for CH and CL soils, respectively. The general trend revealed that the FSI increased linearly with an MMC of up to 85% in the case of CH. Similar trends have been followed in the case of CL also but with significant scatter.
(3) Although with similar MMC, two different FSI values are measured both in CH and CL soils, which is due to the presence of montmorillonite-based mixed minerals.

References

1. IS: 2720-Part XL (1977) Indian standard methods of test for soils: determination of free swell index of soils. Bureau of Indian Standards, New Delhi
2. Sivapullaiah, P. V., Sitharam, T. G., Rao, K. S.: Modified free swell index for clays. Geotechnical Testing Journal 10(2) (1987).
3. Prakash K, Sridharan A (2004) Free swell ratio and clay mineralogy of fine-grained soils. Geotech Test J 27(2):220–225
4. Holtz RD, Kovacs WD (1981) An introduction to geotechnical engineering. Prentice-Hall, Englewood Clifs
5. Mitchell JK (1976) Fundamentals of soil behavior. Wiley, New York
6. Skempton AW (1953) The colloidal activity of clays. In: Proceedings of the third international conference on soil mechanics and foundation engineering, vol 1, Zurich, pp 57–67
7. Reddy PS, Mohanty B, Rao BH (2021) Investigations for chemical parameters effect on swelling characteristics of expansive soils. KSCE J Civ Eng 25:4088–4105
8. Reddy PS, Mohanty B, Rao BH (2021) Influence of Na and Ca contents on swelling behavior of Indian expansive soils. Arab J Geosci 14:2675
9. Rao BH, Reddy PS, Mohanty B, Reddy KR (2021) Combined effect of mineralogical and chemical parameters on swelling behaviour of expansive soils. Sci Rep 11:16562
10. Reddy PS, Yang YL, Mohanty B, Rao BH (2022) Assessment of testing method influence on swelling characteristics of expansive soils of India. Arab J Geosci 15:1132
11. Mitchell JK, Soga K (2005) Fundamentals of soil behavior, 3rd edn. Wiley, New York
12. Tahasildar J, Rao BH, Shukla SK (2017) Mineralogical compositions of some Indian expansive soils and their influence on swelling properties. Int J Geosynth Ground Eng 3:1–10
13. Reddy PS, Mohanty B, Rao BH (2020) Influence of clay content and montmorillonite content on swelling behavior of expansive soils. Int J Geosynth Ground Eng 6:1–12
14. ASTM D1452/D1452M-16 (2016) Standard practice for soil exploration and sampling by auger borings. ASTM International, West Conshohocken
15. Sridharan A, Rao SM, Murthy NS (1986) Liquid limit of montmorillonite soils. Geotech Test J 9(3)
16. Mehta B, Sachan A (2017) Effect of mineralogical properties of expansive soil on its mechanical behavior. Geotech Geol Eng 35:2923–2934
17. Sridharan A, Rao SM, Murthy NS (1988) Liquid limit of kaolinite soils. Geotechnique 38(2):191–198
18. Shi B, Jiang H, Liu Z, Fang HY (2002) Engineering geological characteristics of expansive soils in China. Eng Geol 67:63–71
19. Pandya S, Sachan A (2018) Matric suction, swelling and collapsible characteristics of unsaturated expansive soils. J Geotech Transport Eng 4(1):1–9
20. Sun D, Sun W, Fang L (2014) Swelling characteristics of Gaomiaozi bentonite and its prediction. J Rock Mech Geotech Eng 6:113–118

21. Ali R, Khan H, Shah AA (2014) Expansive soil stabilization using marble dust and bagasse ash. J Sci Res 3(6):2812–2816

3D Numerical Investigation of Stone Columns Simulated in Soft Soil

Pooja Bhatia and Murtaza Hasan

1 Introduction

Stone columns, also known as granular piles, are frequently used as a ground-improving technique in coastal locations with thick layers of soft marine clay [1–3]. The method involves using stone columns to strengthen foundations for various structures, such as liquid storage tanks, earthen embankments, raft foundations, to increase their bearing capacity, and lessen settling. [4, 5]. The many theoretical techniques have provided by various studies to determine the ultimate bearing capacity, such as the settlement of foundations reinforced by stone columns, and the diverse failure modes for stone columns, such as general shear failure, bulging, punching, and sliding [6–8]. The most important design factors for stone columns that improve ground stability are the column's stiffness and the weight distribution between the column and soil [9–11]. The behavior of inner stone columns has been studied using several characteristics, including the distance between the columns, their length, diameter, shear strength of the clay, and the angle of internal friction of the column materials [12–14]. In the current work, stone column performance was modeled using numerical calculations in three dimensions using Plaxis 3D. When modeling the unit cell, the short-term stress on the stone column was taken into account. Among the many ground-improvement methods used in geotechnical engineering, stone columns (also called granular or aggregate columns) stand out as revolutionary and technologically sophisticated. Using them efficiently improves the load-bearing capacity, decreases settlement, and strengthens unstable soils. Stone columns are a game-changer for solving geotechnical problems since they involve replacing the current soil with columns of crushed stone or gravel that have been painstakingly compressed. Incorporating cutting-edge numerical analytic methods in recent years has taken our knowledge and skill at designing stone columns to new heights. Plaxis

P. Bhatia (✉) · M. Hasan
Chandigarh University, Mohali 140301, India
e-mail: Poojabhatia.22222@gmail.com

S. Kolathayar et al. (eds.), *Best Practices in Geotechnical and Pavement Engineering*, Lecture Notes in Civil Engineering 449, https://doi.org/10.1007/978-981-99-8505-0_14

3D Finite Element Method (FEM) analysis is a state-of-the-art method that allows engineers to model the complex dynamics of soil–structure interaction and precisely assess the effectiveness of stone column systems. Plaxis 3D, the gold standard in geotechnical software, uses Finite Element Method (FEM) analysis to build three-dimensional models that faithfully reflect actual conditions. Engineers may experiment with a wide variety of parameters, boundary conditions, and loading scenarios on this sturdy platform to mimic the behavior of earth and stone columns. Based on the intensive literature review conducted, certain lagging was found corresponding to which following objectives were defined to address the research gaps. The objectives are as follows:

1.1 Objective 1

Assess the Impact of Variations in Stone Column Diameter and Length on Load-Carrying Capacity, Objective 1. This paper's initial objective is to examine how altering the diameter and length of stone columns affects their load-bearing capability in comparison to a clay bed. This research seeks to measure the percentage improvement in load-bearing capacity for varied diameter and length modifications of the stone columns by analyzing the load-settling plots acquired from Plaxis 3D simulations.

1.2 Objective 2

Evaluation of the Impact of Shear Strength Variation on Load-Carrying Intensity is Objective 2. The second objective is to examine how shifting the clay bed's shear strength affects the stone columns' load-bearing capacity. This research intends to determine the percentage increase in load-bearing intensity by comparing the load-carrying capacity of stone columns with various shear strength values (14 and 30 kPa) to that of the clay substrate.

1.3 Objective 3

Examine the Effect of Changing Internal Friction Angle on Load-Carrying Capacity is Objective 3. The third objective is to compare the load-bearing capacity of the stone column to the clay substrate as a function of changing the stone column's internal friction angle (ϕ). The goal of this research is to calculate the % increase in load-carrying capacity over the range of internal friction angles (40°, 43°, and 46°) by analyzing the load-settling plots for each.

2 Finite Element Analyses

In this study, the settlement curve pertaining to load was evaluated by its behavior on the stone column which has been studied with Plaxis 3D, a FEM-based program. In geotechnical engineering, the load-bearing capacity of stone columns is a crucial element that strongly affects the success of ground development approaches. This research makes extensive use of the cutting-edge capabilities of Plaxis 3D and 2D analysis to better comprehend the complicated behavior of stone column systems. Numerical simulations are used to assess the effect of modifying the characteristics of stone columns on their load-bearing capability. These factors include the column's length, diameter, shear strength, and angle of internal friction. It is possible to accurately predict soil–structure interaction, stress distribution, settlement, and load transfer processes using both three-dimensional (3D) and two-dimensional (2D) Plaxis models. The purpose of these simulations is to get insight into the best way to construct stone columns for load carrying. Geotechnical engineers may evaluate stone column designs with the help of Plaxis software to improve the load-bearing capacity and assure the stability of buildings built on weak or compressible soils. The validation of the model was done by using the findings of the laboratory model load test on the stone column carried out by Ambily and Gandhi [1]. In a tank with an effective diameter of De = 210 mm, $d = 75$ mm, and a length of 450 mm, the authors conducted laboratory model tests in which the tank was loaded alone. The Mohr–Coulomb failure criteria were applied in the analysis. Table 1 includes the material specifications used in the validation and preset study; and applying the load up to the required displacement. The comparison of the findings revealed that they were generally in accord, as shown in Fig. 1.

Elementary parameters that are crucial for the study were taken and their effect was evaluated which includes the diameter and length of the column, internal friction (ϕ), and Shear strength. The rear end of the tank was not allowed to move in any way, and the vertical boundaries of the tank could only move in the vertical direction. With the supposition that the loading plate would act rigidly, a certain displacement was needed in order to impart the weight to the stone column. The consolidation impact of clay was not considered in the current investigation because of the brief loading duration. The linear elastic completely plastic Mohr–Coulomb model was

Table 1 Physical properties used in the present study [1]

Parameters	Clay	Stone aggregates
Young's modulus, E (kPa)	5500	55,000
Undrained shear strength, Cu (kPa)	30	–
Poisson's ratio, μ	0.42	0.3
The angle of internal friction, ϕ (°)	0	43
Bulk unit weight (kN/m^3)	19.45	–
Dry unit weight (kN/m^3)	15.56	16.62
Dilation angle ψ (°)	–	10

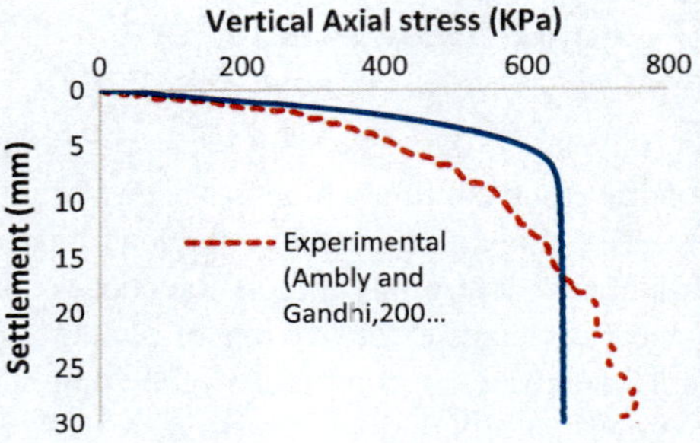

Fig. 1 Plaxis model validation

employed in related studies [1, 7, 14] for both clay and stone aggregates. Table 1 lists the material characteristics utilized in the analysis, including E, Cu, ϕ, γd, and ψ, which were established by lab testing by Ambily and Gandhi [1]. An interface element was not considered since the shear strength qualities of the bed may vary based on the installation method at the mixed zone where the stone column and clay meet. In a volume and connectivity plot for soil treated with stone columns under a medium element distribution, as illustrated in Fig. 2, the created mesh, consisting of nodes and stress points, was exhibited. The data represented in Table 1 for conducting the present study have been taken from the past research data for the evaluation and comparison study conducted in a redefined pattern in the current research study.

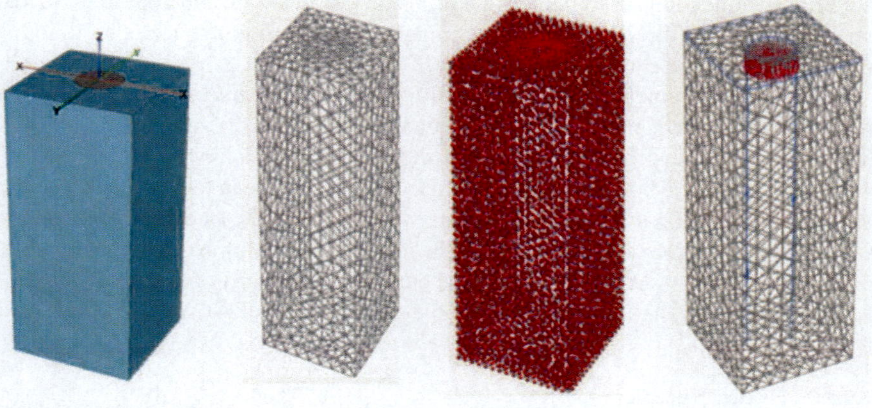

Fig. 2 Unit cell idealization, mesh generation with nodes, stress points

3 Result and Discussion

The findings with regard to clay bed load settlement plots; the effects on modifying column diameter, the longitudinal dimension of a stone column, the impact of changing column's internal friction angle, and the influence of changing shear strength of clay bed discussed in detail. Every single time, it was considered that the stone column could support a load equal to a 30 mm settlement. This study provides an important new understanding of how different factors affect the load-bearing capacity of stone columns by analyzing clay bed load-settling plots. Changes in column diameter, longitudinal dimension, internal friction angle, and shear strength of the clay substrate are all included in the study. First, it was discovered that increasing the diameter of the column increased its load-bearing capacity. Because of the increased surface area of contact between the stone column and the surrounding soil, settling was mitigated because of the bigger diameter. Second, testing columns of varied lengths revealed that the longer the column, the greater its load-bearing ability. The weight was spread out across a broader area, which the longer columns helped accomplish, resulting in less settling.

In addition, there was a noticeable change once the internal friction angle of the stone column was adjusted. The load-carrying capacity was shown to improve with increased internal friction angles. The better shear resistance at the column-soil contact was a direct outcome of the higher internal friction angle, which allowed for less settlement. Finally, research into the effect of varying the shear strength of the clay bed indicated that higher shear strength values resulted in increased load-bearing capability. Because of the clay bed's increased shear strength, it was better able to resist deformation and distribute the applied load evenly throughout the stone column, reducing settling. The analysis accounted for the stone column's ability to bear a weight equal to a 30 mm settlement in all cases. These results show how crucial it is to optimize stone column characteristics to get the required load-bearing capacity while meeting the required settling conditions. Engineers may make better judgments to improve the performance of stone column systems in geotechnical engineering projects if they have a thorough grasp of the consequences of varying column diameter, longitudinal dimension, internal friction angle, and shear strength of the clay substrate.

3.1 Impact of the Stone Column's Diameter

Figure 3 illustrates the results of the FEM study for the load-settlement relationship for stone columns with different column diameters as mentioned in the above section. An increase in the percentage was shown to boost the load-bearing capacity of the columns. As the column's diameter increases, more clayey soil may be replaced with broad granular material. Regarding the clay bed, the stone column's load-carrying capacity increases by 151, 234, and 376% for diameters of $d = 65$ mm, 75 mm,

Fig. 3 Impact of the stone column's diameter

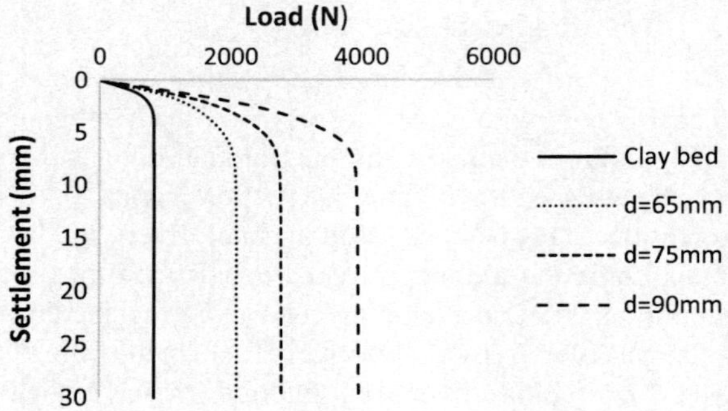

and 90 mm. The rise in the ultimate load of a stone column set upon soft clay has been illustrated through the ultimate load intensity ratio. The ratio expressed as the comparison of the final load of the stone column to that of the clay bed on a settlement of 30 mm efficacy was assessed with reference to the untreated soil sample. The ultimate load ratio has been determined to be 2.51, 3.33, and 4.75 at $d = 65$ mm, 75 mm, and 90 mm, respectively.

3.2 Effect Study of Stone Column on Length

Figure 4 presents a comparing the Finite Element Method result to other results (FEM) analysis for settlement corresponding to load showing a relationship of stone columns with a 75 mm diameter. The analysis shows that when compared to clay bed, the load-bearing capacity of a stone column increases by 226, 230, and 234% for a length-to-diameter (L/D) ratio of 4, 5, and 6 respectively. Additionally, the ultimate load ratio for L/D ratios 4, 5, and 6 is 3.25, 3.29, and 3.33.

Fig. 4 Effect study of stone column on length

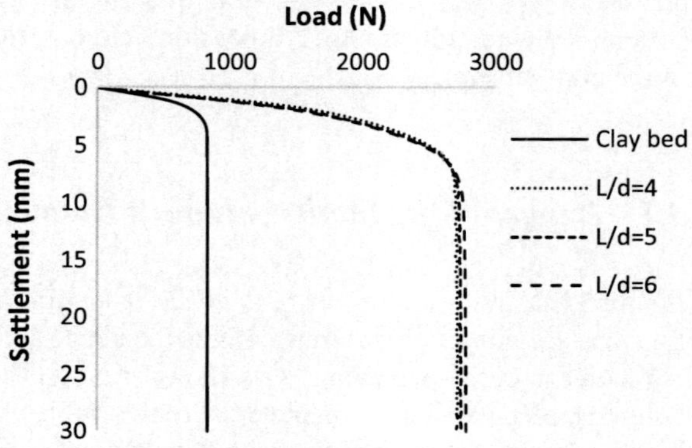

Fig. 5 Impacts of stone column shear strength

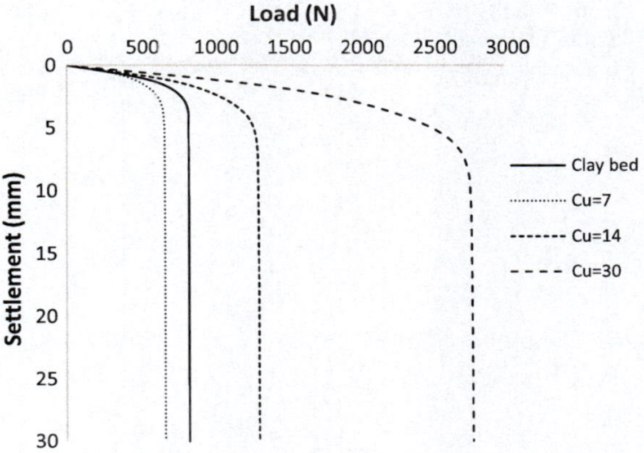

3.3 *Impact of a Stone Column's Shear Strength*

Figure 5 depicts the variance between the stone column's load-bearing capacity and the clay bed's undrained shear strength. The column's diameter was kept fixed at 75 mm, and the angle of internal friction was 43. The observed rise in the safe bearing capacity of the column with increased clay shear strength and column length may be due to higher lateral confining pressure. The stone column's capacity for safe bearing for Cu = 7 kPa is found to be decreased as compared to the clay bed with an undrained shear strength of 30 kPa. Although the strength of the stone column for Cu = 14 kPa and 30 kPa is increased by 57.41% and 234% concerning clay bed, respectively. The ultimate load ratio for Cu = 7, 14, and 30 kPa is 0.80, 1.58, and 3.33.

3.4 *Validity of the Column's Material's Internal Friction Angle*

The investigation of the effects of the internal friction angle of the column material has also increased the use of numerical analysis. For $d = 75$ mm, Cu = 30 kPa, and $l = 6$d, Fig. 6 shows the impact of the (ϕ) of the column material on its capacity to transport loads. The ultimate load ratio for = 40, 43, 46 is 2.92, 3.33, 3.83, and the load-bearing capability of stone columns for clay bed is raised by 193, 234, and 284%.

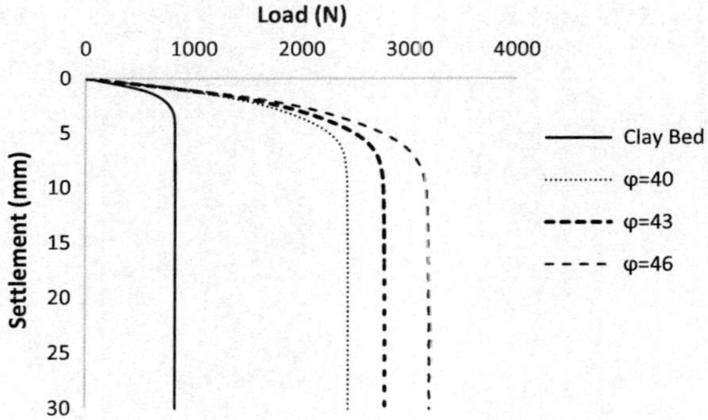

Fig. 6 Validity of the column's material's internal friction angle

4 Conclusion

Using Plaxis 3D simulations, this article examines the behavior and load-bearing capabilities of stone columns in great detail. The research has focused on three main objectives: determining the impact of different diameters and lengths, determining the impact of different shear strengths, and determining the impact of different internal friction angles. According to the results, stone columns are far more effective at supporting weight than a clay base. The load-carrying capacity rose by significant margins after varying the stone column's parameters (diameter, length, shear strength, and internal friction angle). These results help in the further development of optimization and design methods for stone column systems used in ground-improvement initiatives. To improve load-bearing capacity and guarantee the stability and performance of buildings built on weak or compressible soils, the research highlights the need of addressing these characteristics. The effective depth of the column is assumed in the current research to be De = 200 mm, $d = 75$ mm, the L/D ratio is 6, and the column spacing is 2.54 mm, which is within the range specified in IS 15284-1(2003). The area replacement ratio is 14%, and the parametric study has been compared with the clay bed surface. The Plaxis 3D model is used to produce the load-settling plot of various impacts of stone columns with clay beds. With the experimental data extracted and simulation study conducted with the variations of parameters as discussed in the subsequent section, valuable conclusions are made which are noted below:

1. Changing the diameter of the stone column significantly increases its load-bearing capability relative to the clay bed. In terms of load capacity, the increases range from 151 to 234% for a diameter of 65, 75 mm to an incredible 376% for a diameter of 90 mm.
2. The stone column's load-bearing capacity is much enhanced in comparison to the clay substrate as a function of length variations. At 4d in length, the, load-bearing capacity improves by 226%; at 5d, it rises to 230%; and at 6d, it rises to 234%.

3. Significant improvements can be seen in the load-bearing intensity compared to the clay bed, which indicates the difference in ultimate shear strength. For Cu = 14 kPa, the load-bearing intensity rises by 57.41%, and for Cu = 30 kPa, it climbs by a remarkable 234%.
4. The research demonstrates significant enhancements when comparing the load-bearing capability of the stone column to that of the clay substrate. When the internal friction angle is increased from 40 to 46, the stone column's load-bearing capacity improves by 193%, 234%, or 284%, respectively.

References

1. Ambily AP, Gandhi SR (2007) Behavior of stone columns based on experimental and FEM analysis. J Geotech Geoenviron Eng ASCE 133(4):405–415
2. Barksdale RD, Bachus RC (1983) Design and construction of stone columns. Federal Highway Administration, RD-83/026
3. Castro J, Sagaseta C, Costa AD (2013) Consolidation and deformation around stone columns: comparison of theoretical and laboratory results. Comput Geotech 49(1):326–337
4. Greenwood DA (1970) Mechanical improvement of soils below ground surfaces. In: Proceedings of ground engineering conference. Institution of civil engineers, London, pp 11–22
5. Greenwood DA (1975) Vibroflotation: rationale for design and practice. In: Bell FG (ed) Methods of treatment of unstable ground. Newness-Butterworth, London, pp 189–209
6. Hasan M, Samadhiya NK (2016) 3D numerical analysis of granular piles with internal horizontal geogrid strips in layers. In: Proceedings of the Indian geotechnical conference. IIT Madras, Chennai, pp 1–4
7. Hasan M, Samadhiya NK (2017) Performance of geosynthetic-reinforced granular piles in soft clays: model tests and numerical analysis. Computer Geotech 87:178–187
8. IS 1498 (2000) Classification and identification of soils for general engineering purposes. Indian Standards Institution, New Delhi
9. Jayarajan J, Karpurapu R (2020) Bearing capacity and settlement response of ordinary and geosynthetic encased granular columns in soft clay soils: analysis and design charts. Indian Geotech J. https://doi.org/10.1007/s40098-020-00457-9
10. Saxena S, Roy LB (2022) The effect of geometric parameters on the strength of stone columns. Eng Technol Appl Sci Res 12(4):9028–9033
11. Joseph J, Kumar S, Sawant VA, Patel JB (2022) An experimental and numerical comparative study on the uplift capacity of single granular pile anchor and rough pile in sand. Int J Geotech Eng 16(4):499–513. https://doi.org/10.1080/19386362.2021.1999077
12. Shafiee AH, Eskandarinejad A (2022) Bearing capacity of single stone column in clay using finite element limit analysis. Eur J Environ Civil Eng 26(15):7958–7971. https://doi.org/10.1080/19648189.2021.2019616
13. Feng W-Q, Tan D-Y, Yin J-H, Qin J-Q, Chen W-B (2020) Experimental and numerical studies on the performances of stone column and sand compaction pile-reinforced Hong Kong, Marine Clay. Int J Geomech 20(8):06020018
14. Hasan M, Samadhiya NK (2022) Ground improvement by using floating granular piles: experimental studies and numerical investigations. In: Sitharam TG, Kolathayar S, Jakka R (eds) Earthquake geotechnics. Lecture notes in civil engineering, vol 187. Springer, Singapore. https://doi.org/10.1007/978-981-16-5669-9_38

Uncertainty of Measurement of Geotechnical Parameters of a Lateritic Soil

Saisha U. Naik and Nisha P. Naik

1 Introduction

Soil plays a major role in the design of foundation required to keep the building safe. However, in the field of geotechnical engineering, a great number of uncertainties are encountered. It may be due to various sources, such as inherent soil variability present due to natural geological processes, measurement errors facing during testing, errors in soil sampling methods, human error, environmental conditions. As a result, the degree of uncertainty coming from sources is often determined by factors such as the variability of the soil profile at the site, the equipment used, the procedure maintained during testing, and the precision of the correlation model utilized. Too much uncertainty could weaken decision dependability and complicate the situation. As a result, it is essential to assess the uncertainty and minimize its impact on the test results. Bilgin et al. [1] studied the variability associated with the uncertainty due to the measurement error in direct shear test. Kuhinek et al. [2] focuses to determine the combined uncertainty of a rock sample for sample dimensions (diameter and height), uniaxial strength, young's modulus, and Poisson's coefficient using the guide of measurement uncertainty (GUM) method. The repeatability for direct shear testing on compacted alluvial fine-grained soil was investigated by Matteo et al. [3]. The objective of the present study is to investigate the uncertainty associated with geotechnical parameters of a lateritic soil such as specific gravity, liquid limit, plastic limit, maximum dry density, optimum moisture content, angle of internal friction, and cohesion by performing each test five times. The National Accreditation Board for Testing and Calibration Laboratories (NABL) 141 [4] document is used to calculate the uncertainty. And, verify the soil test results with tests result performed in NABL-accredited laboratory.

S. U. Naik (✉) · N. P. Naik
Civil Engineering Department, Goa College of Engineering, Ponda, Goa 403401, India
e-mail: falsesaishanaik26@gmail.com

S. Kolathayar et al. (eds.), *Best Practices in Geotechnical and Pavement Engineering*,
Lecture Notes in Civil Engineering 449, https://doi.org/10.1007/978-981-99-8505-0_15

2 Methodology

The lateritic soil used for tests was sourced from Bandora, Ponda taluka. About 50 kg of the sample was brought from the site to the laboratory. It was essential that the sample used for testing was uniform across all the repeated tests. The conning and quartering procedure was used to obtain test soil sample. In this, soil sample is divided into four parts, and the diagonally opposite portions of the soil are mixed together and the test samples are taken from the mixed portion. All the soil tests such as specific gravity, liquid limit, plastic limit, standard compaction, and direct shear were carried out by following the procedure from the respective IS codes such as IS:2720 (Part 3) (sec-2)-1980, IS:2720 (Part 5)-1985, IS:2720 (Part 5)-1985, IS:2720 (Part 7)-1980, and IS:2720 (Part 13)-1986. To assess the repeatability, five tests of each soil parameter were performed. Care was taken to ensure that the soil sample is uniform across each of the tests and test was performed by the same person and by using the same equipment/machine and same procedure. Photographs of the tests performed are shown in Fig. 1.

The uncertainty of measurement of soil parameters was evaluated based on the procedural guidelines given in NABL 141 [4] document. The steps followed are mentioned below:

(1) Specify the uncertainty measurement procedure by choosing the particular geotechnical or other test, its procedure, number of tests 'n' of the soil as

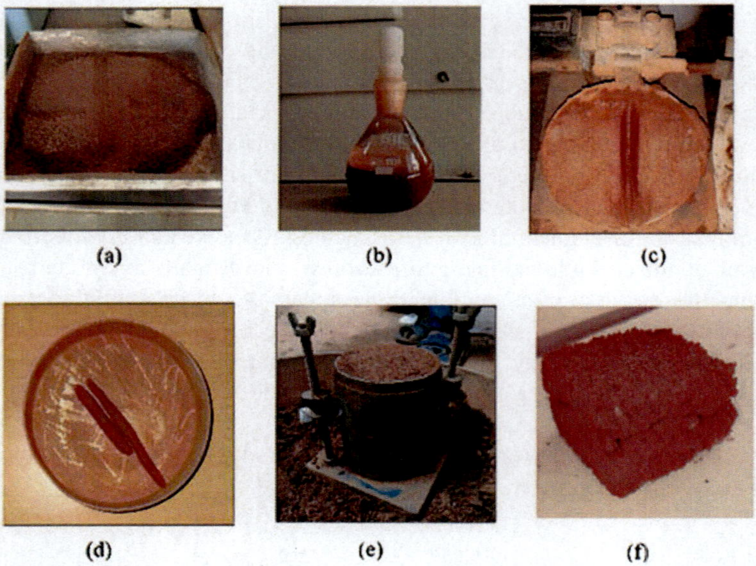

Fig. 1 **a** Sampling of soil, tests performed on soil **b** specific gravity test, **c** liquid limit test, **d** plastic limit test, **e** standard compaction test, **f** direct shear test

repeatability analysis that are going to performed on a control soil sample, etc.

(2) Identify and characterize all the sources of uncertainty that are associated with a particular test. This study considered two sources for all tests i.e., repeatability error and machine error.
(3) Find out the input estimate xi (soil parameter result value), of soil parameter from a particular soil test formula and tabulate these values as tests X_1, X_2, X_3, etc., and determine its mean value or average value $\overline{x}$.
(4) Estimate the standard deviation (SD) of each soil parameter using the formula from MS Excel [SD = STDEVA ($X_1 + X_2 + \cdots X_N$)] or using equation,

$$\mathrm{SD} = \sqrt{\sum_{i=1}^{n} \frac{(x_i - \overline{x})^2}{n-1}}. \tag{1}$$

(5) Now, determine the standard uncertainty $u(x_i)$ of soil parameter. The standard uncertainty is calculated as a Type A evaluation of standard uncertainty or a Type B evolution of standard uncertainty, and both types are based on the sources of uncertainty. In Type A evaluation, the standard uncertainty can be found by dividing the standard deviation by SQRT(n), i.e., $u(x_i)$ = SD/SQRT(n), which is also referred to as the standard deviation about the mean. In Type B evaluation, the standard uncertainty can be found out directly from the calibration certificate.
(6) Estimate the combined standard uncertainty $Uc(y)$ from the standard uncertainties associated with the input estimates using the following equation:

$$Uc(y) = y\sqrt{\left(\frac{u(p)}{p}\right)^2 + \left(\frac{u(q)}{q}\right)^2 + \left(\frac{u(r)}{r}\right)^2 + \cdots}, \tag{2}$$

where y is the measurement result (mean value) derived from measurement of input quantities x_{i1}, x_{i2}, as p, q, r, etc., and $u(p)/p$, $u(q)/q$, $u(r)$, etc., are the standard uncertainties divided by mean value, which is expressed as relative standard deviations or relative standard uncertainty.

(7) Calculate the effective degree of freedom (v_{eff}) to know the degree of parameter uncertainty from equation.

$$V_{\text{eff}} = \frac{uc(y)^4}{\frac{u(x_i)^4}{v_i}}, \tag{3}$$

where v_i is degree of freedom equal to (n – 1).

(8) Determine expanded uncertainty U of soil parameter, whose purpose is to provide an interval $y - U$ to $y + U$, so for this purpose, multiply the combined standard uncertainty $uc(y)$ by a coverage factor $\underline{k}$, i.e.,

$$U = k\ uc(y). \tag{4}$$

Typically, coverage factor k is in the range of 2–3.

Select k on the basis of the level of confidence required for the interval $y - \mathrm{U}$ to $y + U$. Generally, k is assumed to equal 2, at an interval having a level of confidence of approximately 95.45%. In this work, 'k' value is taken 2 where the confidence level is 95.45%.

(9) Report the result of the uncertainty, together with its combined standard uncertainty $Uc(y)$ or expanded uncertainty U along with confidence level.

Giang et al. [5] have used the GUM procedure to find out the uncertainties of shear strength.

3 Results and Discussions

3.1 Results of Uncertainty for Geotechnical Parameters of Lateritic Soil

(A) **Specific gravity**

As a repeatability approach, five tests of specific gravity were performed and two sources of uncertainty are considered, i.e., repeatability as Type A uncertainty and weight machine uncertainty as Type B, and uncertainty for specific gravity was found out to be 2.51 ± 0.02 at coverage factor $k = 2$ (95.45% of confidence level) by following the steps mentioned in methodology which is shown in Table 1.

(B) **Liquid limit**

As a repeatability approach, five tests of liquid limit were performed and two sources of uncertainty are considered, i.e., repeatability as Type A uncertainty and weight machine uncertainty as Type B. The uncertainty for liquid limit parameter was found out to be 33.11 (%) ± 0.79 (%) at coverage factor $k = 2$ (95.45% of confidence level) by following the steps mentioned in above methodology, which is shown in Table 2.

(C) Plastic limit

As a repeatability approach, five tests of plastic limit were performed and two sources of uncertainty are considered, i.e., repeatability as Type A uncertainty and weight machine uncertainty as Type B. The uncertainty for plastic limit parameter was found out to be 23.31 (%) ± 1.98 (%) at coverage factor $k = 2$ (95.45% of confidence level) by following the steps mentioned in above methodology, which is shown in Table 3.

Table 1 Calculation of uncertainty for specific gravity

No. of tests	1	2	3	4	5
Wt. of water (g)	33.06	33.09	33.15	32.98	33.07
Wt. of dry soil (g)	35.04	34.70	35.23	34.04	34.36
Specific gravity	2.49	2.50	2.49	2.51	2.54
Mean	2.51				
Standard deviation, SD	0.02				
Standard uncertainty, $u(x_i)$	0.01				
Sources of uncertainty	Type	Uncertainty	Mean values	Relative std. uncertainty	
(1) Repeatability	A	0.01	2.51	0.0037	
(2) Wt. Machine	B	0.02	33.07 34.67	0.0006 0.0006	
Combine uncertainty, $Uc(y)$	0.001				
Effective degree of freedom	4.42				
Coverage factor, k	2				
Expanded uncertainty, U	0.02				

(D) Optimum moisture content

As a repeatability approach, five tests of optimum moisture content were performed and two sources of uncertainty are considered, i.e., repeatability as Type A uncertainty and weight machine uncertainty as Type B. The uncertainty for OMC parameter was found out to be 16.39 (%) ± 0.66 (%) at coverage factor $k = 2$ (95.45% of confidence level) by following the steps mentioned in above methodology which is shown in Table 4.

(E) Maximum dry density

As a repeatability approach, five tests of maximum dry density were performed and two sources of uncertainty are considered, i.e., repeatability as Type A uncertainty and weight machine uncertainty as Type B. The uncertainty for MDD parameter was found out to be 1.67 (g/cc) ± 0.09 (g/cc) at coverage factor $k = 2$ (95.45% of confidence level) by following the steps mentioned in above methodology which is shown in Table 5.

(F) Angle of internal friction

As a repeatability approach, five tests of direct shear were performed at 0.5, 1, 1.5 kg/cm^2 loads, and two sources of uncertainty are considered, i.e., repeatability

Table 2 Calculation of uncertainty for liquid limit

No. of tests	1	2	3	4	5
Wt. of water (g)	2.12	2.28	2.25	2.21	1.78
Wt. of dry soil (g)	6.41	6.97	6.94	6.33	5.34
Liquid limit %	33.35	32.97	32.33	33.64	33.26
Mean	33.11				
Standard deviation, SD	0.50				
Standard uncertainty, $u(x_i)$	0.22				
Sources of uncertainty	Type	Uncertainty	Mean values	Relative std. uncertainty	
(1) Repeatability	A	0.22	33.11	0.0037	
(2) Wt. Machine	B	0.02	2.13 6.40	0.0094 0.0031	
Combine uncertainty, $Uc(y)$	0.40				
Effective degree of freedom	40.32				
Coverage factor, k	2				
Expanded uncertainty, U	0.79				

Table 3 Calculation of uncertainty for plastic limit

No. of tests	1	2	3	4	5
Wt. of water (g)	0.48	0.54	0.52	0.48	0.48
Wt. of dry soil (g)	2.05	2.32	2.33	2.06	1.99
Plastic limit (%)	23.30	23.23	22.49	23.49	24.03
Mean	23.31				
Standard deviation, SD	0.55				
Standard uncertainty, $u(x_i)$	0.25				
Sources of uncertainty	Type	Uncertainty	Mean values	Relative std. uncertainty	
(1) Repeatability	A	0.25	23.31	0.0106	
(2) Wt. machine	B	0.02	0.50 2.15	0.0400 0.0093	
Combine uncertainty, $Uc(y)$	0.99				
Effective degree of freedom	1011.49				
Coverage factor, k	2				
Expanded uncertainty, U	1.98				

Table 4 Calculation of uncertainty for optimum moisture content

No. of tests	1	2	3	4	5
Wt. of water (g)	1.71	1.41	1.05	1.06	1.58
Wt. of dry soil (g)	10.45	8.14	6.35	6.71	9.69
OMC (%)	16.23	17.20	16.45	15.93	16.14
Mean	16.39				
Standard deviation, SD	0.49				
Standard uncertainty, $u(x_i)$	0.22				
Sources of uncertainty	Type	Uncertainty	Mean values	Relative std. uncertainty	
(1) Repeatability	A	0.22	16.39	0.012	
(2) Wt. machine	B	0.02	1.36 8.27	0.015 0.002	
Combine uncertainty, $Uc(y)$	0.33				
Effective degree of freedom	19.86				
Coverage factor, k	2				
Expanded uncertainty, U	0.66				

as Type A uncertainty and proving ring uncertainty as Type B. The uncertainty for angle of internal friction parameter was found out to be 20.60° ± 1.16° at coverage factor $k = 2$ (95.45% of confidence level) by following the steps mentioned in above methodology, which is shown in Table 6.

(G) **Cohesion**

As a repeatability approach, five tests of direct shear were performed at 0.5, 1, 1.5 kg/cm^2 loads and two sources of uncertainty are considered, i.e., repeatability as Type A uncertainty and proving ring uncertainty as Type B. The uncertainty for cohesion parameter was found out to be 0.12 (kg/cm^2) ± 0.03 (kg/cm^2) at coverage factor $k = 2$ (95.45% of confidence level) by following the steps mentioned in above methodology which is shown in Table 7.

3.2 *Validation of Soil Parameters with NABL-Accredited Laboratory*

To ensure that the lateritic soil properties acquired are correct, all soil tests are performed in NABL-accredited laboratory and the results obtained are verified with the results obtained for uncertainty calculation. Table 8 shows the soil results obtained

Table 5 Calculation of uncertainty for maximum dry density

No. of tests	1	2	3	4	5
Wt. of water (g)	1.71	1.41	1.05	1.06	1.58
Wt. of dry soil (g)	10.45	8.14	6.35	6.71	9.69
MDD (g/cc)	1.77	1.70	1.63	1.68	1.56
Mean	1.67				
Standard deviation, SD	0.08				
Standard uncertainty, $u(x_i)$	0.04				
Sources of uncertainty	Type	Uncertainty	Mean values	Relative std. uncertainty	
(1) Repeatability	A	0.04	1.67	0.012	
(2) Wt. machine	B	0.02	1.36 8.27	0.015 0.002	
Combine uncertainty, $Uc(y)$	0.04				
Effective degree of freedom	8.99				
Coverage factor, k	2				
Expanded uncertainty, U	0.09				

from the NABL laboratory. A plot of mean value of soil parameter obtained in laboratory with respect to that obtained from NABL-accredited laboratory is shown in Fig. 2. It was observed that the soil parameter results obtained from NABL-accredited laboratories vary by 2.79, 7.73, 4.24, 2.01, 5.12, 14.16, 16.67%, respectively, for each soil test.

The uncertainty might be reported along with the soil test results as an application. It is a measure of the lower and higher limits (confidence level: range of estimates for unknown parameters) of the test parameter of the soil sample under the study. The findings for uncertainty of measurement (UOM) for any test parameter can be extended to other types of soils and also to lateritic soil of varying compositions found in the state of Goa and elsewhere. UOM values define the confidence level in the test results.

Table 6 Calculation of uncertainty for angle of internal friction

No. of tests	1	2	3	4	5
Proving ring reading @ 0.5 kg/cm^2	30	31	35	36	33
Proving ring reading @ 1 kg/cm^2	50	42	55	59	51
Proving ring reading @ 1.5 kg/cm^2	67	67	70	76	72
Angle of internal friction (degree)	20	19	21	22	21
Mean	20.60				
Standard deviation, SD	1.14				
Standard uncertainty, $u(x_i)$	0.51				
Sources of uncertainty	Type	Uncertainty	Mean values	Relative std. uncertainty	
(1) Repeatability	A	0.51	20.60	0.025	
(2) Proving ring uncertainty	B	0.34	33.00 51.40	0.010 0.007	
			70.40	0.005	
Combine uncertainty, $Uc(y)$	0.58				
Effective degree of freedom	6.58				
Coverage factor, k	2				
Expanded uncertainty, U	1.16				

Table 7 Calculation of uncertainty for cohesion

No. of tests	1	2	3	4	5
Proving ring reading @ 0.5 kg/cm^2	30	31	35	36	33
Proving ring reading @ 1 kg/cm^2	50	42	55	59	51
Proving ring reading @ 1.5 kg/cm^2	67	67	70	76	72
Cohesion (kg/cm^2)	0.10	0.11	0.15	0.16	0.10
Mean	0.12				
Standard deviation, SD	0.03				
Standard uncertainty, $u(x_i)$	0.01				
Sources of uncertainty	Type	Uncertainty	Mean values	Relative std. uncertainty	
(1) Repeatability	A	0.01	0.13	0.097	
(2) Proving ring uncertainty	B	0.34	33.00 51.40	0.010 0.007	
			70.40	0.005	
Combine uncertainty, $Uc(y)$	0.013				
Effective degree of freedom	4.15				
Coverage factor, k	2				
Expanded uncertainty, U	0.03				

Table 8 Results of lateritic soil parameters obtained from NABL-accredited laboratory

Lateritic soil Parameters	Results	Results variation (%)
Specific gravity	2.44	2.79
Liquid limit	35.88%	7.73
Plastic limit	24.34%	4.24
OMC	17.45%	2.01
MDD	1.82 g/cc	5.12
Angle of internal friction (ϕ)	24 °	14.16
Cohesion (c)	0.1 kg/cm^2	16.67

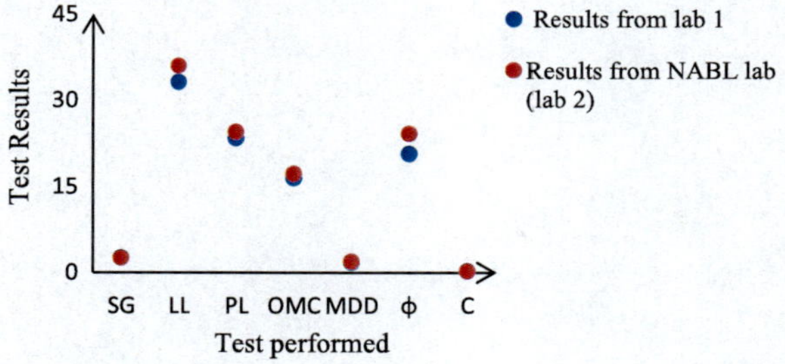

Fig. 2 Graph of showing soil test results from lab 1 and lab 2

4 Conclusion

Uncertainty of measurement was determined for soil parameters of a lateritic soil at 95.45% confidence level of coverage factor $k = 2$. For parameters such as liquid limit, plastic limit, optimum moisture content, maximum dry density, angle of internal friction, and cohesion, the uncertainty was determined to be 2.51 ± 0.02, 33.11% ± 0.79%, 23.31% ± 1.98%, 16.39% ± 0.06%, 1.67g/cc ± 0.09 gm/cc, 20.60° ± 1.16 °, and 0.12 kg/cm^2 ± 0.03 kg/cm^2, respectively. The results of lateritic soil parameters acquired from a NABL-accredited laboratory show a variation of 2.79%, 7.73%, 4.24%, 2.01%, 5.12%, 14.16%, 16.67%, respectively, for each soil parameter. The UOM values define the confidence level in the test results.

References

1. Omer Bilgin PE, Shah B (2010) Statical assessment of repeatability of soil-geomembrane interface shear tests. In: GeoFlorida 2010: advances in analysis, modeling & design (GSP 199). ASCE
2. Kuhinek D, Zoric I, Hrzenjak P (2011) Measurement uncertainty in testing of uniaxial compressive strength and deformability of rock samples. Measure Sci Rev 11(4)
3. Di Matteo L, Valigi D, Ricco R (2013) Laboratory shear strength parameters of cohesive soils; variability and potential effects on slope stability. Bull Eng Geol Environ 72:101–106
4. Guidelines for Estimation and Expression of Uncertainty in Measurement: NABL 141 (2016)
5. Nguyen G, Hrubesova E, Marsalek J, Petpik T, Mohyla M (2015) Uncertainty of shear strength parameter of soil reinforced by plastic waste. In: Applied mechanics and materials, vols 744–746, pp 695–701. ISSN: 1662-7482
6. Husein Malkawi AI, Hassan WF, Abdulla FA (2000) Uncertainty and reliability analysis applied to slope stability. Civil Engineering Department, Jordan University of Science and Technology, Irbid
7. Michael Duncan J (2000) Factors of safety and reliability in geotechnical engineering. J Geotech Geoenviron Eng 307
8. Chalermyanont T, Benson CH (2004) Reliability-based design for internal stability of mechanically stabilized earth walls. J Geotech Geoenviron Eng 130(2)
9. Bareither CA, Benson CH, Edil TB (2008) Reproducibility of direct shear tests conducted on granular backfill materials. Geotech Test J 31(1)
10. Alshibli KA, Okeil AM, Alramahi B, Zhang Z (2009) Statistical assessment of repeatability of CPT measurements. In: 2009 International Foundation Congress and Equipment Expo Contemporary Topics in In Situ Testing, Analysis, and Reliability of Foundations
11. Qasim S, Harahap I (2012) Geotechnical uncertainty and reliability theory applications. Int J Eng Res Technol (IJERT) 1(6). ISSN: 2278-0181
12. Ding D, Likos WJ, Erik Loehr J (2016) Variability and uncertainty in consolidation and settlement parameters from different sampling and testing methods. From Soil Behavior Fundamentals to Innovations in Geotechnical Engineering

Analysis of Slope Stability for Hill-Based Construction in NE India: A Case Study in Guwahati

Anup Kaushik Sharma and Utpal Kumar Das

1 Introduction

In the Indian state of Assam, the city of Guwahati is expanding quickly. It is bounded by hills on three sides. The city is extremely vulnerable to landslides due to its mountainous landscape, which is characterized by steep slopes, rocky outcrops, and deep valleys. The vulnerability of the hill slopes of Guwahati to landslides is affected by a number of factors, including: Geology, Topography, Land use changes, and Climate, which essentially includes the effect of monsoon rains.

As per a Rapid Visual Screening (RVS) [1] project carried out by the Assam Disaster Management Authority, Guwahati has been identified as one of the high landslide susceptible zones in 2012. The ASDMA project identified around 366 landslide-prone areas in the Kamran Metropolitan District, including the study area.

In terms of the number of reported landslide incidents, several unfortunate events have happened in Guwahati during this period, particularly during the monsoon season. In June 2020, a massive landslide in the Lalmati area led to the collapse of several houses, resulting in the death of three people. Several others were injured in the incident. Three months later, in September 2020, another landslide claimed the lives of three people in the Panjabari area of Guwahati and collapse of a residential building. Both the events have been preceded by rainfall, which signifies that the vulnerability of Guwahati's hill slopes towards landslide is remarkably triggered by rainfall, particularly during the monsoon season. The heavy rainfall during this period may saturate the soil and rock layers, leading to increased pore water pressure and reduced shear strength. This, in turn, can make the slopes more susceptible to landslides.

A. K. Sharma (✉) · U. K. Das
Department of Civil Engineering, Tezpur University, Napam, Sonitpur, Assam 784028, India
e-mail: anupksrma@gmail.com

U. K. Das
e-mail: ukrdas@tezu.ernet.in

S. Kolathayar et al. (eds.), *Best Practices in Geotechnical and Pavement Engineering*,
Lecture Notes in Civil Engineering 449, https://doi.org/10.1007/978-981-99-8505-0_16

Also, the likelihood and severity of landslides can be significantly influenced by the geometry of a slope, including its angle of inclination. A slope's stability is influenced by a number of variables, such as the slope's external stresses, groundwater conditions, and soil qualities. However, a slope's vulnerability to sliding is greatly influenced by its geometry, mainly steepness and shape.

2 Hill Soil Characterization

The hill soils in Guwahati and the adjoining areas are predominantly composed of residual soils, which are formed by the weathering of underlying rocks. The top soil is typically reddish-brown in colour, with a fine-to-coarse texture depending on the degree of weathering and the parent rock. This is often underlain by a pale yellowish, sandy rather fragile soil constituting the bulk of the slope body [2].

2.1 Index Property and Geotechnical Property Determination

The present investigative study has been performed on undisturbed as well as remoulded samples collected from three different sites of the Khanapara–Jorabat area of the Greater Guwahati municipal region (26° 6.009′ N, 91° 52.248′ E, 26° 6.147′ N, 91° 51.754′ E, 26°7′ 5.46″ N, 91° 56′ 4.44″ E). Laboratory investigations of the physical and engineering properties of the both the soils have been conducted, namely specific gravity (*G*), particle size determination, Atterberg's limits [i.e. liquid limit (LL), plastic limit (PL)], infiltration characteristics, and shear parameters. The experimentations were carried out at the Geotechnical laboratory of Department of Civil Engineering, School of Engineering, Tezpur University.

Visual and Geological Characteristics

The two soils were taken from the same hillside site at different depth conditions and are visually identified in Fig. 1a, b. Although the soils come from the same location, it can be seen that they exhibit noticeably different visual characteristics. The topsoil is depicted in Fig. 1a contained at depths ranging from 100 cm to around 30 m. When compared to the other variety, it had a brick red tone and felt tougher and more cohesive to the touch. This layer is a lateritic formation of residual silty clay, according to geology. Underlying this layer is another type of soil that is yellowish to pale-yellow in colour. In comparison to the top layer, it is considerably more porous and brittle. Geologically, it is a saprolite deposit made up of residual soil that resulted from the parent bedrock's isovolumetric weathering.

Specific Gravity Determination

Using the standard density bottle (100 ml) method, the specific gravities of the soils were calculated. Table 1 represents the specific gravity values of reddish-type and

(a) (b)

Fig. 1 Samples of **a** reddish soil and **b** pale-yellowish soil

pale-yellow-type soils. The ratio of the mass of dry particles to the mass of water that they displace is known as specific gravity (*G*). The *G*-value for the majority of soils typically ranges from 2.60 to 2.80.

Density Determination

The in situ bulk and dry density of the samples were determined using the standard water displacement method. Table 2 represents the densities obtained for reddish-type and pale-yellow-type soils.

Particle Size Determination

The classification of the grains into several categories of coarse and fine fractions is done using particle size distribution. In light of this, dry sieving was carried out in accordance with standard method in accordance with IS: 2720-1985 [3], and it was found that the reddish soil contains around 77.6% fines, whereas the yellowish soil around 7.6% fines. Wet sieving was also performed for the reddish-type soil. For particles less than 75 μm, a hydrometer study has also been carried out in accordance with IS: 2720 (Part-IV)-1985 [4] to determine the percentage of silt and clay in the soil from the coefficient of curvature and coefficient of uniformity

Table 1 Specific gravity of reddish soil and pale-yellowish soil

Soil type	Reddish soil	Pale-yellowish soil
Specific gravity	2.42	2.63

Table 2 In situ density of reddish soil and dale yellowish soil

Property	Reddish soil	Pale yellowish soil
Bulk Density (g/cc)	1.64	1.60
Dry density (g/cc)	1.48	1.40

Table 3 Coefficient of uniformity (Cu) and coefficient of curvature (Cc)

Property	Reddish soil	Pale-yellowish soil
Cu	4.1	5.2
Cc	1.0	1.4

values represented in Table 3, and the reddish-type soil with a significant and well-distributed representation of silt and clay should be classified as well-graded silty clay. The yellowish soil, however, with negligible amount of clay, but a good fraction of silt and sand fractions, may be classified as gap-graded silty sand.

Liquid Limit and Plastic Limit

The Atterberg limits define the various levels of consistency. Using the static cone penetration method in accordance with IS: 2720 (Part V)-1985 [5], the liquid limit was calculated. The average LL was determined to be 48% for reddish soil and 35% for yellowish soil, respectively.

The plastic limit for reddish soil was found to be 29% using the standard process in accordance with IS: 2720 (Part-V)-1985 [5]. However, because the yellowish soil is relatively sandy, it was not possible to estimate the plastic limit.

The reddish-type soil has been found to be a clay with intermediate plasticity with plasticity index 21%.

Infiltration Characteristics: Permeability Determination

The ability of a porous substance to allow water to seep through the connecting voids is known as permeability. As a result, it is a crucial factor controlling both seepage and infiltration. Because both types of soils mostly contain clay, silt, and sand, with relatively modest proportions of coarse and gravel, the falling head permeability test in accordance with IS2720 (Part 17)-1986 [6] was employed to measure permeability. As a result, it was discovered that the average permeability of the reddish-type soil was 1.59×10^{-7} m/s and that of the yellowish-type soil was 1.44×10^{-6} m/s.

2.2 *Shear Parameters and Engineering Properties*

For determination of shear parameters, direct shear test has been preferred by many researchers in the past, as because of its simplicity and applicability to a wide range of soils [7], and also, it consumes less time, as shorter drainage path is observed [8]. In the present study, particularly in case of the saprolitic soil, undisturbed sampling and testing of the same were not found possible by any other method, which outlines direct shear test as the only alternative. The shear parameters obtained for reddish soil and pale-yellowish soil are represented in Table 4.

The samples were collected from exposed cut surfaces of hill sites and tested for determination of soil properties in the laboratory separately for reddish and yellowish soils under the following conditions.

Table 4 Shear parameters of reddish soil and pale yellowish soil

Property		Reddish soil	Pale-yellowish soil
Cohesion C (kPa)	Undisturbed	17.45	3.85
	Remoulded	16.23	1.53
Angle of friction, $\varphi°$	Undisturbed	27.7°	39.7°
	Remoulded	27.0°	34.11°
Unit weight, $\Upsilon_{unsaturated}$ (kN/m^3)		14.26	16.06
Unit weight, $\Upsilon_{saturated}$ (kN/m^3)		18.05	19.55
In situ moisture content (%)		1.53	2.2

(1) From this study, the shear parameters were determined based on the original Mohr–Coulomb equation

$$\tau' = c' + (\sigma - u_w)\tan\varphi'.$$

Shearing was performed on a predetermined shearing plan under undrained conditions. The experimentation were performed at in situ moisture content. The measurement of pore water pressure u_w was not possible.

(2) Strain rate of 0.25 mm/min was maintained in all the tests.

(3) The method adopted for collection of undisturbed samples of the brittle yellowish soil and detail procedure for shear strength determination in direct shear apparatus are described in Sharma and Das [9].

3 Stability Analysis

Strength characteristics obtained from both the undisturbed and remoulded samples were used in the stability analysis of the hill slope. The typical slope profile utilized in this research is depicted in Fig. 2. The slope's height was estimated at 20 m and span of 35 m. A two-layer system of reddish lateritic soil at the top and pale yellowish at the bottom is considered to constitute the slope body. Based on field observations, it was also assumed that the groundwater table was very deep. On the basis of the hill slopes in study's field conditions, these dimensions were presumptive. It was believed that the test slip surfaces would be circular and would be located above the slope's toe.

Safety factors were calculated for slopes with an inclination of 35°, 45°, 55°, 65°, and 75° with regard to the horizontal, respectively. Soil profiles were considered at in situ moisture content only, and accordingly, the shear and engineering properties have been put into the simulation. The study analysis was completed using Morgenstern-Price method of slope stability analysis using the SLOPE/W software. In Fig. 3a, b, the calculated factors of safety are shown against the slope angle for both the undisturbed and the remoulded examples.

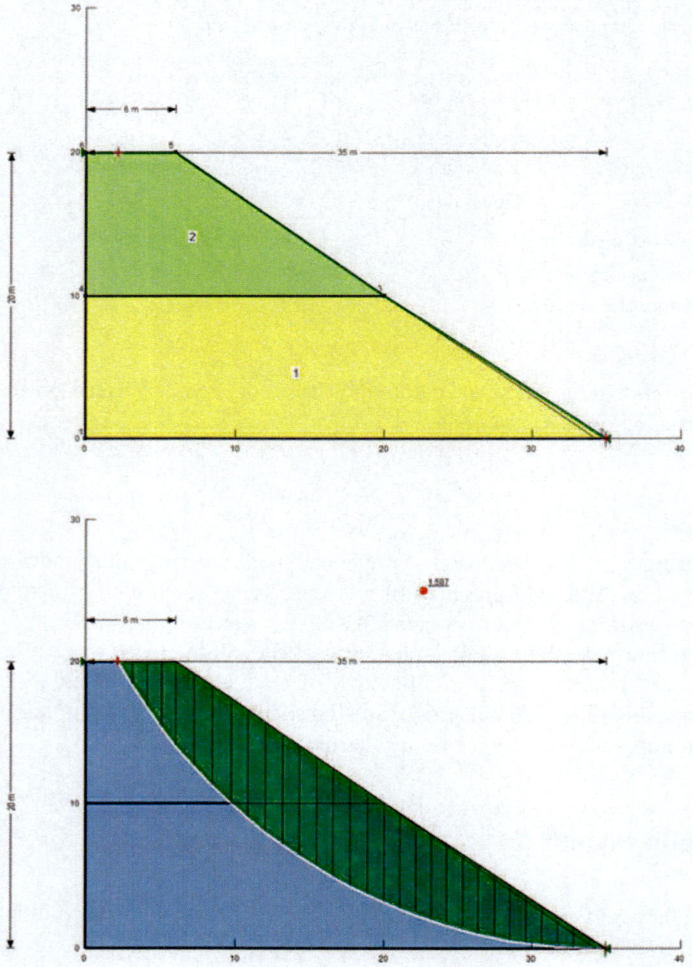

Fig. 2 Typical 35° slope used in the analysis before and after failure

The stability of slopes appears to decrease significantly with increase in steepness, with around 40% decrease in FoS can be attributed for about 30°–40° increase in slope angle.

The gain in stability in terms of critical slope angle in undisturbed soil condition over remoulded soil condition have also been determined and given in Table 5. The undisturbed strength parameters have contributed to higher stability of the slopes.

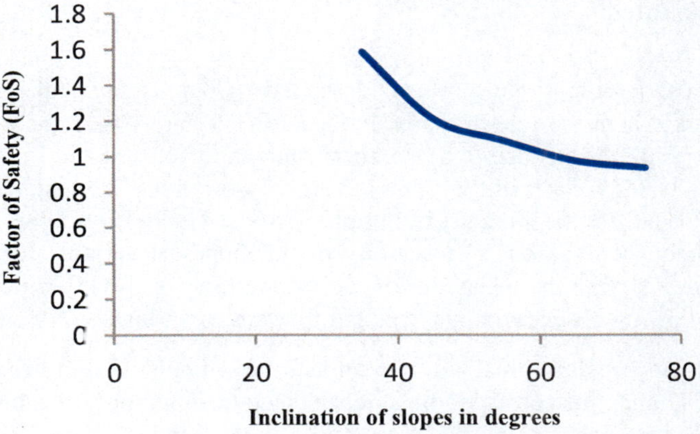

(a)

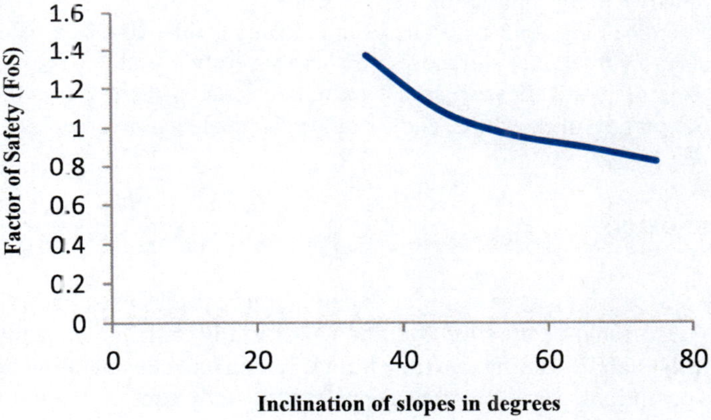

(b)

Fig. 3 Variation of factor of safety with slope angle for **a** Undisturbed and **b** Remoulded slopes

Table 5 Comparison of parameters

Parameters	Undisturbed slope	Remoulded slope
Critical slope angle	58°	47°
Difference in critical slope angle	11°	
% Difference	23.4	
Factor of safety for slope angle 35°	1.587	1.376
Factor of safety for slope angle 75°	0.939	0.83
Loss of FoS in gain in steepness 35°–75°	0.648	0.546
% Loss	41	39.7

4 Discussion

The Northeastern India is rain surplus with around 1500 mm annual rainfall received by the region where the research area is located. The monsoon rains account for most of the rainfall from June to September, while the winter season i.e. November to March, is usually dry with very low levels of saturation. The residual soil that constitutes the bulk of the hill mass is thus often exposed to low saturation, especially in the non-monsoon season. The stability of the slopes at various steepnesses is important for studying the behaviour of such slopes during and after monsoon. As such, the following observations may be put forward as findings.

(1) The stability of slope obtained from undisturbed samples is significantly higher (10–15%) than the corresponding slopes made from remoulded samples.
(2) Critical slope angle is about 23% more for undisturbed soil slope.
(3) As the steepness increases, the FoS decreases. A 35° slope is found to have around 40% more stability than a 75° slope.
(4) Slopes in the range of 30–40° are usually firmly stable, 40–50° are fairly stable, 50–60°are vulnerable, and those > 60° are severely unstable.
(5) The role of pale-yellowish soil is extremely crucial, as it forms the inner or bottom portion of the slope, and its failure is brittle and irreversible.

5 Conclusion

The study was representative of highly steep ground profiles existing in the Greater Guwahati region at low saturation condition particularly prior to monsoon in order to investigate the stability during that time frame. Taking into consideration the possible in situ soil conditions, as well as slope geometry, the study aims at simulating as much ground parameters as possible. The results find significant more resistance offered by the slopes in their in situ conditions over models constituted by remoulded soils collected from the same sites. The study also proposes a viable range of steepness for possible instability and for similar slopes. Further, the actual vulnerability of slopes during rainfall can be assessed by introducing the effect of infiltration in the form of wetting front, for which the current study may provide an effective insight. Overall, the geometry of a slope is an important factor in determining its behaviour and the likelihood of landslides. Understanding the geometry of a slope can help engineers and geologists to design effective slope stabilization measures and reduce the risk of landslides.

References

1. Assam State Disaster Management Authority (2012) Executive summary of rapid visual screening for potential landslide areas of Guwahati. http://sdmassam.nic.in/studies&projects.html
2. Das UK, Saikia BD (2010) Shear strength of unsaturated residual soils of the hills in Guwahati. Indian Geotechnical Conference, Mumbai
3. IS: 2720-1985
4. IS: 2720 (Part-IV)-1985
5. IS:2720 (Part V)-1985
6. IS2720 (Part 17)-1986
7. Simoni A, Houlsby GT (2006) The direct shear strength and dilatancy of sand–gravel mixtures. Geotech Geol Eng 24:523–549
8. Gan JKM, Fredlund DG (1988) Multistage direct shear testing of unsaturated soils. ASTM Geotech Test J 1(2):132–138
9. Sharma AK, Das UK (2022) Stability of undisturbed residual soil hill slopes. Geotech Eng J SEAGS AGSSEA 53(2):29–34

Estimation of Unconfined Compressive Strength of Cohesive Soils in and Around Mysore, South India

P. Nagendra, M. Samarth Urs, K. N. Prakash Narasimha, C. Vinay, and M. Savitha

1 Introduction

Qualitative analysis of the strength of soil and rocks is carried out by the unconfined compressive strength test (UCS) [4, 6]. Subsequently, there are several testing methods available across different standards to determine the UCS for soils [10]. A simple technique for determining the cohesion (C) and internal friction angle (α), of soils and stable materials requiring information on solely unconfined compressive strength, is bestowed. The relationship of linear equations between major and minor principal stresses or normal stress (σ) and shear stress (τ) on the failure plane defines the Mohr–Coulomb (MC) failure theory [1, 9, 11]. Even in the case of large-scale in situ tests, [24] emphasized that test results shouldn't be treated as an absolute property because even the most accurate in situ test is still merely an approximation approach used on a small portion of the geological unit being researched. According to [25], the effective stress for an undisturbed sample is typically different from its in situ effective stress. The variation represents that the level of disturbance the sample had experienced. As in [19], the term "disturbance" is used to describe all external factors that will alter the effective stress, such as mechanical disturbances related to sample collection, transportation, and successive operations for specimen preparation. A procedure for establishing C and α from the MC failure envelope constructed using the UCS method is outlined. In soil mechanics, the UCS test is considered to be the most accepted method to obtain soil strength properties [5, 26, 27].

P. Nagendra · M. Samarth Urs (✉) · K. N. Prakash Narasimha · C. Vinay · M. Savitha
Department of Studies in Earth Science, University of Mysore, Mysuru, India
e-mail: murssamarth@gmail.com

S. Kolathayar et al. (eds.), *Best Practices in Geotechnical and Pavement Engineering*, Lecture Notes in Civil Engineering 449, https://doi.org/10.1007/978-981-99-8505-0_17

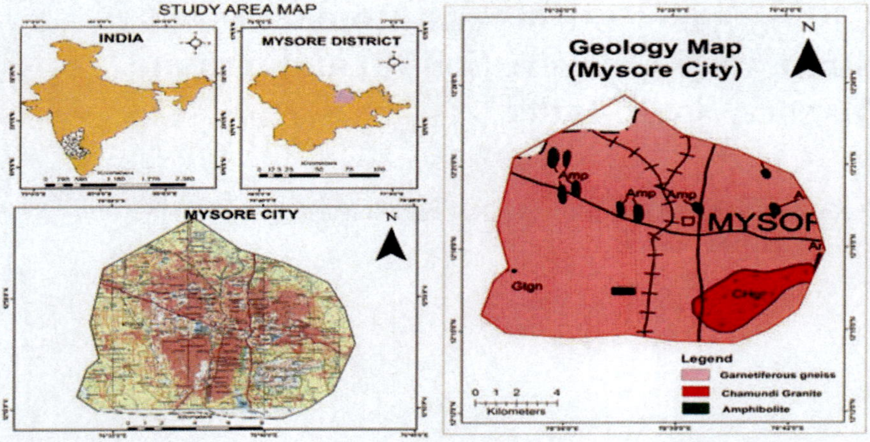

(a) (b)

Fig. 1 **a** Study area map of Mysore city. **b** Geology map of Mysore city [18]

2 Materials and Methods

2.1 Geology of the Study Area

The Mysore city of Karnataka state, South India, is bounded by Latitudes 12° 12′ 00″ N–12° 24′ 00″ N and Longitudes 76° 36′ 00″ E–76° 42′ 00″ E covering an area of 172 km^2 [22]. Most of the rock types of Mysore city belong to the ancient supracrustal rocks of the Sargur group. These rocks have been subjected to three cycles of deformation and two cycles of metamorphism which is greater than 3 B.y of amphibolite facies and 2.5 B.y of old upper amphibolite to granulite facies of metamorphism Fig. 1a, b [15–18]. During Neoproterozoic times, granite batholiths were emplaced and are presently well-known as Chammundi Granites (800 m.y). These granites are situated in the SE direction of Mysore taluk and their extensions are seen in H. D. Kote, T. Narsipura, and K. R. Nagar taluks. Based on the records of the Geological Survey of India, the city comprises many lithological units out of which ultramafic rocks, dolerites, felsite, lamprophyres, metagabbro, etc., are predominantly found as discordant bodies within granitic-gneiss terrain.

2.2 Soil of Study Area

Soil plays a vital role in augmenting the groundwater recharge and also shaping the parameters for the betterment of foundation design. The study area consists of sandy loam, red sandy loam, red loam, and mixed loam. The depth of soil varies from place

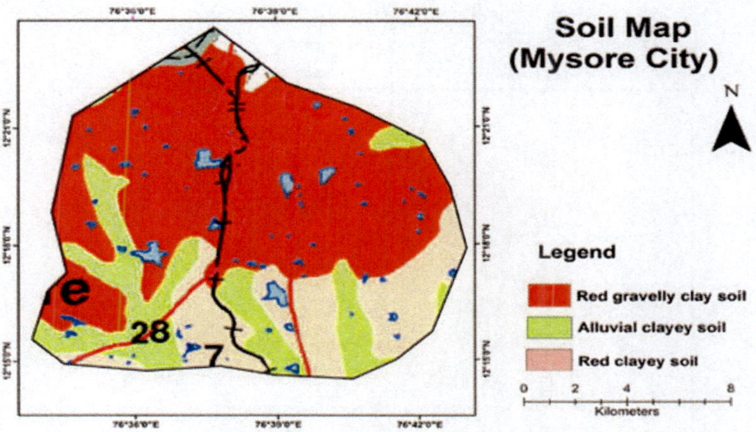

Fig. 2 Soil map of Mysore city (Reference Shiva Prasad et al. [23]. Soils of Karnataka for optimizing land use NBSS Publ. 47)

to place, i.e., from a few centimeters to meters. The description of the soils of the study area is given below (Fig. 2) [15–18].

3 Research Methodology

3.1 Soil Properties

Soil samples comprising clayey sand (SC), silty, clayey sand with gravel (SC–SM), silty sand (SM), and well-graded silty sand (SW–SM) were taken at various locations as shown on the map of Mysore city, Karnataka (Fig. 3) [19]. Table 1 displays the dry density, specific gravity, moisture content, void ratio, plasticity limit, and liquidity limit of those soils. Table 1 also comprises particle size distributions of those soils attained from the sieve analysis; specific iota sizes were calculated using the coefficient of uniformity (C_u), and the coefficient of curvature (C_c) of the soil samples [8, 15, 16]. The coefficients of uniformity of soils vary between 2.0 and 3.0, so these soils are well-graded soil [8].

3.2 Sample Preparation and Testing

According to IS code-2720-10-1991 is followed to determine the unconfined compressive strength of a cylindrical soil specimen required certain equipment's like a compressive device to apply vertical load using appropriate loading device,

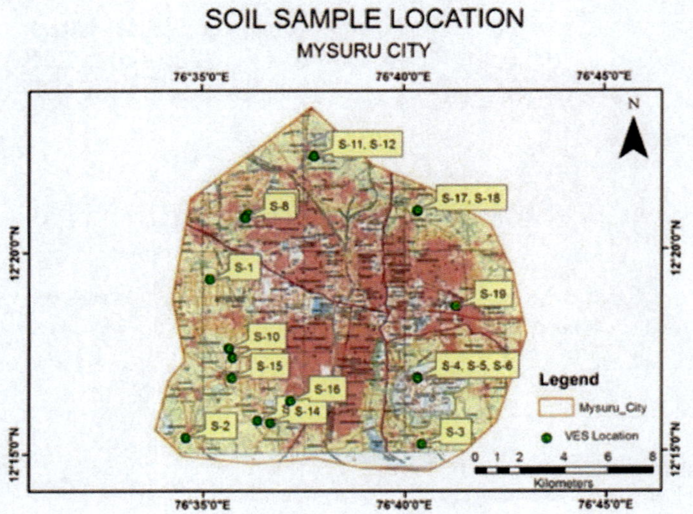

Fig. 3 Soil sample location map of Mysuru city

Table 1 Physical properties of soil samples

Soil	SC	SC–SM	SM	SW–SM
Unit weight (kg/m^3)	1910	1890	1914	1884
Specific gravity (G)	2.65	2.65	2.65	2.65
Water content (%)	9.1	8.5	18.2	10.5
Liquid limit	36.83	35.71	45.66	29.15
Plastic limit	23.16	24.07	31.33	12.88
Void ratio (e)	1.28	1.08	1.19	1.21
Coefficient of uniformity (C_u)	2.58	2.94	2.16	2.87
Coefficient of curvature (C_c)	1.55	1.0	1.14	1.0

sample ejector, dial gauges, vernier calipers, stopwatch, oven, balances, etc. [7]. This test is performed on a cylindrical sample. The sample is subjected to compression until it fails (Fig. 4) [20, 21].

Before conducting the test, the mold is well-greased and later filled with a mixture of soil and water. Before conducting the test, the initial length and the internal diameter of the soil specimen and mold are measured. The mold is unlocked and the sample is removed cautiously. Likewise, two more trials were carried out. The specimen is placed in between the adjustable loading plates and stress is applied until 20% of deformation is observed in the dial gauge, for which a systematic stress–strain curve is obtained. This procedure is repeated for all the samples and the water

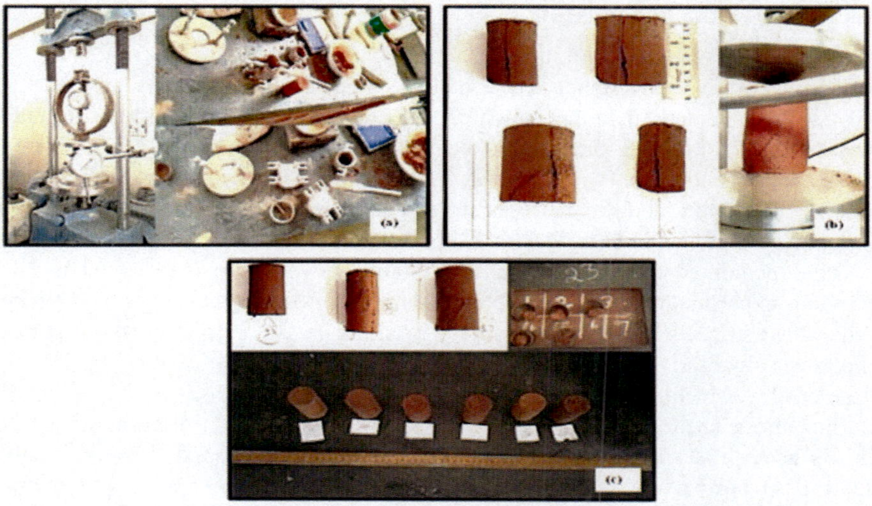

Fig. 4 Unconfined compressive strength test setup. **a** Instrumentation setup **b**, **c** Soil sample after deformation

content of each sample is determined [12–14]. It is possible for cohesive soil only. Initial observations are noted for stress and strain calculations, which are as follows (Table 2).

Table 2 Unconfined compression test values for soil samples

S. No.	Locations	Normal stress σ in kg/cm^2	Shear stress S_u in kg/cm^2
1	S-1	0.230	0.115
2	S-2	0.242	0.121
3	S-3	0.369	0.185
4	S-4	0.089	0.045
5	S-5	0.140	0.070
6	S-6	0.102	0.051
7	S-7	2.453	1.226
8	S-8	9.089	4.544
9	S-9	9.698	4.849
10	S-10	8.374	4.187
12	S-16	4.658	2.329
13	S-17	4.746	2.373
14	S-46	2.249	1.124

4 Result and Discussion

According to the ASTM standard [2, 3, 17], the unconfined compressive strength tests are used to calculate the unconsolidated undrained shear strength of the soil under unconfined conditions. From the test, the following results are plotted between stress (σ in kg/cm^2) and strain (ε in %) (Table 1). Mohr circle for factor of safety is plotted to determine undrained shear stress (Su) in kg/cm^2 and maximum normal stress (σ in kg/cm^2) as shown in Figs. 5, 6, 7, 8, 9, 10, 11, 12, 13, 14, 15, 16, and 17.

The unconfined compressive test results are classified into four types based on previous experimental results. For clayey sand (SC) soil, normal stress is 1.693 kg/cm^2, shear stress is 0.847 kg/cm^2, and internal angle of friction (α) is 57°; hence, the sensitivity of the soil indicates slightly soft due to presence of fines (> 12%). In case of clayey sand to sandy silt soil (SC–SM) which comprises 6.428 kg/cm^2 of normal stress, shear stress is 3.214 kg/cm^2 and internal angle of friction (α) is 62°, this indicates that the soil is hard due to less fines and more sand, whereas in silty sand (SM), normal stress is 4.769 kg/cm^2, shear stress is 2.384 kg/m^2, and internal angle of friction (α) is 58°. Lastly, well-graded sandy silt soil (SW–SM) consists of 4.070 kg/cm^2, shear stress is 2.035 kg/cm^2, and internal angle of friction (α) is 64°.

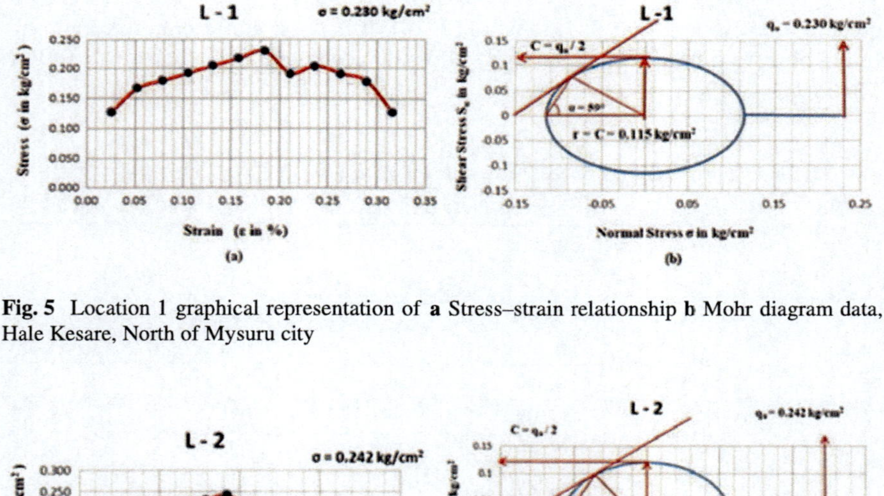

Fig. 5 Location 1 graphical representation of **a** Stress–strain relationship **b** Mohr diagram data, Hale Kesare, North of Mysuru city

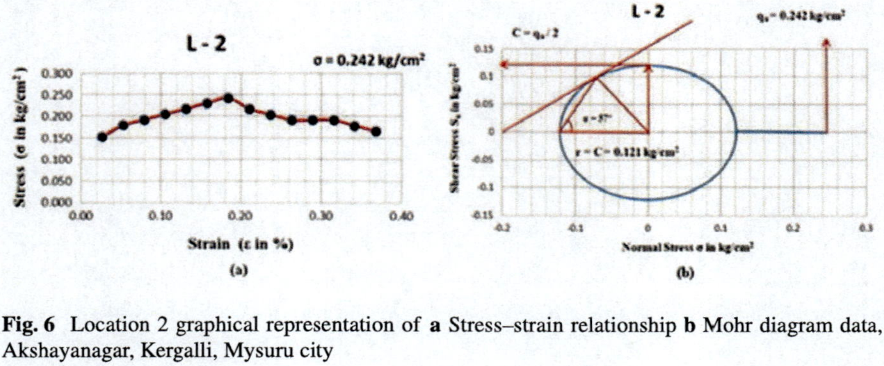

Fig. 6 Location 2 graphical representation of **a** Stress–strain relationship **b** Mohr diagram data, Akshayanagar, Kergalli, Mysuru city

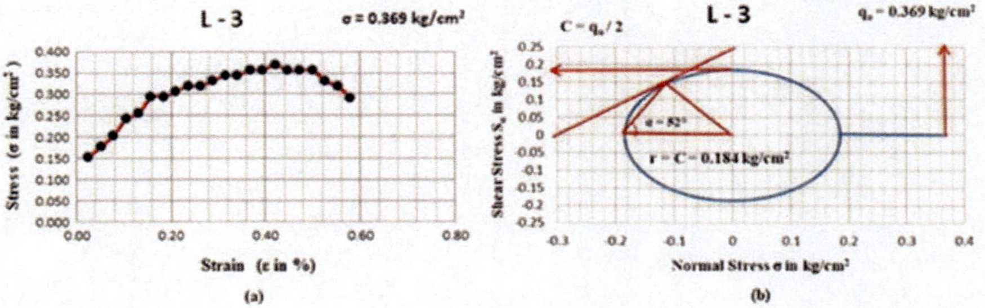

Fig. 7 Location 3 graphical representation of **a** Stress–strain relationship **b** Mohr diagram data, Heggadahalli, Mysuru Taluk

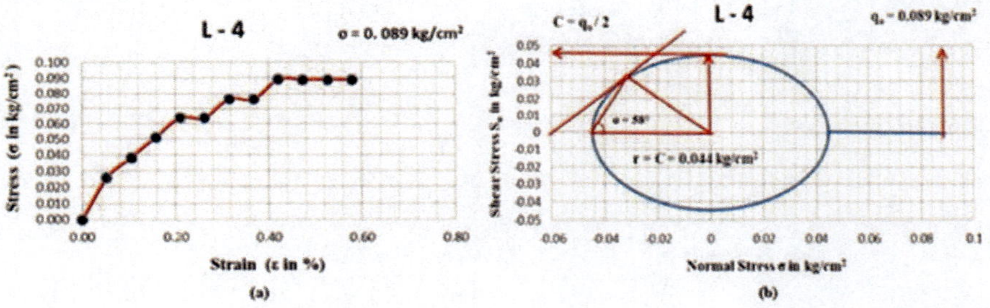

Fig. 8 Location 4 graphical representation of **a** Stress–strain relationship **b** Mohr diagram data, near Nandi temple, Chamundi hill, South of Mysuru city

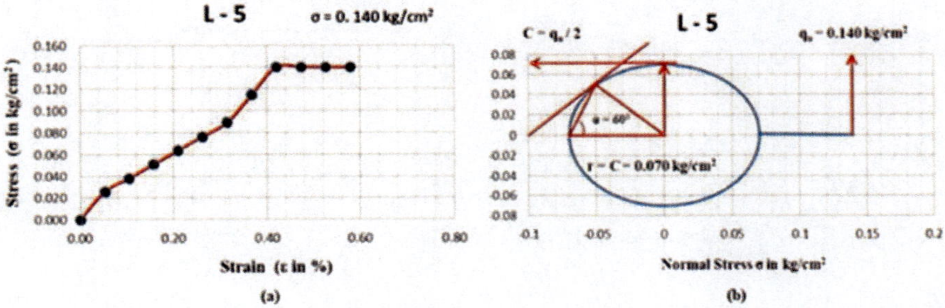

Fig. 9 Location 5 graphical representation of **a** Stress–strain relationship **b** Mohr diagram data, near Nandi temple, Chamundi hill, Mysuru city

From the above results of stress–strain diagram and Mohr circle exhibits average values of normal stress, shear stress and internal angle of friction for 4 major types of soil of study area and the sensitivity of soil of all locations exhibits medium to hard cohesive soil and factor of safety is within the permissible limit; hence, the soil

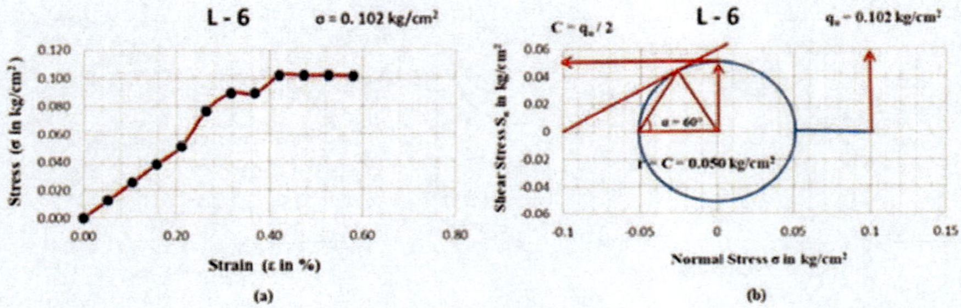

Fig. 10 Location 6 graphical representation of **a** Stress–strain relationship **b** Mohr diagram data, near Nandi temple, Chamundi hill, South of Mysore city

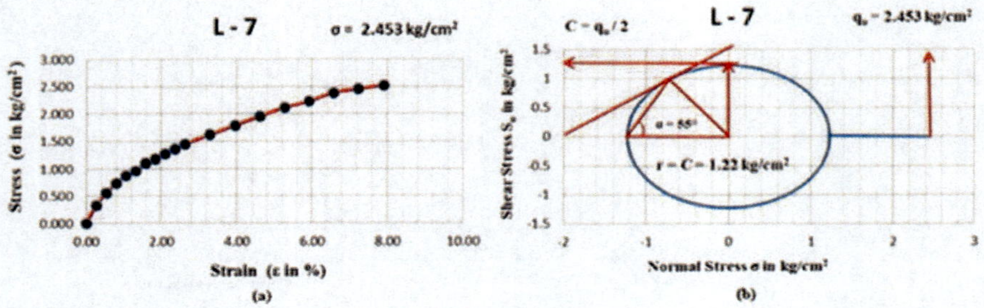

Fig. 11 Location 7 graphical representation of **a** Stress–strain relationship **b** Mohr diagram data, Hebbal 2nd stage, Mysuru city

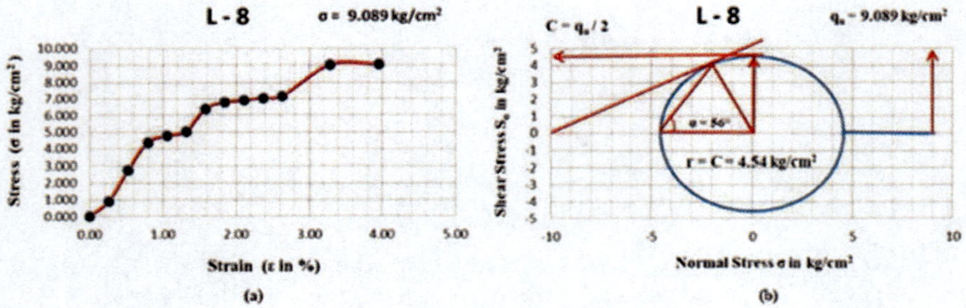

Fig. 12 Location 8 graphical representation of **a** Stress–strain relationship **b** Mohr diagram data, Vijaynagar 4th stage, Mysuru city

exhibits firm to stiffness; therefore, it is predominantly suitable for the construction of multistoried structures (Figs. 5, 6, 7, 8, 9, 10, 11, 12, 13, 14, 15, 16, and 17).

Results

According to the ASTM standard, the unconfined compressive strength tests are used to calculate the unconsolidated undrained shear strength of the soil under unconfined

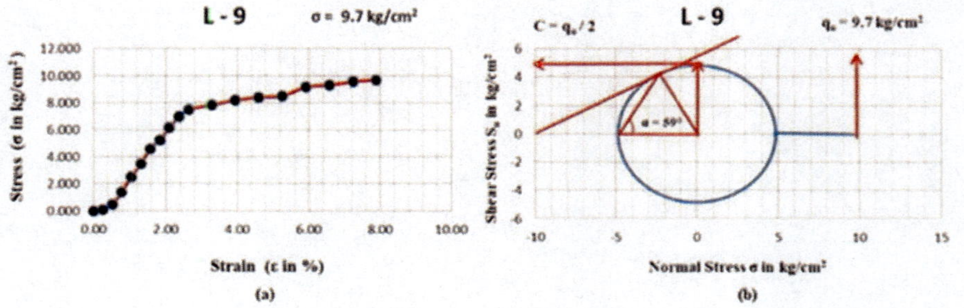

Fig. 13 Location 9 graphical representation of **a** Stress–strain relationship **b** Mohr diagram data, BMEL Layout 2nd stage, Mysuru city

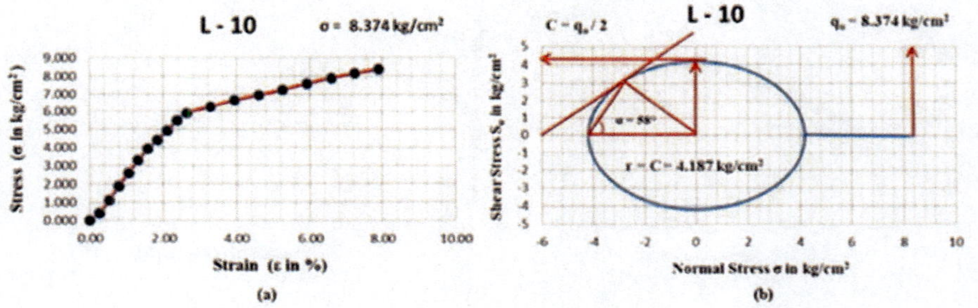

Fig. 14 Location 10 graphical representation of **a** Stress–strain relationship **b** Mohr diagram data, Rajarajeshwari Nagar, Mysuru city

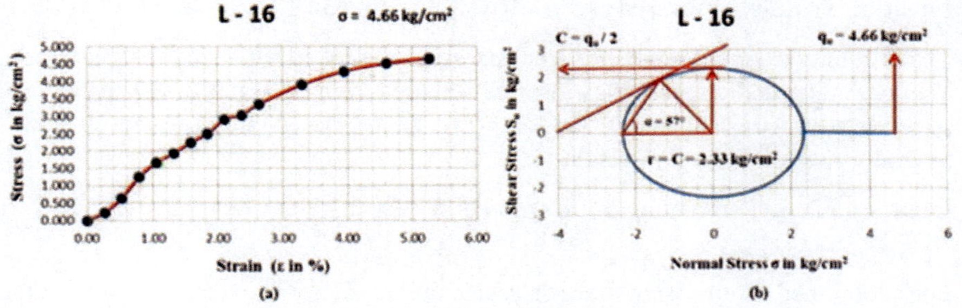

Fig. 15 Location 16 graphical representation of **a** Stress–strain relationship **b** Mohr diagram data, Madhuvana layout, Mysuru city

conditions. From the test, the following results are plotted between stress (σ in kg/cm^2) and strain (ε in %). Mohr circle for the factor of safety plotted to determine undrained shear stress (Su) in kg/cm^2 and maximum normal stress (σ in kg/cm^2).

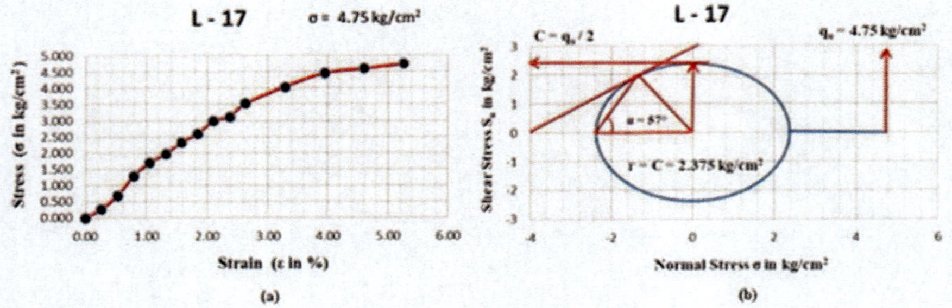

Fig. 16 Location 17 graphical representation of **a** Stress–strain relationship **b** Mohr diagram data, J.P. Nagar 3rd stage, Mysuru city

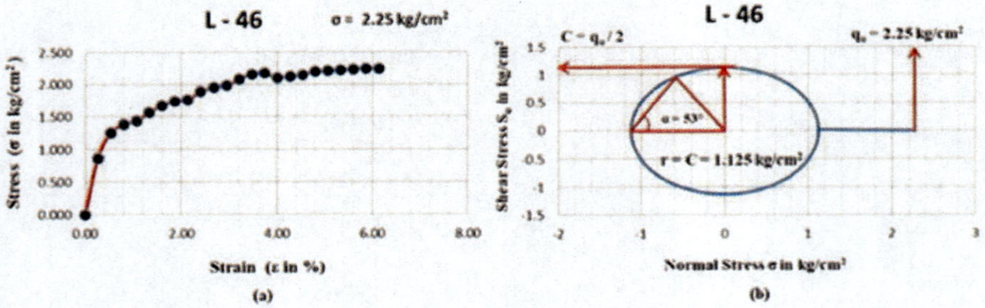

Fig. 17 Location 46 graphical representation of **a** Stress–strain relationship **b** Mohr diagram data, Shankara Nagar, Mysuru city

Location 1: Hale Kesare, Mysuru City (N 12° 20′ 43.2″ E 76° 40′ 36.1″)

The unconfined compressive strength test results are plotted between stress and strain. The maximum undrained stress (σ) value obtained from the test is 0.230 kg/cm^2 and cohesive shear stress Su is 0.115 kg/cm^2 (Fig. 5a, b). Based on the test result, soil is basically soft to medium in condition.

Location 2: Akshayanagar, Kergalli, Mysuru City (N 12° 15′ 23″ E 76° 34′ 35.21″)

The unconfined compressive strength test results are plotted between stress and strain. The maximum undrained stress (σ) value obtained from the test is 0.242 kg/cm^2 and cohesive shear stress Su is 0.121 kg/cm^2 (Fig. 6a, b). Based on the test result, soil is basically soft to medium in condition.

Location 3: Heggadahalli, Mysuru Taluk (N 12° 05′ 12.13″ E 76° 40′ 25.12″)

The unconfined compressive strength test results are plotted between stress and strain. The maximum undrained stress (σ) value obtained from the test is 0.369 kg/cm^2 and cohesive shear stress Su is 0.185 kg/cm^2 (Fig. 7a, b). Based on the test result, soil is basically soft to medium in condition.

Location 4: Near Nandi Temple, Chamundi Hill, Mysuru City (N 12° 16′ 48.5″ E76° 40′ 8.25″)

The unconfined compressive strength test results are plotted between stress and strain. The maximum undrained stress (σ) value obtained from the test is 0.369 kg/cm^2 and cohesive shear stress Su is 0.185 kg/cm^2. From the result, sensitivity of the soil is soft to medium in condition (Fig. 8a, b).

Location 5: Near Nandi Temple, Chamundi Hill, Mysuru City (N 12° 16′ 48.5″ E76° 40′ 8.25″)

The unconfined compressive strength test results are plotted between stress and strain. The maximum undrained stress (σ) value obtained from the test is 0.369 kg/cm^2 and cohesive shear stress Su is 0.185 kg/cm^2. From the result, sensitivity of the soil is soft to medium in condition (Fig. 9a, b).

Location 6: Near Nandi Temple, Chamundi Hill, Mysuru City (N 12° 16′ 48.5″ E76° 40′ 8.25″)

The unconfined compressive strength test results are plotted between stress and strain. The maximum undrained stress (σ) value obtained from the test is 0.102 kg/cm^2 and cohesive shear stress Su is 0.051 kg/cm^2. Based on the test result, soil is basically soft to medium in condition (Fig. 10a, b).

Location 7: Hebbal 2nd Stage, Mysuru City (N 12° 20′ 50.2″ E 76° 36′ 03.5″)

The unconfined compressive strength test results are plotted between stress and strain. The maximum undrained stress (σ) value obtained from the test is 2.453 kg/cm^2 and cohesive shear stress Su is 1.226 kg/cm^2. Based on the test result, soil is basically soft to medium in condition (Fig. 11a, b).

Location 8: Vijaynagar 4th Stage, Mysuru City (N 12° 20′ 56.3″ E 76° 36′ 07.7″)

The unconfined compressive strength test results are plotted between stress and strain. The maximum undrained stress (σ) value obtained from the test is 9.089 kg/cm^2 and cohesive shear stress Su is 4.544 kg/cm^2. Based on the test result, soil is basically soft to medium in condition (Fig. 12a, b).

Location 9: BMEL Layout 2nd Stage, Mysuru City (N 12° 17′ 22″ E 76° 35′ 44″)

The unconfined compressive strength test results are plotted between stress and strain. The maximum undrained stress (σ) value obtained from the test is 9.7 kg/cm^2 and cohesive shear stress Su is 4.85 kg/cm^2. Based on the test result, soil is basically soft to medium in condition (Fig. 13a, b).

Location 10: Rajarajeshwari Nagar, Mysuru City (N 12° 17′ 36″ E 76° 35′ 39″)

The unconfined compressive strength test results are plotted between stress and strain. The maximum undrained stress (σ) value obtained from the test is 8.374 kg/cm^2 and cohesive shear stress Su is 4.187 kg/cm^2. Based on the test result, soil is basically soft to medium in condition (Fig. 14a, b).

Location 16: Madhuvana Layout, Mysuru City (N 12° 16′ 14.8″ E 76° 37′ 09.1″)

The unconfined compressive strength test results are plotted between stress and strain. The maximum undrained stress (σ) value obtained from the test is 4.66 kg/cm^2 and cohesive shear stress Su is 2.33 kg/cm^2. Based on the test result, soil is basically soft to medium in condition (Fig. 15a, b).

Location 17: J.P. Nagar 3rd Stage, Mysuru City (N 12° 15′ 24″ E 76° 38′ 48″)

The unconfined compressive strength test results are plotted between stress and strain. The maximum undrained stress (σ) value obtained from the test is 4.75 kg/cm^2 and cohesive shear stress Su is 2.375 kg/cm^2. Based on the test result, soil is basically soft to medium in condition (Fig. 16a, b).

Location 46: Shankara Nagar, Mysuru City (N 12° 15′ 47.8″ E 76° 36′ 20.4″)

The unconfined compressive strength test results are plotted between stress and strain. The maximum undrained stress (σ) value obtained from the test is 2.25 kg/cm^2 and cohesive shear stress Su is 1.125 kg/cm^2. Based on the test result, soil is basically soft to medium in condition (Fig. 17a, b).

5 Conclusion

Unconfined compressive strength test is predominantly carried for cohesive soil to determine stress–strain relationship, factor of safety, and internal friction angle (α) for major types of soil based on PSD analysis. Unconfined compressive strength test for clayey sand (SC) soil, normal stress is 1.693 kg/cm^2, shear stress is 0.847 kg/cm^2 and internal angle of friction (α) is 57°; hence, the sensitivity of the soil indicates slightly soft, due to presence of fines (> 12%). Unconfined compressive strength test for silty, clayey sand Soil (SC–SM) comprises 6.428 kg/cm^2 of normal stress, shear stress is 3.214 kg/cm^2 and internal angle of friction (α) is 62°, and this indicates that soil is hard, due to less fines (< 12%). Unconfined compressive strength test for well-graded sandy silt soil (SW–SM) consists of 4.070 kg/cm^2, shear stress is 2.035 kg/cm^2, and internal angle of friction (α) is 64°. This indicates that soil is medium to hard, due to less fines (< 12%). From the results of stress–strain diagram and Mohr circle exhibits average values of normal stress, shear stress and internal angle of friction for 4 major types of soil of all locations of study area exhibits medium to hard cohesive soil. From Mohr circle graph, internal angle of friction (α) values falls within the range. Hence, factor of safety of the soil is within the permissible limit and consists of high compressive strength.

References

1. AL-Kinani AM, Ahmed MD (2020) Field study of the effect of jet grouting parameters on strength based on tensile and unconfined compressive strength. IOP Conf Ser Mater Sci Eng 737:012083, 1–10. https://doi.org/10.1088/1757-899X/737/1/012083
2. ASTM D2664-86. Standard test method for triaxial compressive strength of undrained rock core specimens without pore pressure measurements. In: Annual book of ASTM standards, 04.08. American Society for Testing and Materials, Philadelphia
3. ASTM D2938-79. Standard test method for unconfined compressive strength of intact rock core specimens. In: Annual book of ASTM standards, 04.08. American Society for Testing and Materials, Philadelphia, pp 390–391
4. Chae J, Kim B, Park S-W, Kato S (2010) Effect of suction on unconfined compressive strength in partly saturated soils. KSCE J Civ Eng 14(3):281–290. https://doi.org/10.1007/s12205-010-0281-7
5. Correia AG, Tinoco J, Cortez P, Lamas L (2020) Information technology in geo-engineering. Springer International Publishing Science and Business Media LLC. https://doi.org/10.1007/978-3-030-32029-4
6. Dasaka SM, Sumesh KS (2011) Effect of coir fiber on the stress-strain behavior of a reconstituted fine-grained soil. J Nat Fibers 8(3):189–204. https://doi.org/10.1080/15440478.2011.601597
7. Ede K, Thummala SR (2022) Soil stabilization with ortho phosphoric acid and micro steel fiber. Mater Today Proc 52:1576–1582
8. Getahun E, Qi S, Guo S, Shi Z (2020) Experimental evaluation of the shear behavior of large landslide slip zone using multistage—multiphase shearing method. Res Square 1–20. https://doi.org/10.21203/rs.3.rs-37863/v1
9. George KP, Uddin W (2000) Subgrade characterization for highway pavement design, final report. Mississippi Department of Transportation, Jackson, MS
10. Guneyli H, Rusen T (2016) Effect of length-to-diameter ratio on the unconfined compressive strength of cohesive soil specimens. Bull Eng Geol Environ 75:793–806. https://doi.org/10.1007/s10064-015-0835-5
11. Harshavardhan S, Rajendra Kumar P (2018) Study of strength variation in cohesive soil with moisture content and time. Int Res J Eng Technol (IRJET) 05(06):759–762. p-ISSN: 2395-0072
12. Lee W, Bohra NC, Altschaeffl AG (1997) Resilient characteristics of dune sand. J Transp Eng ASCE 121(6):502–506
13. Lee W, Bohra NC, Altschaeffl AG, White TD (1997) Resilient modulus of cohesive soils. J Geotech Geoenviron Eng 123(2):131–136
14. Mohammad LN, Gaspard K, Herath A, Nazzal MD (2007) Comparative evaluation of subgrade resilient modulus from non-destructive, in-situ, and laboratory methods. Louisiana Transportation Research Center, LA, USA, p 30
15. Nagendra P, Suresh Kumar BV, Prakash Narasimha KN, Thanmaya BM, Samarth Urs M (2020) Grain size measurement of varied soil particles by Sieve analysis—a case study in Mysore District, South India. J Eng Sci 11(2):686–691. I F-6.54. ISSN NO: 0377-9254.
16. Nagendra P, Thanmaya BM, Samarth Urs M, Nibiya NT, Prakash Narasimha KN, Suresh Kumar BV (2020) Geotechnical analysis of the soils in and around Mysore, South India. J Eng Sci 11(6):127–134. I F-6.54. ISSN NO: 0377-9254.
17. Nagendra P, Ganesh AV, Hema HC, Samarth Urs M, Prakash Narasimha KN, Suresh Kumar BV (2022) The Casagrande method for determining the consistency of granitic soil. Mater Today: Proc. https://doi.org/10.1016/j.matpr.2022.04.992
18. Nagendra P, Samarth Urs M, Thanmaya BM, Nibiya NT, Prakash Narasimha KN, Suresh Kumar BV (2021). Application of VES techniques for building foundations in Mysuru city. Bull Pure Applied Sci Geol (Geol Sci) 40F(1):68–85. ISSN 0970 4639. https://doi.org/10.5958/2320-3234.2021.00007.X
19. Nakase A (1979) Overview: sampling disturbance in cohesive soils. Tsuchi-To-Kiso. Proc JSSMFE Ser 255:7–10

20. Nigerian Geological Survey Agency (2006) Geological and mineral map of Akwa-Ibom State, Nigeria; Olsen RS, Farr JV (1986) Site characterization using the cone penetration test. In: Proceedings, in-situ'86. ASCE specialty conference, Blacksburg. Virginia Pavementinteractive.org 2007. www.pavementinteractive.org/article/resilientmodulus
21. NIRM Annual Report (2008)
22. Ozdemir A (2020) A comparative study of the frequency ratio, analytical hierarchy process, artificial neural networks and fuzzy logic methods for landslide susceptibility mapping: Taşkent (Konya), Turkey. Geotech Geol Eng 38:4129–4157
23. Shiva Prasad CR, Reddy PSA, Sehgal J, Velayutham M (1998) Soils of Karnataka for optimizing land use, vol 47. National Bureau of Soil Survey and Land Use Planning (NBSS LUP) Publ
24. Sun G (1983) Fundamentals of rock mechanics. China Science Publisher, Beijing, p 241
25. Skempton AW, Sowa VA (1963) The behaviour of saturated clays during sampling and testing. Geotechnique 13(4):269–290
26. Thuro K, Plinninger RJ, Zah S, Schutz S (2001) Scale effects in rock strength properties part 1: unconfined compressive test and Brazilian test. In: Eurock 2001 rock mechanics a challenge for society ISRM regional symposium, pp 169–174
27. Vandre B, Budge A, Nussbaum S (1998) DCP-a useful tool for characterizing engineering properties of soils at shallow depths. In: Proceedings of 34th symposium on engineering geology and geotechnical engineering. Utah State University, Logan, UT

Impact of Plastic Waste on the 3D Consolidation Characteristics of Sandy-Silt with Clay Soil

Soumi Koley, Abhijit Debnath, and Sujit Kumar Pal

1 Introduction

Increased urbanization and human activities produce large amounts of waste worldwide. The existing construction materials, like bricks and cement, are natural resources prone to depletion and, thus, costlier. There are many alternatives to these materials, which are economical and environmentally friendly. Therefore, effectively recycling waste contributes to the well-being of the environment and serves the construction purpose. Chauhan et al. [1] investigated the effectiveness of coir fibre and synthetic fibre reinforcement in subgrade soil from the consideration of strength. An experimental program has been performed to find the influence of WPF with different sizes and contents on maximum dry density (MDD) and optimum moisture content (OMC) [2]. Okoro et al. [3] studied the consolidation characteristics of fly ash, lime and plastic waste stabilized soil. A laboratory study has been executed to investigate the impact of WPF on reinforced soil's consolidation and compaction behaviour [4]. Zhang et al. [5] performed a 3D consolidation study of a vertical drain established on continuous drainage boundary. Ibraim and Fourmont [6] studied the influence of randomly oriented discrete crimped polypropylene fibres on the mechanical response of very fine sand. An experimental study has been performed to measure the c_v and hydraulic conductivity of improved Tunis soft soil when imposed vertical and horizontal drainage by geodrains [7]. Babu and Chouksey [8] deliver an approach for using plastic waste as a soil reinforcement material. The influence of rice husk ash and cement mixtures on the engineering properties of cohesive soil has been studied experimentally [9]. Koley et al. [10] performed an experimental study to investigate the consolidation characteristics of sandy-silt with clay soil using a fabricated 3D consolidation apparatus.

S. Koley (✉) · A. Debnath · S. K. Pal
National Institute of Technology Agartala, Agartala, Tripura 799046, India
e-mail: skoley.nita2021@gmail.com

S. Kolathayar et al. (eds.), *Best Practices in Geotechnical and Pavement Engineering*, Lecture Notes in Civil Engineering 449, https://doi.org/10.1007/978-981-99-8505-0_18

Table 1 Soil properties used in the present study

Property	Values
Soil type	Sandy-silt with clay
Specific gravity (G)	2.55
Liquid limit (%)	38.76
Plastic limit (%)	22.57
Plasticity index (%)	16.19
MDD (kN/m^3)	18.24
OMC (%)	10.17
Coefficient of permeability (m/s)	1.326×10^{-8}
Sand (%)	19.76
Silt (%)	61.27
Clay (%)	18.97

The aim of the current study is to perform a laboratory 3D consolidation test to assess the impact of WPF on soil consolidation behaviour. It is clear from the literature that the plastic strips significantly impact the OMC and MDD of the soil. In this study, scrap plastic bottle fibre was employed as a reinforcing element to reduce post-settlement by increasing the rate of soil consolidation.

2 Soil Profiles

In the current study, an experiment is conducted on the fabricated 3D consolidation apparatus to assess the impact of WPF on soil consolidation behaviour. Table 1 displays the characteristics of the sandy-silt with clay soil employed in the current experimental study.

3 Materials

Using waste plastic bottles, this experiment is carried out on the sandy-silt with clay soil (sand = 19.76%, silt = 61.27%, and clay = 18.97%). With these raw plastic bottle fibres, two different aspect ratios (AR) were used: 2 (8 mm 4 mm) and 4 (8 mm 2 mm), as shown in Table 2 and Fig. 1a, b. These varied plastic strip sizes are used in four different percentages with the local sandy-silty with clay soil: 0.00, 0.20, 0.50, and 1.00%.

Table 2 Properties of plastic strips

Property	Values
Length (mm)	8
Width (mm)	2 and 4
Thickness (mm)	0.20
Specific gravity	1.29

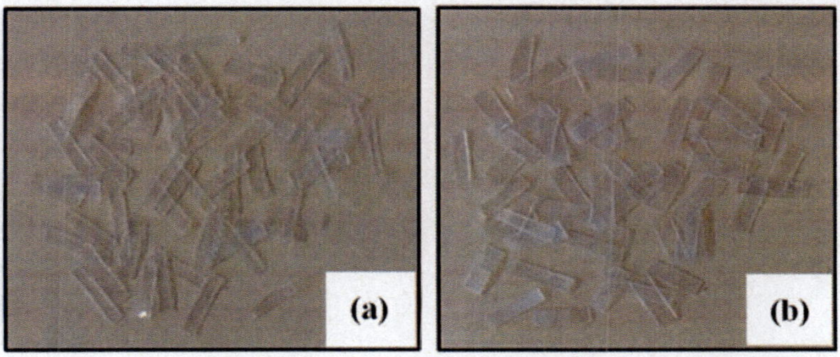

Fig. 1 Plastic strips with **a** AR 4 and **b** AR 2 (not to scale)

4 Experimental Work

This study has made an effort to examine the impacts of including WPF (with AR, l/b = 2 and 4) on the behaviour of soil consolidation. The experiments are conducted on sandy-silt with clay soil. The 3D consolidation equipment consists of a cast iron box with outlets to allow the drainage of water, as shown in Fig. 3. To apply the vertical load on top of the soil sample, a 3.70 cm diameter hole is cut out from a cast iron plate and a 3.60 cm diameter perforated loading plate is inserted, where the load is transferred through a plunger. The internal dimension of the consolidation box is 12 cm × 12 cm × 12 cm. To regulate the flow condition, steel plates are employed along the boundaries of soil to block pore water drainage and porous plates are utilized to allow open drainage passage. The test sample is kept saturated throughout the experiment. In this 3D consolidation test, a surcharge pressure of magnitude 20 kPa is positioned on top of the soil, and the flow condition has porous stone plates that allow drainage on all sides. The surcharge resembles the presence of pavements, roads, and surrounding structures on the soil in actual field conditions. The appropriate quantities of air-dried sandy-silt with clay soil and plastic strips were manually blended under dry conditions. To achieve the MDD, the soil-plastic waste mixture was combined with the OMC, as shown in Fig. 2, then put into the 3D consolidation device and compacted into three layers. The fibres were sliced into two different widths, 2 and 4 mm, and lengths of 8 mm. The soil had three different amounts of fibre added: 0.20, 0.50, and 1.00% of the soil's dry weight. To compare the consolidation features of the soil, tests were performed with and without plastic

Fig. 2 Soil sample mixed with waste plastic strips

strips added, as well as by altering the fibre content and AR. Various studies on soil alone and soil combined with WPF were carried out in accordance with the testing program.

5 Validation of the Experimental Study

The proposed small-scale laboratory model has been validated with the large-scale numerical model test results. Several researchers are extensively used PLAXIS 3D finite element-based software for the numerical analysis [11]. The numerical model has been prepared in PLAXIS 3D finite element-based software, as shown in Fig. 4. The validation has been done considering the soil profile used in the laboratory test under a surcharge pressure of 20 kPa, with drainage permitted along all sides. The Mohr–Coulomb constitutive model has been used to model soil nonlinearity as various researchers successfully use it [12–16]. The model dimensions are fixed conferring to the scaling laws suggested by Wood [16]. To simulate the proposed experimental model with the large-scale numerical model, a scaling factor of 100 (prototype/model) has been chosen while choosing the sizes of the numerical model. The c_v observed in the small-scale model, corresponding to a vertical pressure of 3.20 kg/cm^2 is found as 8.20×10^{-2} cm^2/s and that for the large-scale numerical model test is 7.01×10^{-2} cm^2/s. The minor difference in the values of the c_v of the above models is due to the difference in the height of the soil samples and analysis approach. Hence, the experimental result of the small-scale model is validated with that of the large-scale numerical model test result.

Fig. 3 Laboratory-fabricated 3D consolidation apparatus

6 Results and Discussions

6.1 Change of Void Ratio (e) with Pressure Range (log p) for Different Plastic Sizes and Percentages in 3D Consolidation Test

It is illustrated in Figs. 5 and 6 that as the plastic percentage increases, the reduction in the void ratio decreases. This happens because due to the incorporation of plastic strips in the soil, the void spaces of the soil sample are occupied by these strips, hence lowering the amount of reduction in the void spaces. Therefore, for a particular aspect

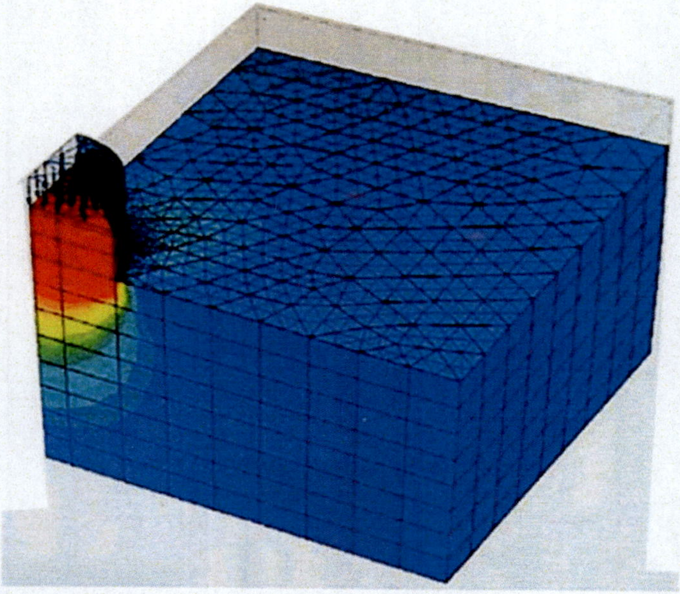

Fig. 4 Model geometry of the 3D consolidation test

ratio, as the amount of plastic percentage added to the soil increases, the settlement of the soil sample decreases.

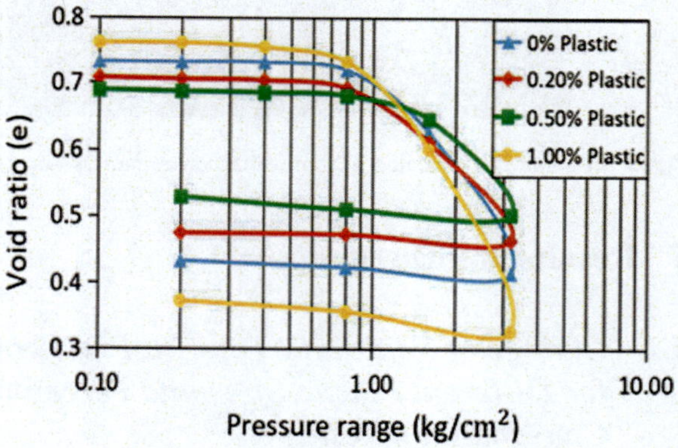

Fig. 5 *e* versus log *p* curve for the increasing plastic percentage for AR 4

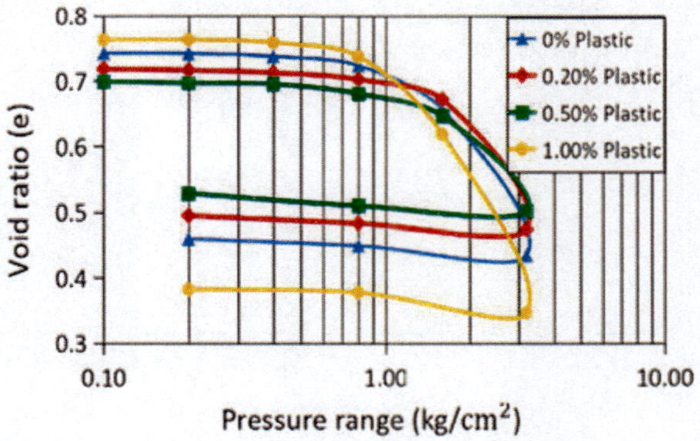

Fig. 6 *e* versus log *p* curve for the increasing plastic percentage for AR 2

6.2 Variation of Compression Index (C_c) for Different Plastic Sizes and Percentages in 3D Consolidation Test

The C_c decreases up to 0.50% with an increase in plastic strip quantity but increases further up to 1.00% for AR 4 and 2. Figure 7 shows how the C_c value decreases for increments of 0.00, 0.20, and 0.50% before increasing until 1.00% of plastic strips are added. The soil voids are occupied with these plastic wastes as the amount of plastic strips introduced increases. This results in an overall decrease in the void ratio. The C_c is shown to decrease in this study with the addition of WPF to the soil up to a point and then increase further because, when added up to a certain amount, the plastic strips act as a reinforcing material that binds the soil particles together, increasing strength, which results in increased strength due to increased confining pressure. Thus, the settlement of the soil is reduced. These plastic strips experience strains from the already strained soil mass, which causes tensile loads to be generated in them. Once the plastic strips added start acting as a reinforcing material, these inhibit the soil movements, improving the soil's capacity to settle. Thus, it illustrates that the compressibility of the soil reduces with the increase in WPF added to it. A similar pattern is also observed by Kar and Pradhan [17].

6.3 Variation of c_v for Different Plastic Sizes and Percentages in 3D Consolidation Test

Figure 8 indicates that for ARs 2 and 4, the c_v values rise with the increased plastic percentage in the soil sample. Including these plastic fibres increases the soil's overall permeability by creating passageways in the soil sample to expel extra pore water. The c_v values of soil alone and soil mixed with WPF are calculated in this study using Taylor's (1948) method. Thus, the c_v values increase with the increase of plastic fibres in the soil. Therefore, as per Terzaghi's one-dimensional consolidation theory, the

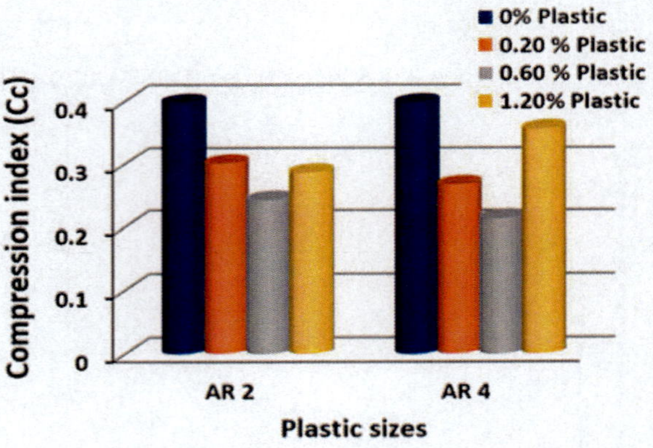

Fig. 7 C_c versus plastic sizes plot for varying plastic percentages

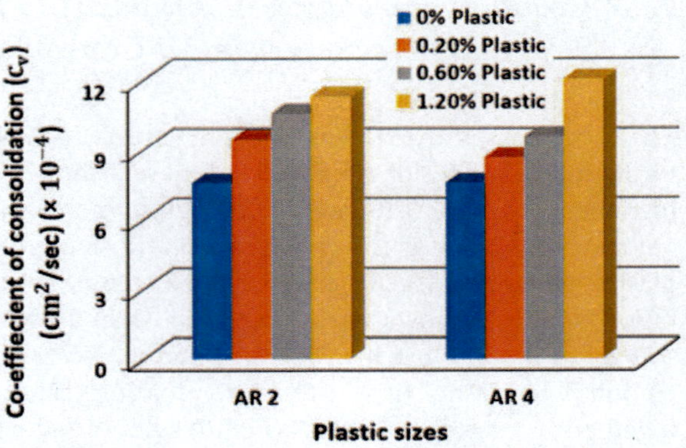

Fig. 8 c_v versus plastic sizes plot for varying plastic percentages

consolidation rate rises as the amount of plastic fibres increases. Thus, enhancing the rate at which pore water is expelled (as permeability increases).

7 Conclusion

This study serves as an example of a laboratory investigation done to assess the impact of WPF on the consolidation phenomenon of reinforced soil. The major conclusions are drawn from the current investigation:

- With the increase of plastic fibres in soil, compression index of soil decreases up to 0.50% WPF content. However, the values increase with further inclusion of WPF of 1.00% in soil.
- Increases in plastic waste percentage in soil, more voids are occupied with WPF, leading to an overall reduction in the void ratio. A similar trend has been observed by Laskar and Pal [4].

- The coefficient of consolidation values increases with the increase of WPF in soil for AR 2 and 4.
- It specifies that soil compressibility reduces as the WPF content increases.
- The WPF-reinforced soil used in this study encourages potential use in geotechnical engineering construction by increasing the strength and reducing the settlement.

References

1. Chauhan MS, Mittal S, Mohanty B (2008) Performance evaluation of silty sand subgrade reinforced with fly ash and fibre. Geotext Geomembr 26(5):429–435
2. Meftahi M, Hamidzadeh Y (2019) Influence of plastic waste reinforcement on compaction and consolidation behavior of silty soil. Int J Geotech Geolog Eng 13(6):467–470
3. Okoro CC, Vogtman J, Yousif A, Agnaou M, Khoury N (2011) Consolidation characteristics of soils stabilized with lime, coal combustion product, and plastic waste. In: Geo-frontiers: advances in geotechnical engineering, pp 1202–1209
4. Laskar A, Pal SK (2013) Effects of waste plastic fibres on compaction and consolidation behavior of reinforced soil. Electron J Geotech Eng 18:1547–1558
5. Zhang Y, Wu W, Mei G, Duan L (2019) Three-dimensional consolidation theory of vertical drain based on continuous drainage boundary. J Civ Eng Manag 25(2):145–155
6. Ibraim E, Fourmont S (2006) Behaviour of sand reinforced with fibres. In: Soil stress-strain behavior: measurement, modeling and analysis. Springer Nature, pp 807–818
7. Jebali H, Frikha W, Bouassida M (2017) 3D consolidation of Tunis soft clay improved by geodrains. Geotech Test J 40(3):361–370
8. Babu GS, Chouksey SK (2011) Stress–strain response of plastic waste mixed soil. Waste Manage 31(3):481–488
9. Prasad D, Borthakur N, Das S (2017) Effect of rice husk ash and cement mixtures on engineering properties of cohesive soil. In: Workshop on sustainable geotechnics. IGS Kanpur Chapter, IIT Kanpur, India
10. Koley S, Debnath A, Pal SK (2023) A comparative study on consolidation of clay using laboratory oedometer and fabricated 3D consolidation apparatus. Indian J Environ Prot
11. Debnath P, Debnath A, Pal SK (2022) A Numerical Study on Laterally Loaded Single CFG Pile Embedded in Layered Soil. In: Proceedings of Indian geotechnical conference, Kochi
12. Debnath A, Pal SK (2022) Influence of surcharge strip loads on the behaviour of cantilever sheet pile walls: a numerical study by ABAQUS. In: Proceedings of Indian geotechnical conference, Kochi
13. Debnath A, Pal SK (2023) Influence of surcharge strip loads on the behavior of cantilever sheet pile walls: a numerical study. J Eng Res 11(1):100029
14. Debnath A, Pal SK (2023) A numerical analysis on anchored sheet pile wall subjected to surcharge strip loading. J Eng Res: 100088
15. Debnath A, Pal SK (2023) Behavior of cantilever concrete diaphragm wall under sequential excavation of front-fill soil: a numerical study. In: Earth retaining structures and stability analysis. Springer Nature, Singapore, pp 395–407
16. Wood DM (2004) Geotechnical modelling. CRC Press Taylor & Francis Group/Books, Boca Raton
17. Kar R, Pradhan P (2011) Strength and compressibility characteristics of randomly distributed fiber-reinforced soil. Int J Geotech Eng 5(2):235–243

Evaluation of Geotechnical Properties of Pond Ash-Bentonite Mixture as a Potential Landfill Liner Material

Naman Kantesaria, Dhvanil Chotani, and Harsh Ganvit

1 Introduction

The purpose of landfill liners is to avert the leachate leakage from the landfill. They prevent groundwater and soil pollution through the ingress of leachate by acting as a barrier system. Hence, the key requirements of any liner system are low permeability to prevent pollutants migration, adequate shear strength, less swelling-shrinkage, and high erosion resistance [1]. As per the specifications of the US Environmental Protection Agency (EPA/600/R-02/099) [4] and Central Pollution Control Board, GoI (HAZWAMS/17/2000-01) [2], the acceptable value of the hydraulic conductivity for a liner system must be below 10^{-9} m/s. The most regularly used liner materials are compacted clay, geomembranes, bitumen materials, soil sealants, and compacted shale. Out of the above-listed liner materials, Compacted Clay Liners (CCL) are one of the most widely employed materials due to their extremely low permeability, resistance to puncture damage, higher cost-effectiveness, adequate strength, and good sorption capacity. However, CCL also have certain limitations. Bentonite which is used in CCL can develop the desiccation crack networks during an unsaturated state due to very high suction. This led to an increase in the risk of leachate migration through CCL. Hence, there is a requirement to search for an alternative material that has all the advantages of CCL along with low desiccation crack formation capabilities. The addition of a significant amount of coarse-sized particle fraction can fulfil this requirement. Hence, Bentonite mixed with pond ash can become a viable substitute material. Pond ash is generated in a thermal power plant as a waste by-product of coal combustion. It is a mixture of coarse-grained bottom ash particles and fine-grained fly ash particles. The utilization of these by-products is one of the big issues. Hence, it will be beneficial if these by-products can be utilized in liner construction. Researchers conducted experimental studies focusing mainly on fly

N. Kantesaria (✉) · D. Chotani · H. Ganvit
Department of Civil Engineering, SoE, IoT, Nirma University, Ahmedabad, Gujarat 382421, India
e-mail: naman.kantesaria@nirmauni.ac.in

S. Kolathayar et al. (eds.), *Best Practices in Geotechnical and Pavement Engineering*, Lecture Notes in Civil Engineering 449, https://doi.org/10.1007/978-981-99-8505-0_19

ash or sand mixture with Bentonite to use in geotechnical and geo-environmental engineering applications. Edil et al. [3], Mollamahmutoğlu and Yilmaz [12], Sankar and Niranjan [16], Gupt et al. [6], and Kantesaria et al. [11] examined the various hydromechanical properties of Bentonite-fly ash mixture. They found that the addition of Bentonite in fly ash significantly reduced permeability of the mixture. They suggested an optimum percentage of Bentonite (> 8%) be added to the fly ash so as to satisfy the permeability and strength criteria. A similar type of research was conducted on a sand-Bentonite mixture by Haug and Wong [7], Gueddouda et al. [5], and Oren et al. [13, 14]. They observed that due to the large void spaces in coarser sand particles, the distribution of Bentonite at a lower percentage was non-uniform. Hence, the amount of Bentonite required to bring down the permeability level below the accepted range was a little higher (> 10–12%) in sand-Bentonite mixtures. Rout and Singh [15] studied the particle shape and morphology impact on various geotechnical properties of pond ash-Bentonite mixture. A linear regression-based equation to predict the permeability of the mixture from basic geotechnical parameters was developed.

Very limited studies have been found on pond ash to ascertain its applicability as a liner material. The objective of the present research is to determine an optimum percentage of Bentonite which is required to be added to the pond ash for making the mixture suitable as a liner material. Bentonite content (0–18%) was added to the pond ash in various percentages and experiments such as consistency limits, consolidation tests, swell pressure tests, and unconfined compression tests were performed on the samples. The ideal Bentonite percentage was determined by considering the geotechnical properties of mixture as hydraulic conductivity, shear strength, and swell pressure criteria.

2 Material Properties

Table 1 displays the geotechnical properties of Bentonite and pond ash. Pond ash was procured from Sabarmati thermal power plant, Ahmedabad. Pond ash majorly contained 73% fine sand-sized particles with a specific gravity of 2.42. Compaction parameters of pond ash were determined from the Standard Proctor test. Commercially available Bentonite with mainly 63% clay content was used in the current research.

3 Experimental Program and Specimen Preparation

0, 2, 4, 6, 8, 10, 12, 14, 16, and 18% of dry Bentonite was mixed in the pond ash by the weight percentage to conduct the series of experiments. The amount of Bentonite was restricted to < 20% to prevent the adverse effects of excessive swelling and shrinkage. Oven-dried Bentonite powder was mixed with oven-dried pond ash to

Table 1 Basic geotechnical properties of Bentonite and pond ash

Geotechnical parameters	Pond ash	Bentonite
Gravel (%)	6	0
Coarse sand (%)	4	0
Medium sand (%)	9	0
Fine sand (%)	73	0
Silt (%)	8	37
Clay (%)	0	63
Specific gravity, G_s	2.43	2.72
Liquid limit, w_L (%)	34	361
Plastic limit, w_p	Non-plastic	48%
Maximum dry density	1.60 g/cc	–
Optimum moisture content	17%	–

prepare PAB-0 to PAB-18 specimens. Here the name of the specimen specifies the amount of Bentonite added, e.g. PAB-18 specifies that 18% of Bentonite by weight was added to the pond ash to form the specimens. Liquid and plastic limit experiments were done according to the specifications of IS: 2720-5 [10]. All the specimens for and consolidation tests and unconfined compression (UC) tests were prepared at maximum dry density (1.60 g/cc) and optimum moisture content (17%) of raw pond ash using a moist tamping technique. UC tests were performed on the cylindrical specimens of 50 mm diameter and 100 mm height as per the specifications of IS: 2720 (Part 10) [8] at a constant rate of deformation of 1.25 mm/min. For each Bentonite content, three UC tests were conducted to certify repeatability. Consolidation tests were conducted as per the specifications of IS: 2720 (Part 15) [9] on the specimens of size 20 mm height and 60 mm diameter. The samples were allowed to swell freely at the start of the test under 5 kPa seating pressure to determine swell pressure later on from the same consolidation series of experiments. Permeability was calculated based on the result analysis of consolidation tests. Table 2 displays all the specimens and results of different experiments.

4 Result and Discussion

4.1 Consistency Limits

Figure 1 displays the variation in the plastic limit (W_P), liquid limit (W_L), and plasticity Index (I_P) with different Bentonite content. Liquid limit increased from 34 to 84% with the addition of Bentonite from 0 to 18%. Bentonite increased the water holding capacity of pond ash mixture by creating a diffused-double layer around clay particles. This resulted in creating plasticity among non-plastic pond ash particles.

Table 2 Geotechnical properties of pond ash-Bentonite mixture

Geotechnical properties	Bentonite content									
	0%	2%	4%	6%	8%	10%	12%	14%	16%	18%
Specimen name	PAB-0	PAB-2	PAB-4	PAB-6	PAB-8	PAB-10	PAB-12	PAB-14	PAB-16	PAB-18
Liquid limit, W_L (%)	34	41	50	58	62	70	74	76	82	84
Plastic limit, W_P (%)	NP	25	32	38	41	48	52	54	57	58
Plasticity index, I_P (%)	–	16	18	20	21	22	22	22	25	26
Unconfined compressive strength, UCS (kPa)	–	351	398	435	579	571	544	506	482	467
Compression index (C_c)	0.032	0.039	0.043	0.049	0.058	0.074	0.111	0.146	0.193	0.254
Permeability (m/s)	3.0E−04	3.1E−07	2.0E−08	5.2E−09	8.3E−10	4.8E−10	4.0E−10	3.0E−10	2.5E−10	2.2E−10
Swell pressure (kPa)	0	0	1	4	8	11	15	19	25	37

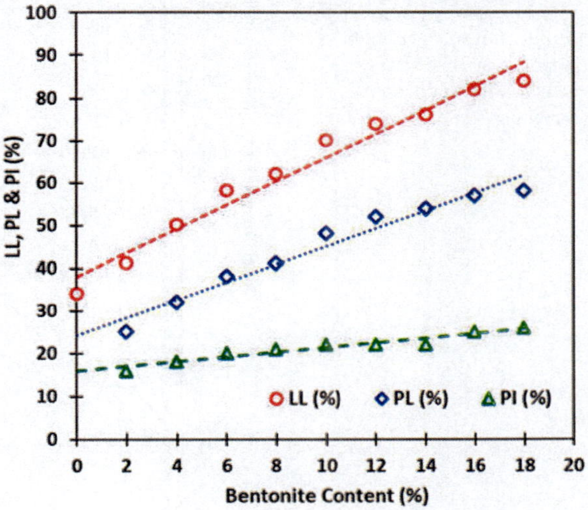

Fig. 1 Impact of Bentonite on consistency limits of pond ash-Bentonite mixture

The plasticity index increased up to 58% for PAB-18 specimen from non-plastic response of PAB-0.

4.2 *Shear Strength*

Figure 2 displays the result of UC tests conducted on PAB-2 to PAB-18 specimens. Figure 2a shows axial stress versus axial strain response during UC tests, and Fig. 2b represents a comparable bar chart of UCS of PAB-2 to PAB-18. UC test for PAB-0 was not conducted as this specimen wasn't stable without any support due to raw pond ash's non-plastic nature. The unconfined compressive strength (UCS) increased by 64% from 351 to 579 kPa by adding Bentonite content from 2 to 8%. Further, Bentonite addition resulted in a gradual reduction in the UCS. UCS decreased to 467 kPa for PAB-18 specimens. This behaviour can be explained through two phenomena: (a) pore filling, and (b) lubrication [11]. At lower Bentonite concentrations ($\leq 8\%$), the tiny Bentonite particles can settle between the larger void skeleton formed by coarser pond ash particles. The presence of Bentonite then stiffens the entire mixture mass and results in higher stiffness and strength. This is called the pore-filling phenomenon. Whereas in the lubrication phenomenon, the excess amount of Bentonite particles can come among the contact points of pond ash particles. Due to this, the lubrication between the pond ash particles increases and that weakens the whole skeleton structure. The initial stiffness response also indicated a similar trend as shown in Fig. 4a. All the specimens failed suddenly after achieving peak and for most of the specimens' failure strain was observed around 1.8–2%.

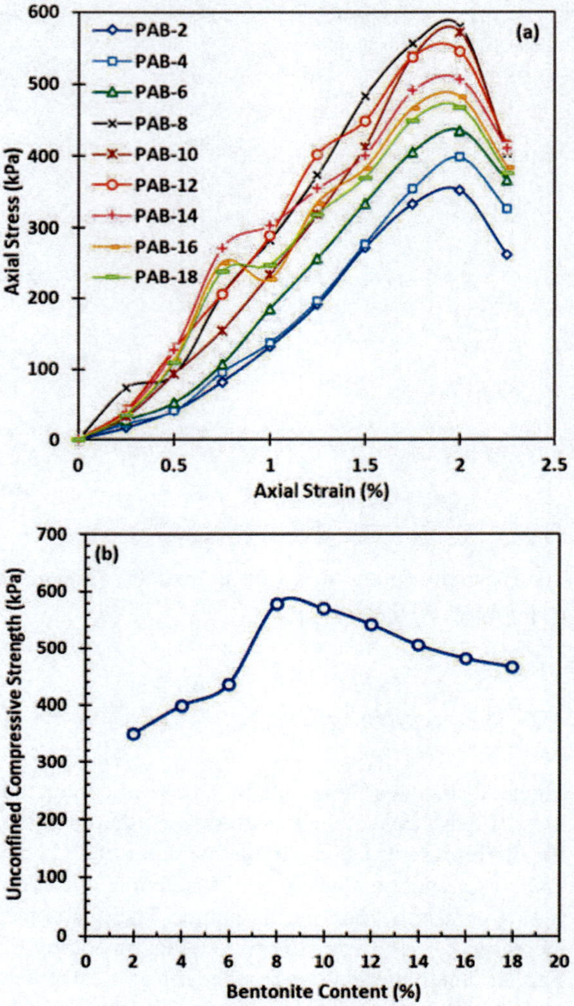

Fig. 2 Impact of Bentonite addition on shear strength response: **a** axial stress versus axial strain curves, **b** comparison of UCS values

4.3 Consolidation, Permeability, and Swell Pressure

Figure 3 demonstrates the consolidation tests' results in terms of permeability and compression index (C_c) variation with Bentonite percentages. The permeability and C_c values are displayed in Table 2. The addition of Bentonite made the pond ash structure more compressible as C_c value was found to increase from 0.032 to 0.254 for raw pond ash to 18% Bentonite added pond ash. Very high C_c values are not recommended for liners as they can be vulnerable to higher differential settlement and damage. Permeability (k) of the specimens was decreased significantly from 3×10^{-4} m/s to 8.3×10^{-10} m/s with an addition of only 8% Bentonite to raw pond

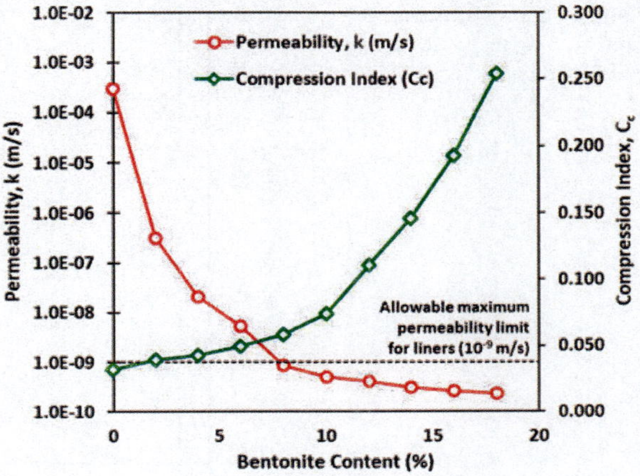

Fig. 3 Impact of Bentonite on permeability and consolidation behaviour of pond ash-Bentonite

ash. The reduction in permeability afterwards with further addition of Bentonite got stabilized and the minimum permeability value of 2.2×10^{-10} m/s was found for the PAB-18 specimen. As observed from experimental results, raw pond ash cannot be used directly as a liner material owing to its high permeability value (range of 10^{-4} m/s). However, with the addition of a small Bentonite concentration, the permeability of the mixture material reduced drastically below the acceptable limit of liner material ($< 10^{-9}$ m/s) specified by the US Environmental Protection Agency (EPA/600/R-02/099) [4] and Central Pollution Control Board (CPCB), GoI (HAZWAMS/17/2000-01) [2]. Figure 4 illustrates the change in swell pressure with a rise in Bentonite percentage. The swell pressure increased with an addition of Bentonite, and a maximum of 37 kPa swell pressure was obtained for the PAB-18 specimen. These values of swell pressures may be ignored in comparison to the swell pressure detected in the raw Bentonite used in CCL.

5 Conclusions

The following important conclusions can be obtained through the present experimental research:

- Optimum content of Bentonite recommended to make pond ash suitable for liner material is 10% based on the criteria of permeability, shear strength, and swell pressure.
- Shear strength of pond ash-Bentonite mixture increased by 64% until the addition of 8% Bentonite content due to pore filling phenomena and subsequently it reduced with further Bentonite addition due to increased lubrication between particles.

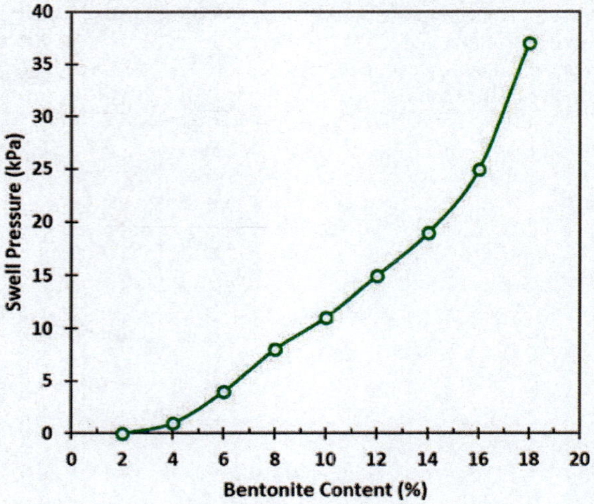

Fig. 4 Impact of Bentonite on swelling response of pond ash-Bentonite

- The hydraulic conductivity of raw pond ash reduced from the order of 10^{-4}–10^{-10} m/s by adding 8% Bentonite. Maximum swell pressure of 37 kPa was attained for the maximum 18% Bentonite mixture.
- Water holding ability and plasticity of pond ash enhanced by the addition of Bentonite.

These experimental results confirmed the possible utilization of a pond ash-Bentonite as a liner material. This cost-effective solution however needs further research in environmental sustainability in terms of leachate properties and long-term stability.

References

1. Brandl H (1992) Mineral liners for hazardous waste containment. Geotechnique 42(1):57–65
2. Central Pollution Control Board (2001) Criteria for hazardous waste landfills, Hazardous waste management series, HAZWAMS/17/2000-01. Ministry of Environments and Forests, New Delhi, India
3. Edil TB, Sandstrom LK, Berthouex PM (1992) Interaction of inorganic leachate with compacted pozzolanic fly ash. J Geotech Eng 118(9):1410–1430
4. EPA/600/R-02/099 (2002) Assessment and recommendations for improving the performance of waste containment systems. United States Environmental Protection Agency, Office of Research and Development
5. Gueddouda MK, Lamara M, Aboubaker N et al (2008) Hydraulic conductivity and shear strength of dune SB mixtures. Electron J Geotech Eng 13:1–15
6. Gupt CB, Bordoloi S, Sahoo RK, Sekharan S (2021) Mechanical performance and micro-structure of bentonite-fly ash and bentonite-sand mixes for landfill liner application. J Clean Prod 292:126033
7. Haug MD, Wong LC (1992) Impact of molding water content on hydraulic conductivity of compacted sand-bentonite. Can Geotech J 29:253–262

8. IS: 2720-10 (1991) Methods of test for soils, part 10: determination of unconfined compressive strength. Bureau of Indian Standards, New Delhi
9. IS: 2720-15 (1986) Methods of test for soils, part XV: determination of consolidation properties. Bureau of Indian Standards, New Delhi
10. IS: 2720-5 (1985) Methods of test for soils, part 5: determination of liquid and plastic limit. Bureau of Indian Standards, New Delhi
11. Kantesaria N, Chandra P, Sachan A (2021) Geotechnical behaviour of fly ash-bentonite mixture as a liner material. In: Proceedings of the Indian geotechnical conference 2019: IGC-2019, vol II. Springer, Singapore, pp 237–247
12. Mollamahmutoğlu M, Yilmaz Y (2001) Potential use of fly ash and bentonite mixture as liner or cover at waste disposal areas. Environ Geol 40:1316–1324
13. Oren AH, Durukan S, Kayalar AS (2014) Influence of compaction water content on the hydraulic conductivity of sand bentonite and zeolitebentonite mixtures. Clay Miner 49:109–121
14. Oren AH, Kaya A, Kayalar AS (2011) Hydraulic conductivity of zeolitebentonite mixtures in comparison with sand-bentonite mixtures. Can Geotech J 48:1343–1353
15. Rout S, Singh SP (2020) Characterization of pond ash-bentonite mixes as landfill liner material. Waste Manage Res 38(12):1420–1428
16. Sankar VSR, Niranjan DV (2015) Effect of compaction conditions on the hydraulic and compressibility behaviour of fly ash-cement. Int J Eng Res Technol 4(7):365–378

2D Analysis of Slope Stability Using Limit Equilibrium Analysis and Finite Element Analysis

S. Sravya, Angshuman Das, Koteswaraarao Jadda, and Dinesh Gundavaram

1 Introduction

Slopes play a vital role in the construction of highways and railway embankments, earth dams, levees, and canals. In hilly areas, slopes along the highway are more susceptible to landslides. Slope failure refers to the sudden movement or collapse of soil, rock, or debris on a sloping surface. It is a common geotechnical problem that can occur due to various factors, including natural phenomena such as rainfall, earthquakes, and erosion, or human activities like excavation, mining, and construction. Slope failure can have severe consequences, including environmental deterioration, property damage, and loss of life. Some of the significant instances where slope failures occurred in India include Agumbe ghat in Karnataka (2019), two landslides along the National Highway 1-A in Ladakh Region of Jammu and Kashmir (2023), Malin landslide in Maharashtra (2014), Kedarnath landslide at Uttarkhand (2013), etc. Therefore, it is crucial to perform slope stability analysis to identify potential failure modes and implement appropriate measures to prevent slope failure.

Slope stability analysis can be performed using empirical or analytical approaches for evaluation the stability of excavated or natural slopes in soil and rock, rock-fill or earthen dams, and embankments. 2D approaches were widely adopted by several engineers for slope stability analysis due to their simplicity [1–3]. 2D slope stability analysis use simple geometry of 3D slope by considering the plain strain or axisymmetric conditions. In past decades the 2D slope stability analyses have

S. Sravya · A. Das (✉)
Chaitanya Bharathi Institute of Technology, Hyderabad, Telangana 50075, India
e-mail: dasangshuman5@gmail.com

K. Jadda
Dr B R Ambedkar National Institute of Technology Jalandhar, Jalandhar, Punjab 144011, India
e-mail: Jaddak@nitj.ac.in

D. Gundavaram
Indian Institute of Technology Hyderabad, Hyderabad, Telangana 502285, India

S. Kolathayar et al. (eds.), *Best Practices in Geotechnical and Pavement Engineering*, Lecture Notes in Civil Engineering 449, https://doi.org/10.1007/978-981-99-8505-0_20

been advanced through a series of development. The modifications are made in the assumptions to simplify the analysis and different equilibriums equations were developed for obtaining the factor of safety (FOS). These methods can be categorized into non-circular methods (e.g. log-spiral procedure), circular methods (e.g. friction circle and Swedish circle methods), and methods of slices (e.g. Bishop [4]; Janbu [5]; Morgenstern and Price [6]; Spencer [7] and Fellenius [8]). The method of slices become a popular 2D approaches due to its ability to deal with complex geometrics, variation in soil and water conditions, and the effects of external forces [9].

In recent years, finite element (FE) methods, are becoming increasingly popular [10]. The literature review indicates that both finite element method (FEM) as well as limit equilibrium method (LEM) have been widely utilized in slope stability analysis, and numerous studies have been conducted to compare their effectiveness [11–14]. Some studies have found that FEM is more accurate in analysing complex geometries and can consider the effects of soil-structure interaction [11–13]. However, FEA requires more computational resources and is more time-consuming than LEM. On the other hand, LEM is simpler, faster, and is suitable for analysing simple slope geometries. Therefore, the objective of the present study became to compare the widely used limit equilibrium (LE) and identify the most accurate LE method compared to the more advanced finite element (FE) method. The study revealed that the FOS obtained from Bishop and Fellinius methods of slices is similar to that obtained from FEM. In addition to that parametric study has been conducted using LEA to investigate the effect of the design parameters of the anchor such as anchor inclination angle, number of anchors on the slope stability analysis.

2 Study Area

In Karnataka, National Highway NH-169A passes through the region of Agumbe ghat (as shown in Fig. 1). Due to the presence of 14 hairpin bends in this specific Ghat section, hence it become one of the deadliest and complex Ghat sections in Karnataka. Among all the hairpin, the bend in the 7th hairpin has the smallest curve having a radius of merely 3.5 m, resulting in challenging for guiding the vehicle. Therefore, in this study, the site between 7 and 8th hairpin bend has been selected for analysis.

Another study area is the Dhobighat Section of Banjarahills is bounded by 17° 27′ 23.4″ N to 17° 41′ 5″ N latitudes and 78° 26′ 31″ E to 78° 29′ 26″ E longitudes. The area is situated in Banjarahills, Telangana (as shown in Fig. 1). Previous reports suggested that this slope is prone to landslides. The slope angle of the study area is about 30°.

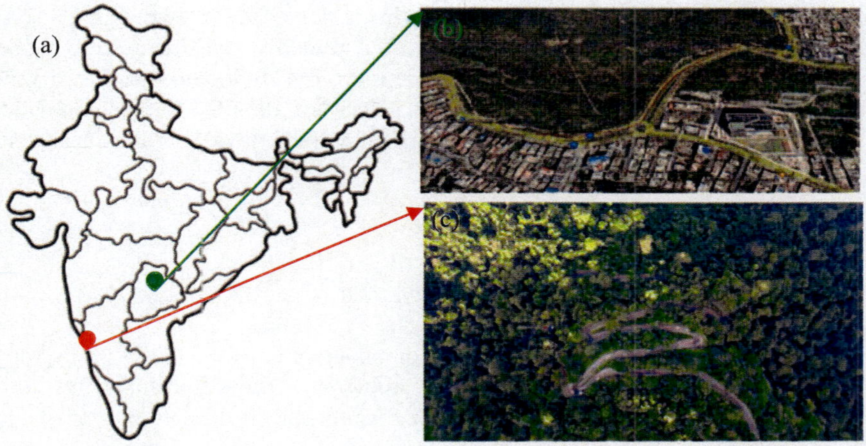

Fig. 1 **a** Location of study areas in India; arial view of **b** Dhobighat region and **c** Agumbe ghat where slope stability analysis has been carried out (*Courtesy* Google)

3 Limit Equilibrium Analysis

It is assumed in the limit equilibrium method that the slope will fail along a well-defined failure surface. Therefore, this method of analysis does not deal with the stress–strain relationship or the subsequent deformation behaviour within the slope body. In the method of slice of the limit equilibrium analysis the slope is divided into number of slices. Afterward the forces acting on every single slice is determined. Eventually, the FOS is calculated by involving the resisting and driving forces. The following methods of slices are considered in the present study for limit equilibrium analysis.

Bishop [4] proposed a simple limit equilibrium analysis approach by assuming that the failure surface is circular. The analysis involves determining the forces acting on each slice of the slope, and calculating the FOS by relating the resisting forces to the driving forces.

Janbu [5] also proposed another simple LEA approach assuming that the failure surface is polygonal. The analysis involves dividing the slope into a series of polygons, and determining the forces acting on individual polygon. The FOS is then calculated by equating the resisting forces to the driving forces.

Spencer [7] proposed a method based on the assumption that the failure surface is a composite of planar and circular surfaces. The analysis involves determining the forces acting on a slice of the slope by considering a "wedge" and calculating the FOS by comparing the resisting forces to the driving forces.

Fellenius [8] proposed a method for slope stability analysis which become popular as ordinary method of slices. In this method all the slices are considered to be act individually and interactions between slices are neglected.

Morgenstern–Price method [6] assume that a slope can be divided into a series of slices, where all individual slice have a different FOS against failure. The method investigates under the effective stress condition. The development of pore water pressure in each slice is being considered. Afterward, the FOS for each individual slice can be obtained by relating the shear strength of the soil to the shear stress acting on the slice.

4 Finite Element Analysis

Finite element analysis is a widely used computational method that is proficient of solving intricate geotechnical engineering problems. In finite element analysis, first the actual geometry of an object is divided using several discrete portions called "finite elements". Each element is connected with adjacent elements and the point between connected elements is called nodes. The finite elements and nodes altogether known as the mesh. In the present study, quadratic 6-node (T6) triangular elements and Mohr-Coulomb criteria have been used to effectively approximate the variables over the region of interest and analysing the stress and strain distribution within each element [15]. The FOS is then calculated by relating the shear strength of the slope material to the applied stresses.

5 Results and Discussion

5.1 Slope Stability Analysis Using LEA

A numerical model of the Agumbe ghat slope has been prepared based on the literature [16]. The slope mainly consists of stiff sandy clay along with a layer of well graded gravel near the toe (as shown in Fig. 1a). The properties of the soils used in the numerical models are detailed in Table 1. The LEA has been performed for the natural slope as well as slope stabilized with five number of anchors. The design parameters for the anchors are taken from the literature [16] and detailed in Table 2.

Table 1 Soil properties assigned for Agumbe ghat slope (as obtained from the literature [16])

Materials	Angle of internal friction (φ)	Cohesion (C) kPa	Unit Weight (γ) kN/m^3	Stress state	Assigned soil
Sandy clay (Stiff consistency)	16	37.26	18.82	Effective	
Well graded gravel	30	10,000	25.39	Effective	
Sandy soil	35	90.00	18.00	Effective	

Table 2 Anchor parameters used in present study

Number of anchors	Free length l (m)	Root length l_k (m)	Slope α (°)	Anchor spacing b (m)	Force F (kN)
Anchor details for Agumbe ghat slope [16]					
5	7	0.50	165.0	2.00	100
Anchor details for Dhobighat slope					
2	6	0.50	145.0	2.00	100

The first set of analysis has been performed using the method of slices in LEA. The factor of safety for all five methods were obtained and compared with the results available in literature [16]. The slope stability analysis of Agumbe ghat indicates that the slope become unsafe for all the methods follows for LEA. The factor of safety (FOS) obtained for the natural slope from all the methods are below 1.5 as detailed in Table 3. Table 3 is also showing the comparison of the FOS value from the literature [16] and present study. The FOS value below 1.5 indicate the slope is either unstable or critically stable [17]. Consequently, further the LEA has been performed for the slope stability analysis of Agumbe ghat with slope stabilization technique. Similar factor of safety value has been obtained as compared to the literature [16]. The comparison of FOS for the stabilized slope is also provided in Table 3. It can be observed from the table that, under any condition the Fellenius method and Bishop method are providing more conservative value than other methods of slices. The reason behind that could be the simplified assumptions, consideration of conservative soil parameters, such as angle of internal friction, cohesion, and simple failure mechanism in these method of slices [18]. Figure 2 shows the slip surface obtained from limit state method.

Similarly, another slope has been considered from the Dhobighat region of Hyderabad and LEA has been carried out. As depicted in Fig. 3, the slope is mainly consisting of well graded sandy soils. However, the density of soil is different for two layers as detailed in Table 4. The slope stability assessment at the Dhobighat region employed two different method of slices. The results, presented in Table 5, indicates that the FOS for the natural slope is 1.08 and 1.05 consequently for Bishop

Table 3 Comparison between factor of safety (FOS) obtained from present study and literature for natural slope and slope stabilized with anchors

Method of analysis	Without slope protection		With anchorage	
	FOS obtained from present study	FOS obtained from literature	FOS obtained from present study	FOS obtained from literature
Bishop	1.4	1.35	1.65	1.56
Fellenius	1.37	1.32	1.6	1.58
Janbu	1.42	1.4	1.85	1.74
Spencer	1.44	1.36	1.86	1.8
Morgenstern–Price	1.44	1.47	1.86	1.8
FEM	1.3		1.5	

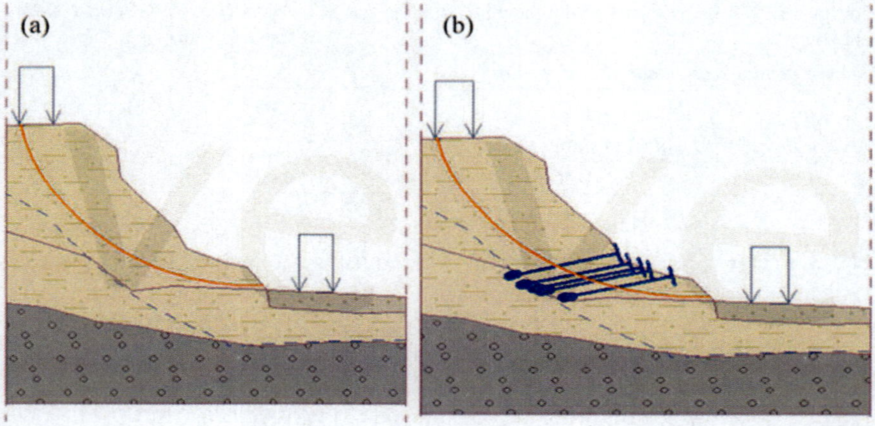

Fig. 2 Slip surface obtained from limit state method for **a** Agumbe ghat slope and **b** Agumbe ghat slope stabilized with anchors

and Fellenius method. Therefore, the natural slope is found to be critically stable [17]. Further, considering the anchor design parameters provided in Table 2 the LEA was performed for stabilized slope. It was found that the factor of safety increases from 1.08 to 2.38 using Bishop method and from 1.05 to 2.35 using Fellenius method. For the verification of the numerical model the following mathematical formulation has been carried out considering slope angle (φ) = 32°, and slope angle (i) = 30.5°.

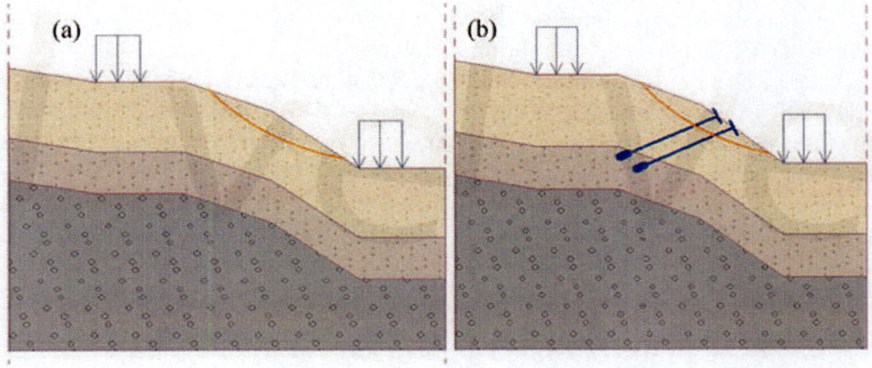

Fig. 3 Slip surface obtained from limit state method for **a** the slope at Dhobighat region and **b** the slope at Dhobighat region stabilized with anchors

Table 4 Soil properties assigned for Agumbe ghat slope

Materials	Angle of internal friction (φ)	Cohesion (C) kPa	Unit weight (γ) kN/m^3	Stress state	Assigned soil
Well graded sand	30	0.00	15.19	Effective	
Well graded sand	32	0.00	16.19	Effective	
Weathered rock	30	10,000	25.39	Effective	

Table 5 Comparison between factor of safety (FOS) obtained from different method of analysis for natural slope and slope stabilized with anchors

Method of analysis	Without slope protection	With slope protection
Bishop	1.08	2.38
Fellenius	1.05	2.35
Mathematical formulation considering no flow	1.06	–
FEM	0.9	1.2

For cohesion less soils in dry state,

$$\text{FOS} = \frac{\tan \varphi}{\tan i} \tag{1}$$

This mathematical formulation also provide almost similar value as compared to Bishop and Fellenius method as shown in Table 5.

It can be also depicted from Tables 3 and 5 that the natural flatter slope consists sandy soil exhibit lower FOS value compared to the steeper stiff clayey slope. However, the introduction of anchor highly affected the sandy slope compared to stiff clayey slope. The literature also indicates that the alteration in the slope conditions have a greater influence the FOS value for sandy slope compared to the clayey slope [19].

5.2 Slope Stability Analysis Using FEA

FEA has been carried out on the prepared models for both the Agumbe ghat and Dhobighat regions. The potential slip surface along with localized deviatoric plastic strain are obtained for both natural slopes as well as slopes stabilized using anchors as shown in Figs. 4 and 5. The potential slip surface indicates the weak soil in the overall slope. After placing the anchors in those week zone the potential slip surface changes as well as the factor of safety increases as presented in Tables 3 and 5. In all conditions the FEA provides less FOS value compared to all the methods for LEA. As depicted from Table 3, for the natural slope at Agumbe ghat region FEA provides 7 and 5% lower FOS value consequently compared to Bishop and Fellenius method of slices. Similarly, as depicted from Table 5, for the natural slope at Dhobighat region FEA provides 16 and 14% lower FOS value consequently compared to Bishop and Fellenius method of slices. This may be due to the discretization of single structure into several number of nodes, element, and regions in FEA. Whereas the classical LEA depend on the direction of the forces acting on each slice. So, FEA gives more conservative results than the methods for LEA as observed in earlier studies [20, 21]. However, even after using the anchors for the stabilization the FOS obtained from FEA as 1.5 and 1.2 for Agumbe ghat slope and Dhobighat slope consequently. Therefore, the limit equilibrium method has been considered further for the parametric study of both slopes.

5.3 Effect of Anchor Design Parameters on Slope Stability

The Bishop method of slices has been employed in this present study for the parametric study using LEA to analyse the stability of slopes both the Agumbe ghat and Dhobighat regions. The inclination of anchor and number of anchors are two important design parameters for the slope stabilization using anchor [22, 23].

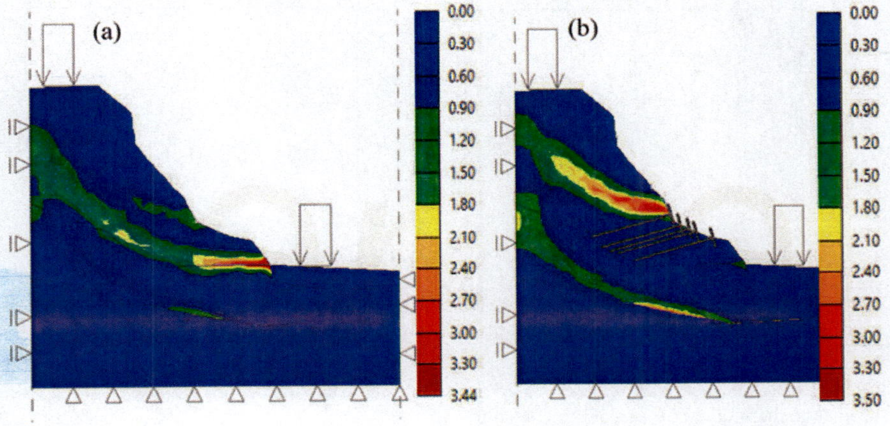

Fig. 4 Plastic deformation behaviour of **a** Agumbe ghat slope and **b** Agumbe ghat slope stabilized with anchors obtained from finite element method analysis

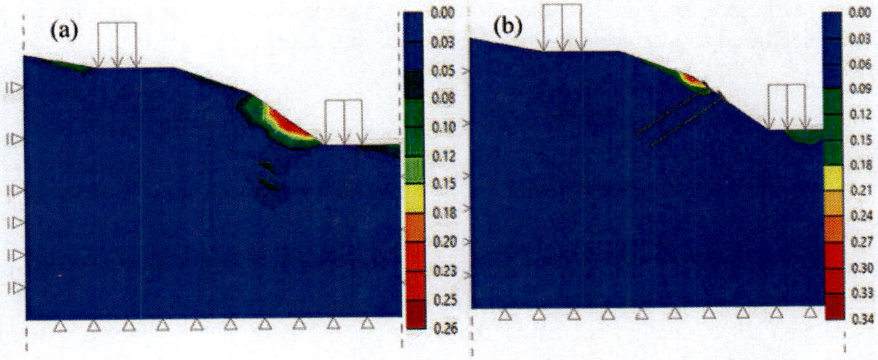

Fig. 5 Plastic deformation behaviour of **a** the slope at Dhobighat region and **b** the slope at Dhobighat region stabilized with anchors

It can be found from Fig. 6a, b that increasing in the inclination of the anchors resulted in higher FOS value for both slopes. This can be attributed to the fact that as inclination of the anchor increases, the stabilizing axial force in anchor cables increases [24]. This results in an enhanced holding capacity of the anchors, enabling them to better resist external loads acting on the slopes. It can also be observed from Fig. 6 as the inclination of anchors increases FOS increases linearly.

The following equations, Eqs. (2) and (3) are representing the relationship between the FOS and inclination angle of anchors consequently for Agumbe ghat and Dhobighat region. It can be observed from the equation that the FOS increases

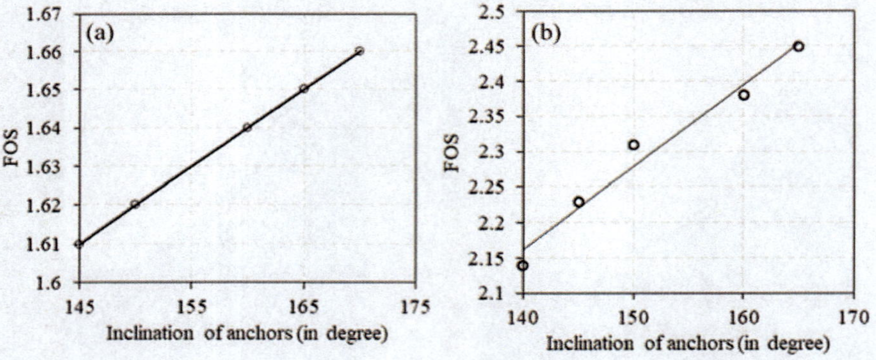

Fig. 6 Change in the factor of safety (FOS) at **a** Agumbe ghat slope and **b** Dhobighat slope with the change the inclination angle of the anchor

almost 6 times more in the case of studied sandy slope compared to the studied stiff clay slope with the alteration in the inclination angle of the anchors. This is similar to the earlier observation that the natural flatter slope consists sandy soil are more influenced due to alteration in the slope conditions compared to stiff clayey slope [19].

$$\text{FOS} = 0.002 * \text{Inclination of anchors} + 1.32, \quad R^2 = 1 \tag{2}$$

$$\text{FOS} = 0.012 * \text{Inclination of anchors} + 0.54, \quad R^2 = 0.97 \tag{3}$$

Similarly, from Fig. 7a, b, it can be observed that the number of anchors is directly proportional to the FOS value for both slopes. The reason behind it could be that as the number of anchors increases, the holding capacity of the soil slope improved due to an increase in the stabilizing axial force. Consequently, the FOS value also increases.

Furthermore, the correlations in Eqs. (4) and (5) indicate that in case of studied sandy slope the FOS increases almost ten times compared to studied stiff clay slope. The reason is similar as already stated in the case of the alteration of inclination of anchors for both slopes.

$$\text{FOS} = 0.05 * \text{Number of anchors} + 1.4, \quad R^2 = 1 \tag{4}$$

$$\text{FOS} = 0.517 * \text{Number of anchors} + 1.16, \quad R^2 = 0.99 \tag{5}$$

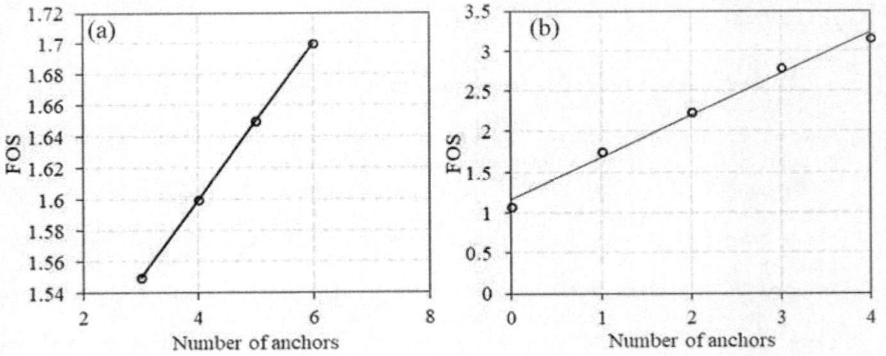

Fig. 7 Change in the factor of safety (FOS) of **a** Agumbe ghat slope and **b** Dhobighat slope due to use of different number of anchors

6 Conclusions

Based on the overall study, several conclusions can be drawn regarding the slope stability analysis using limit equilibrium analysis (LEA) and finite element analysis (FEA) for both the Agumbe ghat and Dhobighat slopes.

The FOS value obtained from LEA is more than that obtained from FEA for both slopes. The FOS for the normal Agumbe ghat slope using Bishop method is about 1.40, while for normal Dhobighat slope, it is found to be 1.08. This suggests that the Dhobighat slope is relatively more unstable compared to Agumbe ghat slope.

Further when the slopes are stabilized with anchorages, the FOS increases. For the stabilized Agumbe ghat slope (using 5 anchors), the FOS become 1.65. Thus the stabilized slope is 17% safer than the normal slope. In the case of the stabilized Dhobighat slope (using 2 anchors), the FOS become 2.38, indicating it is 120% safer compared to the normal slope. Therefore it can be said that the natural flatter slope consists sandy soil are more influenced due to alteration in the slope conditions compared to stiff clayey slope.

It can be also concluded from the study that the FOS obtained from LEA is more compared to that from FEA. Also, compared to the other method of slices used in the study the Bishop and Fellenius method provides similar FOS compared to FEA. The reason behind it could be the discretization of single structure into several number of nodes, element, and regions in FEA. Whereas the classical LEA depend on the direction of the forces acting on each slice.

It can be concluded from the parametric study that FOS is directly proportional to the studied two anchor design parameters inclination of anchors and number of anchors. Also the parametric study indicates that the sandy slope in the Dhobighat region is more influenced by the alteration in anchor design parameters compared to the stiff clayey slope in the Agumbe ghat region.

References

1. Albataineh N (2006) Slope stability analysis using 2D and 3D methods. Doctoral dissertation. University of Akron
2. Cheng YM, Lansivaara T, Wei WB (2007) Two-dimensional slope stability analysis by limit equilibrium and strength reduction methods. Comput Geotech 34(3):137–150
3. Chakraborty R, Dey A (2019) Effect of toe cutting on hillslope stability. In: Geotechnical applications: IGC 2016, vol 4. Springer Singapore, pp 191–198
4. Bishop AW (1955) The use of the slip circle in the stability analysis of earth slope. Geotechnique 5(1):7–17
5. Janbu N (1954) Application of composite slip surface for stability analysis. In: Proceeding of the conference on stability of Earth slopes, part 3. Stockholm, Sweden, pp 43–49
6. Morgenstern N, Price V (1965) The analysis of the stability of general slip surfaces. Geotechnique 15(1):79–93
7. Spencer E (1967) A method of analysis of the stability of embankments assuming parallel inter-slice forces. Geotechnique 17(1):11–26
8. Fellenius W (1936) Calculation of the stability of Earth Dams. In: Proceeding of the second congress on large Dams, Washington, DC, pp 445–463
9. Fredlund DG, Krahn J (1977) Comparison of slope stability methods of analysis. Can Geotech J 16:121–139
10. Nordal S, Glaamen MG (2004) Some examples of slope stability evaluations from Norwegian geotechnical practice. In: Brinkgreve RBJ, Schad H, Schweiger HF, Willand E (eds) Geotechnical innovations, pp 347–363
11. Mebrahtu TK, Heinze T, Wohnlich S, Alber M (2022) Slope stability analysis of deep-seated landslides using limit equilibrium and finite element methods in Debre Sina area, Ethiopia. Bull Eng Geol Env 81(10):403
12. Griffiths DV, Lane PA (1999) Slope stability analysis by finite elements. Geotechnique 49(3):387–403
13. Zein AKM, Karim WA (2017) Stability of slopes on clays of variable strength by limit equilibrium and finite element analysis methods. GEOMATE J 13(38):157–164
14. Vinod BR, Shivananda P, Swathivarma R, Bhaskar MB (2017) Some of limit equilibrium method and finite element method based software are used in slope stability analysis. Int J Appl Innov Eng Manag (IJAIEM) 6(9):5
15. Poklopová T, Pavelcová V, Šejnoha M (2021) Comparing the Hoek-Brown and Mohr-Coulomb failure criteria in FEM analysis. Acta Polytech CTU Proc 30:69–75
16. Keshava V, Patil NN, Venkatesh BR (2021) Slope stability analyses at Agumbe ghat section NH-169A with and without slope protection. IOP Conf Ser Mater Sci Eng 1070(1):012030
17. Kamal ASM, Hossain F, Ahmed B, Rahman, MZ, Sammonds P (2023) Assessing the effectiveness of landslide slope stability by analysing structural mitigation measures and community risk perception. Nat Hazards 1–26
18. Zolkepli MF, Ishak MF, Zaini (2019) MSI slope stability analysis using modified Fellenius's and Bishop's method. IOP Conf Ser Mater Sci Eng 527(1):012004
19. Shiferaw HM (2021) Study on the influence of slope height and angle on the factor of safety and shape of failure of slopes based on strength reduction method of analysis. Beni-Suef Univ J Basic Appl Sci 10(1):31
20. Kadakci Koca T, Koca MY (2020) Comparative analyses of finite element and limit-equilibrium methods for heavily fractured rock slopes. J Earth Syst Sci 129(1):49
21. Rabie M (2014) Comparison study between traditional and finite element methods for slopes under heavy rainfall. HBRC J 10(2):160–168

22. He D, Yang W, Cheng Y, Chen B (2019) Effect of anchor layouts on the safety factor and slip surface of slope. Geotech Geol Eng 37:1073–1078
23. Liao J, Li J, Hao R, Liu B (2014) Stability of consequent rock slopes and analysis of prestressed anchors reinforcement. J Central South Univ 45:231–236
24. Zhang J, Li M, Yi J, Liu Z (2021) Investigation on the stability of fissured slopes reinforced with anchor cables under seismic action. Math Probl Eng 2021:1–14

Sustainable Usage of Calcium Carbide Residue for Soil Stabilization: A Review

Mahesh Endait and Tejashri Sambre

1 Introduction

Soil stabilization techniques play a crucial role in civil engineering, particularly in the construction and maintenance of infrastructure. Over the years, researchers and engineers have explored various methods to enhance soil properties, including incorporating stabilizing agents. Among these agents, calcium carbide residue (CCR) has gained substantial attention due to its potential for sustainable utilization and ability to improve soil characteristics [1]. Calcium carbide residue is a by-product generated during the production of acetylene gas from calcium carbide. Traditionally considered a waste material, CCR has been disposed of in landfills, posing environmental concerns, and occupying valuable landfill space. However, recent studies have highlighted the beneficial properties of CCR when employed as a soil stabilizer, showcasing its potential for sustainable usage and waste management [2, 3]. The sustainable utilization of CCR for soil stabilization offers several advantages. Because CCR is a commonly accessible by-product of industrial operations, it lessens the demand for the exploitation of natural resources. By repurposing CCR, the environmental impact associated with extracting and processing conventional stabilizers, such as lime and cement, can be significantly reduced. Additionally, incorporating CCR into soil matrices can enhance their geotechnical properties, including improved compaction, increased shear strength, reduced swelling potential, and enhanced load-bearing capacity [4–6]. The study provides a thorough assessment of the research on the subject. It aims to offer an overview of the methodology used, a summary of the research's findings, and the identification of possible research areas in the future. The evaluation covers the impact of several parameters, including CCR content,

M. Endait · T. Sambre (✉)
Department of Civil Engineering, Sandip University, Nasik, Maharashtra 422213, India
e-mail: kulkarniatejashri@gmail.com

M. Endait
e-mail: mahesh.endait@sandipuniversity.edu.in

S. Kolathayar et al. (eds.), *Best Practices in Geotechnical and Pavement Engineering*, Lecture Notes in Civil Engineering 449, https://doi.org/10.1007/978-981-99-8505-0_21

curing time, and environmental conditions, and considers how CCR affects diverse soil types. It also assesses CCR's long-term performance. The utilization of CCR as a soil stabilizer not only offers environmental benefits but also presents economic advantages. Its availability as a waste material at minimal or no cost, coupled with the potential reduction in construction costs due to improved soil properties, makes it an attractive alternative to conventional stabilizers. However, it is crucial to understand the limitations and challenges associated with CCR utilization to ensure its successful implementation in real-world scenarios. By providing a comprehensive review of the sustainable usage of calcium carbide residue for soil stabilization, this paper aims to contribute to the knowledge base and promote further research in this area. The findings of this review can be valuable for researchers, engineers, and policymakers involved in soil stabilization projects, offering insights into the potential of CCR as a sustainable and cost-effective solution for improving soil performance while reducing environmental impact.

2 Sustainable Development of CCR Utilization

The sustainable utilization of calcium carbide residue (CCR) for soil stabilization is supported by existing literature, highlighting its potential benefits and positive environmental impact. Several studies have investigated the properties and performance of CCR when used as a soil stabilizer, shedding light on its sustainability and waste management advantages. CCR is an alkaline material with a high calcium oxide concentration and other minerals. The utilization of CCR in sustainable development has been studied and researched in recent years. CCR, which is generated as a by-product during the production of acetylene gas, has traditionally been considered waste and disposed of in landfills. However, researchers have recognized its potential as a valuable resource for soil stabilization, thus promoting its sustainable utilization [2, 7]. By repurposing CCR, the need for extracting and processing conventional stabilizers, such as lime and cement, can be reduced, decreasing resource consumption and associated environmental impacts. Furthermore, the sustainable utilization of CCR aligns with circular economy and waste management principles. Treating CCR as a valuable resource rather than waste reduces the demand for landfill space and mitigates potential environmental hazards related to its disposal. This approach helps turn waste into valuable input, fostering a more sustainable and resource-efficient approach to soil stabilization practices. Research has shown that CCR can effectively improve the geotechnical properties of soils, enhancing their stability and performance. Adding CCR has been found to increase soil compaction, shear strength, and load-bearing capacity [4, 8, 9]. CCR can be particularly beneficial in construction projects where the stabilization of weak or problematic soils is required. By stabilizing soils with CCR, the need for extensive excavation, replacement, or more resource-intensive stabilizers can be minimized, resulting in significant cost savings and reduced environmental burden [10, 11]. Literature on the sustainable utilization of CCR for soil stabilization indicates its successful application and performance.

Various case studies, laboratory experiments, and field trials have been conducted to assess the effectiveness of CCR in different soil types, climatic conditions, and geotechnical contexts [12, 13]. Overall, the existing literature supports the sustainable utilization of CCR for soil stabilization by highlighting its potential environmental, economic, and waste management advantages. By repurposing this industrial by-product, significant strides can be made toward reducing resource consumption, minimizing environmental impact, and improving the sustainability of soil stabilization practices. Continued research and application of CCR in soil stabilization projects can further enhance our understanding of its effectiveness, enabling its widespread adoption as a sustainable solution in the construction industry. Another example is the use of CCR in the production of building materials. CCR can substitute traditional materials such as cement and aggregates in producing bricks, concrete, and other building materials. As a result, using CCR in building materials can decrease the demand for natural resources and reduce the environmental impact of construction activities. Additionally, using CCR in building materials can provide economic benefits by reducing material costs and improving structures' durability and longevity—using CCR in producing autoclaved aerated concrete (AAC) blocks. The study found that using CCR as a partial replacement for sand improved the compressive strength and reduced the environmental impact of AAC blocks [14]. There are many reasons to increase the amount of calcium carbide residue (CCR) being re-utilized. Here are some of the main ones: Environmental benefits: CCR is an alkaline material that can improve soil quality and reduce erosion. Using CCR as a soil amendment, we can eliminate the need for chemical stabilizers, which can cause harm the environment [15]. Economic benefits: CCR can be used as a substitute for traditional materials such as cement and aggregates to produce building materials. Using CCR in building materials can reduce material costs and improve the durability and longevity of structures [14]. Waste reduction and Resource conservation: CCR is a by-product of the production of acetylene gas and is often disposed of in landfills. By re-utilizing CCR, we can reduce the total waste that goes to landfills, which can help conserve space and reduce environmental pollution. Using CCR in various applications can help conserve natural resources. Using CCR as a substitute for traditional materials, we can decrease the demand for virgin materials such as sand, gravel, and cement, which are finite resources [16, 17]. Increasing the amount of CCR re-utilized can provide economic, environmental, and social benefits. However, proper testing and evaluation of CCR are necessary to ensure their safe and effective use in various applications. Regulatory frameworks and guidelines for managing and utilizing CCR have also been established to promote sustainable development.

3 Characterization of CCR

3.1 Physical

Calcium carbide residue (CCR) is produced when acetylene is burned. Despite being dangerous by nature, lime is plentiful. Consequently, it can be applied to alter the characteristics of expansive soils. Stabilization with CCR yields superior results than stabilization with lime from an economic and environmental standpoint. They classified the strength development of CCR-stabilized soils into active, inert, and degrading zones. Only the first two zones help boost power. The pozzolanic reaction can be initiated solely using natural pozzolanic materials. In the initial zone where the CCR is below 7%, the strength enhances naturally. However, it results in a substantial increase in strength within the inert zone (CCR ranging from 7 to 11%) [4, 17, 19]. Somna et al. [18] observed a 22% increase in UCS (Ultimate Compressive Strength) between 28 and 180 days of curing by employing a CCR and Rice Husk Ash mixture. Moreover, these minerals are employed in creating high-strength.

3.2 Chemical

Some waste products that are Ca $(OH)_2$ rich can be combined with pozzolanic elements found naturally in clay to create cementitious materials with minimal negative economic and environmental effects. When calcium carbide (CaC_2) is hydrolyzed to produce acetylene (C_2H_2), calcium carbide residue (CCR) is produced as a by-product.

The CCR is produced as an aqueous slurry and consists primarily of calcium hydroxide $Ca(OH)_2$, with trace amounts of unreacted carbon, calcium carbonate ($CaCO_3$), and silicates. During the manufacturing of acetylene, processing conditions impact the sludge's properties. Although not considered dangerous or hazardous, the high base sludge ($pH > 12$) might contain metals such as Mg, Br, Sr, Cd, Cu, Pb, Fe, Mn, Ni, and Zn [20, 21]. Thailand needs 18,500 tons of CaC_2 per year to produce acetylene gas. The supply of calcium carbide residue is 21,500 tons per year, while the annual demand is rising. Soil stabilization, pozzolan activation, asphaltic paving mixtures, and concrete are all examples of how the CCR is used as building, construction, and pavement materials. The strength development in the CCR-stabilized clay was examined in [4, 7, 18]. They demonstrated that, up to a specific content level, the plasticity index of the stabilized clay declines with the CCR concentration before becoming practically constant. This certain content level is designated as the CCR fixation point. It is helpful as a gauge for determining the ideal CCR concentration for increasing strength. Although there has been a prior study on soil stabilization using CCR [7], that research concentrated on strength development and is insufficient for pavement and earth structure construction. It is necessary to consider other engineering characteristics, such as bearing capacity, swelling, and collapse behavior.

3.3 Mineral

Calcium carbide residue (CCR) is an alkaline material that contains a high concentration of calcium oxide (CaO) and other minerals. The mineral characteristics of CCR have been studied and reported in various research studies. Here are some of the main mineral characteristics of CCR. The chemical composition of CCR contained a high CaO, SiO_2, and Al_2O_3 concentration. CCR has a low heavy metal concentration, making it appropriate for various applications [21]. Table 1 describes various mineral and chemical components of CCR and their application. Various studies found that adding CCR improved the soil's pH, increased the availability of nutrients, and improved plant growth. Overall, the mineral characteristics of CCR make it a potentially valuable resource for various applications, including soil stabilization, building materials, and steelmaking. However, proper testing and evaluation of CCR are necessary to ensure its safe and effective use in numerous applications; regulatory frameworks and guidelines for managing and utilizing CCR have also been established to promote sustainable development.

Table 1 Chemical and mineral composition present in CCR

Chemical and mineral composition present in CCR	% Range CCR	Use for various application
CaO	50–90	The main mineral is in CCR, used for soil stabilization, cement production, and steelmaking [22, 23]
SiO_2	4.2–4.8	Important for soil structure and can improve soil water-holding capacity [14, 22]
Al_2O_3	033–0.50	Affect soil acidity [14]
Fe_2O_3	0.10	Affect soil color and nutrient availability [14]
MgO	0.22	Affect soil color [14]
$CaCO_3$	13.06	Important for soil structure and can improve soil water-holding capacity [14, 22]

4 Effect on Geotechnical Properties with Emphasis on Chemical Processes

Calcium carbide residues can significantly impact the geotechnical properties of soil due to the chemical processes that occur when the residue comes into contact with the soil. Calcium carbide residue contains a high concentration of calcium hydroxide.

($Ca(OH)_2$) and calcium carbide (CaC_2). When mixed with soil, calcium carbide residue can cause several chemical reactions that can alter the geotechnical properties of the soil.

The first reaction is the dissolution of calcium hydroxide in water, which leads to an increase in the pH of the soil. This increase in pH can cause the clay particles in the soil to become more cohesive; it may improve the soil's shear strength. This is because the hydrolysis of calcium carbide releases alkaline compounds, which can increase the pH of the surrounding environment. This can have significant implications for the area, as changes in pH can affect the availability of nutrients and other resources.

The second reaction is the production of acetylene gas when calcium carbide comes into contact with water. This gas can create voids in the soil, which can decrease the soil density and, therefore, the strength of the soil. The voids can also cause the soil to become more compressible, leading to settlement issues. In addition to these chemical reactions, CCR can affect the soil structure by altering the pore water pressure and changing the soil's hydraulic conductivity. The high concentration of calcium ions in the residue can increase the pore water pressure, which can lead to an increase in the effective stress of the soil. This can source the soil to become more compact and increase its strength.

The chemical theory involved in the reaction between CCR and silicate aluminate constituents of the expansive soils is complex. The primary factors include cation exchange, flocculation and agglomeration, pozzolanic reaction, and carbonation [24, 25]. The reaction mechanisms can be further classified into two groups: modification for plasticity reduction and solidification [26]. Modification may be reversible, but solidification results in irreversible changes in the soil characteristics Modification results through flocculation and cation exchange activity, whereas solidification results through a pozzolanic reaction [27]. These reactions contribute to the treated soils' physical, chemical, mineralogical, and microstructural changes. However, the chemical effects of calcium carbide residues are more complex. Various factors, including the type and concentration of the residues, the nature of the soil or rock, and the environmental conditions, can influence them. One important chemical process that can occur is the release of calcium ions from the residues into the soil or rock. These ions can react with various minerals in the soil, such as clay minerals, to form new compounds [21–25]. CCR has the potential to be used as a soil stabilizer. It discusses the chemical composition of the residue and its effects on soil properties, including compaction, permeability, and shear strength. The study examines the effects of calcium carbide residue on the engineering properties of cemented paste backfill, a commonly used material in underground mining operations. It discusses the role of the residue in influencing the strength, stiffness, and permeability of the backfill. The study also explores the chemical processes involved in the residue and backfill interaction.

As summarized below, CCR can significantly affect geotechnical properties through various chemical processes. When calcium carbide residues come into contact with water, they can undergo hydrolysis, producing acetylene gas and calcium hydroxide. The calcium hydroxide can then react with the surrounding soil or rock, potentially altering its properties. Calcium released from the residues can react with various minerals in the soil, such as clay minerals, to form new compounds, which has implications for soil strength and stability [18, 19]. Calcium carbide residues can affect the pH of the surrounding soil or water. The hydrolysis of calcium carbide releases alkaline compounds, which can increase the pH of the surrounding environment. This can have significant implications for the biota in the area, as changes in pH can affect the availability of nutrients and other resources [28, 29]. Calcium carbide residues have been shown to have a high capacity for adsorbing heavy metals such as copper, zinc, and lead. It can have implications for soil and water quality [28]. Recently 2019, a researcher [30] studied the addition of RHA-CCR on expansive soil stabilization and observed that it reduces the proportion of clay particles in the expansive soil. The low plasticity of RHA and CCR will correspondingly reduce expansive soil's liquid limit and plasticity index. Due to the lack of montmorillonite and illite, RHA and CCR are non-expansive materials; they can lower the expansion and contraction of expansive soil. In addition, the particles of CCR are very fine; they can also have a good filling effect on soil particles. The effects of CCR on the compressibility of clayey soil were investigated. The results showed that the residues could significantly increase soil compressibility, affecting the stability of slopes and embankments. The authors attributed this effect to the filling of soil pores by the residues [31]. Various studies suggest that calcium carbide residues can significantly affect geotechnical properties through physical and chemical processes. However, the exact nature and extent of these effects can depend on various factors, such as the type and concentration of the residues, the nature of the soil or rock, and the environmental conditions.

5 Effect on Physicochemical and Microstructure Interaction

The physicochemical and microstructure interaction of CCR can positively and negatively affect different materials and systems. Here are some possible effects:

Cementitious systems: CCR can be a supplementary cementitious material in concrete production. When blended with cement, CCR can improve the compressive strength, reduce the porosity, and enhance the durability of concrete. This is due to the pozzolanic reaction between the calcium hydroxide in the cement and the silicates and aluminates in the CCR, leading to the formation of cementitious compounds [17]. Soil improvement: CCR can also be used as a soil amendment due to its alkaline properties. The calcium hydroxide in the CCR can react with the soil acidity and neutralize it, improving the soil's pH and fertility. In addition, the porosity of the soil can be improved by forming cementitious compounds between the CCR and soil particles.

Environmental impact: CCR can hurt the environment if disposed of improperly. The acetylene in the CCR can react with water to produce acetylene gas, a potent greenhouse gas. The calcium hydroxide in the CCR can also leach into the surrounding soil and water, increasing alkalinity and potentially harming aquatic life [24]. Microstructure interaction: The microstructure of materials can be affected by the addition of CCR due to the formation of new compounds and the modification of existing ones. For example, adding CCR to cement can lead to forming calcium silicate hydrate (C–S–H) gels with different morphologies and characteristics than those formed in pure cement, affecting the resulting concrete's mechanical properties [32]. The physicochemical and microstructure interaction of CCR can have different effects depending on the application and the specific characteristics of the CCR. While it can positively affect cementitious systems and soil improvement, its environmental impact and potential effects on microstructure should also be considered. Proper disposal and handling of CCR can minimize its negative environmental impact.

6 Discussion

The performance of CCR was evaluated against several traditional and nontraditional stabilizing agents reported for the same clays and the literature.

CCR can have a complex and significant impact on the geotechnical properties of soil.

When CCRs are deposited on soil or rock, they can have physical and chemical effects on the surrounding environment. Regarding physical effects, the residues can cause compaction and reduce the permeability of the soil or rock. It is because the residues tend to fill the voids between soil particles, effectively reducing the pore space and limiting water and air movement through the soil.

The microstructural characteristics indicate that PI is significantly increased with 8% CCR due to the formation of C–S–H and C–A–H gels which rise in the calcium-to-silica ratio and fill the pores in clays with discontinuous structure.

According to Dash and Hussain, these studies concur with those seen for the microstructural behavior of expansive soils. Thus, depending on the availability of the appropriate component, CCR stabilization is possible for soils rich in SiO2, soils containing gypsum, and soils containing iron with differing optimum CCR percentages for each. Lime is not suggested in SO4-rich soils because there isn't any hydrated calcium oxide. Thus, it may be concluded that the primary method of lime stabilization of expansive soils is a function of the environment of lime-soil reactions and the type of clay minerals. For cement versus lime stabilization, it is often misinterpreted that both stabilizers are identical in yielding C–S–H and C–A–H formation results. Still, CCR proves to be a relatively better choice while modifying the CCR is also proven to increase the strength of expansive soils, such as by 10% more. However, further research that could result in the development of multipurpose CCR is still needed to determine the optimal dosage for combined use, such as one that simultaneously increases UCS and decreases expansion.

7 Conclusions

Considering the importance and impact of CCR in physical, chemical, and mineralogical aspects, it is significant to understand any new developments regarding CCR. Over time CCR has moved from a critical waste to an important industrial by-product in many countries [5, 6, 32] A better understanding and push toward sustainable development have shifted the research focus from high-volume recycling to high-value utilization [1, 14]. One should be able to appreciate the role of new technological developments in the use of CCR as a sustainable stabilizer and apply them in the industry. The authors hope that readers will be able to apply this knowledge and move from the waste minimization phase to sustainable development in the CCR and, in time, apply the lessons of sustainable development learned from using CCR toward developing other such industrial wastes.

8 Conflicts of Interest

There are no conflicts of interest, according to the authors.

References

1. Makusa GP (2012) Soil stabilization methods and materials in engineering practice. Journal 1:1–35
2. Endait M, Wagh S, Kolhe S (2019) Stabilization of black cotton soil using calcium carbide residue. In: Proceedings of the Indian geotechnical conference (IGC 2019)
3. Horpibulsuk S, Rachan R, Chinkulkijniwat A, Raksachon Y, Suddeepong A (2010) Analysis of strength development in cement-stabilized silty clay from microstructural considerations. Constr Build Mater 24(10):2011–2021. https://doi.org/10.1016/j.conbuildmat.2010.03.011
4. Horpibulsuk S, Phetchuay C, Chinkulkijniwat A, Cholaphatsorn A (2013) Strength development in silty clay stabilized with calcium carbide residue and fly ash. Soils Found 53(4):477–486. https://doi.org/10.1016/j.sandf.2013.06.001
5. Vichan S, Rachan R (2013) Chemical stabilization of soft Bangkok clay using the blend of calcium carbide residue and biomass ash. Soils Found 53(2):272–281. https://doi.org/10.1016/j.sandf.2013.02.007
6. Liu Y et al (2019) Stabilization of expansive soil using cementing material from rice husk ash and calcium carbide residue. Constr Build Mater 221:1–11. https://doi.org/10.1016/j.conbuildmat.2019.05.157

7. Horpibulsuk S, Phetchuay C, Chinkulkijniwat A (2012) Soil stabilization by calcium carbide residue and fly ash. J Mater Civ Eng 24(2):184–193. https://doi.org/10.1061/(ASCE)mt.1943-5533.0000370
8. Chindaprasirt P, Kampala A, Jitsangiam P, Horpibulsuk S (2020) Performance and evaluation of calcium carbide residue stabilized lateritic soil for construction materials. Case Stud Constr Mater 13. https://doi.org/10.1016/j.cscm.2020.e00389
9. Phetchuay C, Horpibulsuk S, Arulrajah A, Suksiripattanapong C, Udomchai A (2016) Strength development in soft marine clay stabilized by fly ash and calcium carbide residue based geopolymer. Appl Clay Sci 127–128:134–142. https://doi.org/10.1016/j.clay.2016.04.005
10. Du YJ, Jiang NJ, Liu SY, Horpibulsuk S, Arulrajah A (2016) Field evaluation of soft highway subgrade soil stabilized with calcium carbide residue. Soils Found 56(2):301–314. https://doi.org/10.1016/j.sandf.2016.02.012
11. Joel M, Edeh JE (2013) Soil modification and stabilization potential of calcium carbide waste. In: Advanced materials research, pp 29–36. https://doi.org/10.4028/www.scientific.net/AMR.824.29
12. Estabragh AR, Rafatjo H, Javadi AA (2014) Treatment of an expansive soil by mechanical and chemical techniques. Geosynth Int 21(3):233–243. https://doi.org/10.1680/gein.14.00011
13. Raot KSS (2000) Swell-shrink behaviour of expansive soils-geotechnical challenges
14. Wang Q, Guo H, Yu T, Yuan P, Deng L, Zhang B (2022) Utilization of calcium carbide residue as solid alkali for preparing fly ash-based geopolymers: dependence of compressive strength and microstructure on calcium carbide residue, water content and curing temperature. Materials (Basel) 15(3). https://doi.org/10.3390/ma15030973
15. Akinwumi II, Ajayi OO, Agarana MC, Ogbiye AS, Ojuri OO, David AO (2019) Investigation of calcium carbide residue as a stabilizer for tropical sand used as pavement material. In: WIT transactions on the built environment. WIT Press, pp 285–294. https://doi.org/10.2495/UT180261
16. Cardoso FA, Fernandes HC, Pileggi RG, Cincotto MA, John VM (2009) Carbide lime and industrial hydrated lime characterization. Powder Technol 195(2):143–149. https://doi.org/10.1016/j.powtec.2009.05.017
17. Zhu X, Niu F, Ren L, Jiao C, Jiang H, Yao X (2022) Effect of calcium carbide residue on strength development along with mechanisms of cement-stabilized dredged sludge. Materials (Basel) 15(13). https://doi.org/10.3390/ma15134453
18. Somna K, Jaturapitakkul C, Kajitvichyanukul P (2011) Microstructure of calcium carbide residue-ground fly ash paste. J Mater Civ Eng 23(3):298–304. https://doi.org/10.1061/(asce)mt.1943-5533.0000167
19. Vipulanandan C, Mohammed A (2016) XRD and TGA, swelling and compacted properties of polymer treated sulfate contaminated CL soil. J Test Eval 44(6):2270–2284. https://doi.org/10.1520/JTE20140280
20. Graduate ME, Horpibulsuk S, Raksachon Y (2009) Role of fly ash on strength and microstructure development in blended cement stabilized silty clay. Japanese Geotechnical Society
21. Jaturapitakkul C, Roongreung B (2003) Cementing material from calcium carbide residue-rice husk ash. https://doi.org/10.1061/ASCE0899-1561200315:5470
22. Latifi N, Vahedifard F, Ghazanfari E, Rashid ASA (2018) Sustainable usage of calcium carbide residue for stabilization of clays. J Mater Civ Eng 30(6). https://doi.org/10.1061/(asce)mt.1943-5533.0002313
23. Lin B, Cerato AB, Andrew SM, Madden MEE (2013) Effect of fly ash on the behavior of expansive soils: microscopic analysis. Environ Eng Geosci 19(1):85–94. https://doi.org/10.2113/gseegeosci.19.1.85
24. Al-Mukhtar M, Khattab S, Alcover JF (2012) Microstructure and geotechnical properties of lime-treated expansive clayey soil. Eng Geol 139–140:17–27. https://doi.org/10.1016/j.enggeo.2012.04.004
25. Mosa AM, Taher AH, Al-Jaberi LA (2017) Improvement of poor subgrade soils using cement kiln dust. Case Stud Constr Mater 7:138–143. https://doi.org/10.1016/j.cscm.2017.06.005

26. Cheshomi A, Eshaghi A, Hassanpour J (2017) Effect of lime and fly ash on swelling percentage and Atterberg limits of sulfate-bearing clay. Appl Clay Sci 135:190–198. https://doi.org/10.1016/j.clay.2016.09.019
27. Chindaprasirt P, Kampala A, Jitsangiam P, Horpibulsuk S (2020) Performance and evaluation of calcium carbide residue stabilized lateritic soil for construction materials. Case Stud Constr Mater 13:e00389. https://doi.org/10.1016/j.cscm.2020.e00389
28. Zhang LM, Xu Y, Tang WH (2008) Calibration of models for pile settlement analysis using 64 field load tests. Can Geotech J 45(1):59–73. https://doi.org/10.1139/T07-077
29. Mishra S, Sachdeva SN, Manocha R (2019) Subgrade soil stabilization using stone dust and coarse aggregate: a cost effective approach. Int J Geosynth Gr Eng 5(3). https://doi.org/10.1007/s40891-019-0171-0
30. Ying Z, Cui YJ, Benahmed N, Duc M (2020) Changes in mineralogy and microstructure of a lime-treated silty soil during curing time. In: E3S web of conferences, EDP Sciences. https://doi.org/10.1051/e3sconf/202019503044
31. He S, Yu X, Banerjee A, Puppala AJ (2018) Expansive soil treatment with liquid ionic soil stabilizer. Transp Res Rec 2672(52):185–194. https://doi.org/10.1177/0361198118792996
32. Juneja A, Shinde ST (2020) Undrained yielding of black cotton soil treated with calcium carbide residue. Indian Geotech J 50(3):319–329. https://doi.org/10.1007/s40098-019-00380-8

Mechanical Strength of Local Soil Enhanced by Hybrid Saw Dust Ash

Sagar D. Turkane, Arti A. Wagh, Rau N. Dohale, Komal N. Shinde, Dinesh M. Pandhure, and Abhay A. Bhagat

1 Introduction

In the recent years, demand for good quality road is increasing day by day due rapid development of industrial and urbanization growth. For construction of good road requires adequate strength of various layers of road pavements especially the subgrade and subbase layer. These layers can be constructed using locally available soil that may not have desired strength, hence that soil may be replaced with good soil or their geo-characteristics is enhanced by various ground improvement techniques. In India, almost 20% of land has inadequate soil strength properties that suggest improvement in geotechnical properties [1, 2]. The ground improvements techniques involve the soil stabilizations by means of mechanical enhancement or chemical enhancement. The chemical soil stabilization is quite quick and cheap comparing to other soil stabilization techniques. This chemical stabilization involves the addition of chemical reagents or material that initiate chemical reaction inside the material or will leach cementitious product that hold the soil particles closely [3].

Also, any developing country needs huge energy to fulfil the required demand of society. This energy can be fulfilled throught the renewable sources such ash biomass (forestry and agricultural wastes) [4]. Power plants that run on biomass have low operational costs and a steady supply of renewable fuel in the current trends of energy production. Additionally, using wastes produced by the biomass industries as fuel (including sawdust, woodchips, wood bark, sawmill scraps, and hard chips) provides a safe and effective method for their disposal. Typically, factories that make wood products that uses wood waste, they produce as their primary fuel to generate heat energy for their numerous operations, such drying the final goods. The substantial amount of ash created during the burning of such wastes is a serious

S. D. Turkane (✉) · A. A. Wagh · R. N. Dohale · K. N. Shinde · D. M. Pandhure · A. A. Bhagat
Structural Engineering Department, Sanjivani College of Engineering, Kopargaon, Maharashtra 423603, India
e-mail: sagarturkane24@gmail.com

S. Kolathayar et al. (eds.), *Best Practices in Geotechnical and Pavement Engineering*, Lecture Notes in Civil Engineering 449, https://doi.org/10.1007/978-981-99-8505-0_22

issue associated with the use of forest and wood waste products as fuel ashes are often produced in greater quantities by hardwood than by softwood, and ashes are typically produced in greater quantities by the bark and leaves of trees than by their interiors. The usual amount of ash produced by burning raw wood is in the range of 6–10% by weight, and its makeup can vary greatly depending on the environment and industrial operations [5]. Wood ash is potentially hazardous since it is largely made up of fine particle matter that is easily transported into the air by winds and might cause respiratory problems to the nearby living organisms or groundwater contamination by leaching toxic compounds into the water [6, 7]. A sustainable and effective saw dust ash management disposal strategy that needs to define without destroying natural cycles, as disposal of this material, is very expensive since the amount of ash is growing rapidly.

Several studies have been conducted for improvement of geotechnical properties and that showed the significant enhancement and could be suitable material for road construction [8–13]. In India, huge amount of waste generated from various sectors such as agriculture, industry, per year. This suggests proper utilization or disposal in enviro-friendly manner. Therefore, in view of all this issues with saw dust ash it can be effectively used for soil stabilizer, and further, it can be used for road construction material. When weak soils are stabilized using undesired products like sawdust ash (SDA), there are several technical, economic, and environmental advantages. Utilizing such agricultural waste and by-product ashes thereby addresses the dual issue of how to dispose of them while also offering a practical substitute for various soil stabilizers. Therefore, using wood ash as a substitute for various soil stabilizers is advantageous from an environmental standpoint as well as providing affordable building entities, leading to a sustainable connection.

A several research has been carried out on stabilization of expansive soil or black cotton soil using SDA along with binders such cement and lime [8, 11, 14] and still date no research find for SDA mixed with sodium hydroxide (NaOH) and sodium silicate (Na_2SiO_3, water glass) as alkaline activators. This paper aims to study the feasibility of SDA for road construction material as subgrade layer via unconfined compression strength test. Soil stabilization using alkaline activators is very effective technique shown by various researcher in past such as [2, 15–22]. This research may help to engineers, practitioners, and contractors to understand the feasibility of saw dust ash as soil stabilizer, subjected to requirement of further investigation.

2 Materials and Methods

2.1 *Expansive Soil*

The expansive (black cotton) soil used in this study, which is depicted in Fig. 1, was obtained from a natural deposit close to Shirdi, Maharashtra, India and is well known as "Black cotton soil". After removing the top 500 mm of soil, the soil was collected

Fig. 1 Black cotton soil

Table 1 Geotechnical properties of soil and SDA

Geotechnical engineering properties	Soil	SDA
Sand size fractions %	09	11
Silt size fractions %	30	53
Clay size fractions %	61	36
G (specific gravity)	2.67	2
LL (liquid limit) %	58.8	42.2
PL (plastic limit) %	22.4	–
PI (plasticity Index) %	36.4	–
IS classification	CH	–
MDD (maximum dry density) gm/cc	1.81	1.1

using the disturbed sampling technique and transported in bags to the laboratory. To determine the sample's natural moisture content, a little portion of the substance was enclosed in a polythene bag. The soil was air dried, the surface, and 4.75 mm sieved before drying. For laboratory experiments, materials that passed through a 4.75 mm sieve were employed. Table 1 provides a geotechnical characterization of the sample.

2.2 *Sawdust Ash*

The saw dust utilized in this investigation, as seen in Fig. 2, originated from sawmill locations at the Kopargaon wood market in India. To prevent combining sand and sawdust, the sample was carefully taken. The sample was collected and open-burned in a metal container until it was reduced to ashes. After cooling, the sawdust ash (SDA) was crushed in a mortar and pestle. To ascertain the physical characteristics of the saw dust ash (SDA), tests were performed and a yield calculation was made,

Fig. 2 Sawdust ash

as indicated in Table 1. For this investigation, the sawdust ash (SDA) particles that passed through the sieve with an aperture of 600 m were employed.

2.3 *Alkaline Activator*

Here, sodium hydroxide (NaOH) and sodium silicate (Na_2SiO_3) solutions (Fig. 3a, b, respectively) were utilized in a proportional ratio to activate fly ash using an alkaline solution. 20% of the total soil mass is made up of an evenly mixed liquid of NaOH and Na_2SiO_3 in a 1:1.5 ratio. NaOH should be present at a concentration of between 4.5 and 18 molars, according to the study carried by [2, 23, 24]. Therefore, low concentrated NaOH, or 5 molars, was used for the experimental inquiry from a cost and safety point of view.

(a) (b)

Fig. 3 **a** NaOH solution and **b** Na_2SiO_3 solution

Studies have revealed that a number of factors, including the activator concentration, activator ratio, and liquid to solid ratio, among others, affect how well a geopolymer binder performs. Most geopolymer mixtures have good compressive strength at a ratio of 1.5 alkaline activator.

The methodology of the investigation was completed in following stages: soil and SDA characterization, selection of proportion of stabilizer, casting and curing of samples, and testing of the samples. Firstly, soil was dried for two weeks and investigated for basic properties. For the compaction testing, soil samples were separated into five layers in a mould and exposed to 25 blows of 44.5 N rammer descending from a height of 0.46 m to ascertain its impact on the soil's strength attributes, saw dust ash replacements of 5, 10, 15, and 20% were made after the first process had been completed without the addition of any saw dust ash. To further determine if soil is a viable option for use as a building material for pavement, unconfined compressive strength tests were carried out using conventional techniques in accordance with the applicable IS code IS 2720-10 (1991). For the experimental testing (UCS), a known quantity of soil and SDA mixed together and alkaline activator equal to OMC was added and mixed homogeneously. After stabilizing a soil sample with alkaline activated saw dust ash, its durability was tested after 7, 14, and 28 days of curing. The soil sample was poured into UCS mould of 38 mm diameter and 76 mm in height and compacted to prepare a cylindrical sample. After compaction, sample was placed in plastic air tight bag according to methods suggested by Kaniraj and Gayathri 2003 [2, 25].

3 Result and Discussion

3.1 Compaction Characteristics

With varied amounts of saw dust ash, i.e. 0, 5, 10, 15, and 20% by total weight of the soil, compaction behaviour was observed in order to understand the impact of the saw dust ash on the black cotton soil. The findings demonstrate that as saw dust is replaced with ash, the maximum dry density gradually declines and the ideal moisture content rises. Due to the diversity of non-plastic material and the lowering of the specific gravity of the soil mass, which may weaken the cohesive link between soil particles, the dry density may have decreased. The increased ideal moisture content may be brought on by the sawdust ash's increased water needs. Figure 4 illustrates the variations in dry density and ideal moisture content relate to sawdust ash contents. The increased water demand for SDA is what causes the rise in optimal moisture content as a result of the increase in SDA%. The compaction of soil with SDA% without any binding material or adhesive material reduction the maximum dry density was noted, this suggests that requirement of any binding materials.

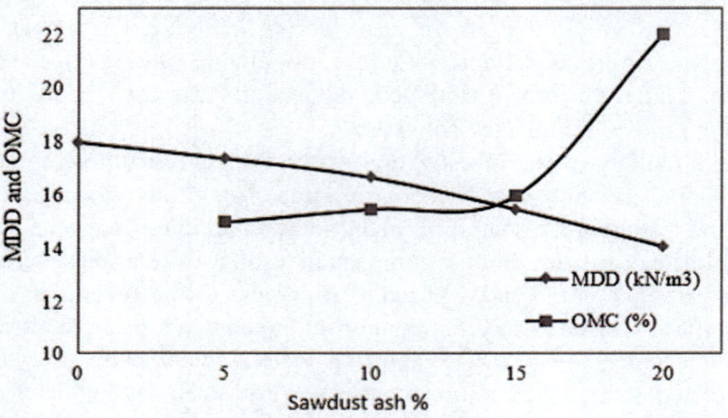

Fig. 4 Compaction characterization of soil-saw dust ash

3.2 *Unconfined Compression Strength Test*

The unconfined compression strength test is the most frequent, well-liked, and often applied method of evaluating the durability of cohesive and stabilized soils. For calculating the amount of additive required for soil stabilization, it is the main test that is recommended. Generally speaking, the higher the compressive strength for a specific kind of stabilization, the higher the calibre of the stabilized material. For each sample, Fig. 5 shows the stress–strain curve.

Figure 5 depicts the fluctuation in unconfined compressive strength with SDA concentration but no admixtures. According to the findings, the soil's maximum strength was 330 kPa at 5% SDA. Comparing UCS values reveals that 180 kPa serves as SDA's substitute for 0% SDA. The gradation of soil changes that leads to

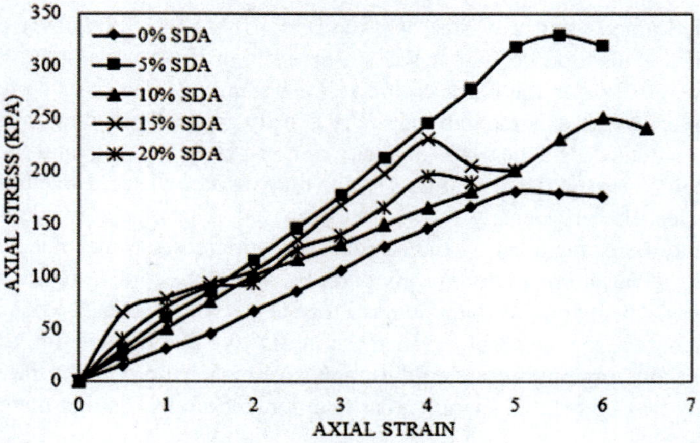

Fig. 5 Stress–strain curve for various percentage of SDA

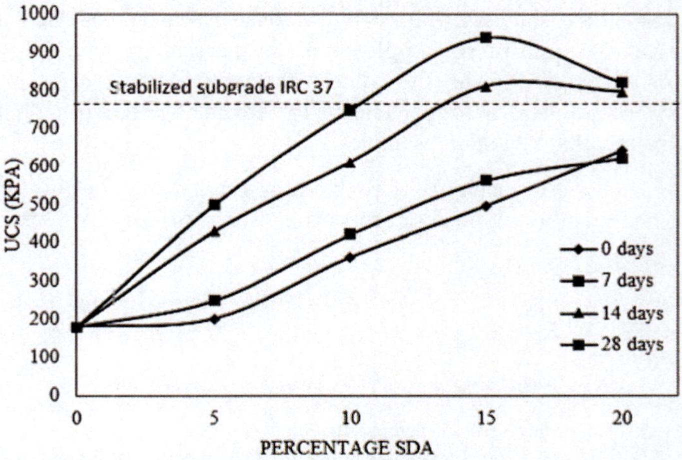

Fig. 6 UCS with addition of alkaline activators for various curing days

an increase in soil's UCS values. Figure 6 displays the expected UCS values for SDA with alkaline activators added for various curing times. With the addition of 15% SDA, the continuous rise was seen in the soil, going from 230 to 900 kN/m^2. The pozzolanic interactions of SDA with the soil's NaOH and the pozzolana in SDA to produce cementitious products may be the cause of the increased strength. Additional SDA replacement at 20% UCS value decrease may be caused by inadequate alkaline liquid or by the establishment of weak links between the soil and cementitious compounds.

The curing of soil sample at ambient temperature propagates the geopolymeric reaction that may lead to enchantments of strength of soil. The stabilized soil may be suitable for construction of rural roads or low volume roads as it satisfied the minimum strength requirement of for subgrade layer of flexible pavement as per IRC 37:2012 [26].

4 Conclusion

Following conclusion can be drawn from the experimental investigation for stabilized expansive soil with alkaline activated saw dust ash.

- The use of SDA for black cotton soil stabilization using alkali activated SDA at varied percentages, i.e. 5, 10, 15, and 20%, has been proven to be a good stabilizer, and their disposal problem may be reduced.
- In terms of the effect of SDA on density and compaction, test findings reveal that SDA lowers compacted dry density while increasing soil optimal moisture content.

- The black cotton soil is stabilized by sawdust ash and alkali activators. The compressive strength of soil improves as the percentage of SDA with alkali activators increases, allowing the soil to withstand maximal stress.
- The 15% is the ideal SDA percentage for enhancing the strength properties approximately 900 kPa of soil samples.

Utilizing such waste materials while building rural roads not only reduces the amount of natural resource material in our environment but also strengthens the subgrade layer.

References

1. Das BM (2014) Advanced soil mechanics, 4th edn
2. Turkane SD, Chouksey SK (2022) Design of flexible pavement thickness using stabilized high plastic soil by means of fly ash-based geopolymer. Int J Pavement Eng 1–15. https://doi.org/10.1080/10298436.2022.2044035
3. Garcia-Lodeiro I, Palomo A, Fernández-Jiménez A (2015) An overview of the chemistry of alkali-activated cement-based binders. In: Handbook of alkali-activated cements, mortars and concretes. Elsevier, pp 19–47
4. Onyeka FC (2019) Application of industrial waste (saw-dust ash) in the production of self-compacting concrete. Int Res J Innov Eng Technol 3(11):1–9
5. James J (2019) Strength benefit of sawdust/wood ash amendment in cement stabilization of an expansive soil. Rev Fac Ing 28(50):44–61. https://doi.org/10.19053/01211129.V28.N50.2019.8790
6. Misra MK, Ragland KW, Baker AJ (1993) Wood ash composition as a function of furnace temperature. Biomass Bioenerg 4(5):103–116
7. Marthong C (2012) Sawdust ash as partial replacement of cement. Int J Eng Res Appl 2(4):1980–1985. (Online). Available: https://www.ijera.com/papers/Vol2_issue4/LW2419801985.pdf
8. Butt WA, Gupta K, Jha JN (2016) Strength behavior of clayey soil stabilized with saw dust ash. Int J Geo-Eng 7(1). https://doi.org/10.1186/s40703-016-0032-9
9. Ogunribido THT (2012) Geotechnical properties of saw dust ash stabilized southwestern Nigeria lateritic soils. Environ Res Eng Manag 60(2):29–33. https://doi.org/10.5755/j01.erem.60.2.986
10. Owamah H, Atikpo E, Oluwatuyi O, Oluwatomisin A (2017) Geotechnical properties of clayey soil stabilized with cement-sawdust ash for highway construction. J Appl Sci Environ Manag 21(7):1378–1381. https://doi.org/10.4314/jasem.v21i7.29Copyright
11. Elahi TE, Shahriar AR, Alam MK, Abedin MZ (2022) Engineering characteristics of soil stabilised with saw dust ash and cement. Geomech Geoengin 17(6):1923–1940. https://doi.org/10.1080/17486025.2021.1981464
12. Mohamed AAMS, Yuan J, Al-Ajamee M, Dong Y, Ren Y, Hakuzweyezu T (2023) Improvement of expansive soil characteristics stabilized with sawdust ash, high calcium fly ash and cement. Case Stud Constr Mater 18(2022):e01894. https://doi.org/10.1016/j.cscm.2023.e01894
13. Oyedepo OJ, Oluwajana SD, Akande SP (2014) Investigation of properties of concrete using sawdust as partial replacement for sand. Civ Environ Res 6(2):35–42
14. Ikeagwuani CC, Obeta IN, Agunwamba JC (2019) Stabilization of black cotton soil subgrade using sawdust ash and lime. Soils Found 59(1):162–175. https://doi.org/10.1016/j.sandf.2018.10.004
15. Arioz E, Arioz O, Mete Kockara O (2012) Leaching of F-type fly ash based geopolymers. Proc Eng 42:1114–1120. https://doi.org/10.1016/j.proeng.2012.07.503

16. Miao S, Shen Z, Wang X, Luo F, Huang X, Wei C (2017) Stabilization of highly expansive black cotton soils by means of geopolymerization. J Mater Civ Eng 29(10):04017170. https://doi.org/10.1061/(ASCE)MT.1943-5533.0002023
17. Mazhar S, GuhaRay A, Kar A, Avinash GSS, Sirupa R (2018) Stabilization of expansive black cotton soils with alkali activated binders. In: Construction and building materials, vol 31, no 2. Elsevier Ltd., pp 826–829
18. Khadka SD, Jayawickrama PW, Senadheera S (2018) Strength and shrink/swell behavior of highly plastic clay treated with geopolymer. Transp Res Rec J Transp Res Board 2672(52):174–184. https://doi.org/10.1177/0361198118797214
19. Abdullah A et al (2021) The effects of various concentrations of NaOH on the inter-particle gelation of a fly ash geopolymer aggregate. Materials (Basel) 14(5):1–11. https://doi.org/10.3390/ma14051111
20. Phummiphan I, Horpibulsuk S, Sukmak P, Chinkulkijniwat A, Arulrajah A, Shen S-L (2016) Stabilisation of marginal lateritic soil using high calcium fly ash-based geopolymer. Road Mater Pavement Des 17(4):877–891. https://doi.org/10.1080/14680629.2015.1132632
21. Turkane SD, Chouksey SK (2022) Design of low volume road pavement of stabilized low plastic soil using fly ash geopolymer. Mater Today Proc 1–7. https://doi.org/10.1016/j.matpr.2022.04.167
22. Turkane SD, Chouksey SK (2022) Partial replacement of conventional material with stabilized soil in flexible pavement design. Int J Eng 35(5):908–916. https://doi.org/10.5829/IJE.2022.35.05B.07
23. Nematollahi B, Sanjayan J (2014) Effect of different superplasticizers and activator combinations on workability and strength of fly ash based geopolymer. Mater Des 57:667–672. https://doi.org/10.1016/j.matdes.2014.01.064
24. Mustafa Al Bakri, AM, Kamarudin H, Bnhussain M, Rafiza AR, Zarina Y (2012) Effect of Na_2SiO_3/NaOH ratios and NaOH molarities on compressive strength of fly-ash-based geopolymer. ACI Mater J 109(5):503–508. https://doi.org/10.14359/51684080
25. Kaniraj SR, Gayathri V (2003) Factors influencing the strength of cement fly ash base courses. J Transp Eng 129(5):538–548. https://doi.org/10.1061/(ASCE)0733-947X(2003)129:5(538)
26. IRC: 37-2012 (2012) Guidelines for the design of flexible pavements

Behavior of Reinforced Soil

Seismic Stability Analysis of Road Embankment Resting on Geotextile Reinforced Soft Soil

Uzma Azim and Siddhartha Sengupta

1 Introduction

Disruption in transportation corridor, that is, roadways, railways or airways, either due to man-made or due to natural causes, can increase the damages incurred at minor to catastrophic scales. The problem is enhanced if the strength of the foundation soil is not compatible to withstand the load, as in the case of soft soil. Soft soils are characterized as highly under-consolidated soils bearing high water content, low strength and holding the potential to undergo large deformation. Under the action of seismic forces, these soils are highly susceptible to liquefaction, under which the loosely packed saturated geomaterial undergoes sudden loss in their inherent shear strength in response to strong ground motions. Liquefaction beneath the foundation of buildings and transportation embankment are a major cause of damage during seismic activities, and hence, remediation measures are an urgent requirement. This paper attempts to analyze the stability and effect on horizontal and vertical stresses on reinforcing the soft soil foundation underlying a road embankment with geotextile under the influence of seismic forces.

Several studies have been performed on stabilization of road embankments subjected to earthquake. Koseki [1] studied the effect of geotextile inclusion with consideration of factors like rigidity, arrangements and properties of reinforcements and subsoil conditions. In comparison with unreinforced simulations, the ductile behavior of geotextile enhanced the load taking capacity of the structure with its effectiveness increasing with increase in the rigidity, pullout resistance and proper arrangement of the geotextile layers. Sharma et al. [2] performed seismic stability

U. Azim · S. Sengupta (✉)
Department of Civil and Environmental Engineering, Birla Institute of Technology, Mesra, Ranchi 835215, India
e-mail: siddhartha@bitmesra.ac.in

U. Azim
e-mail: phdcee10001.20@bitmesra.ac.in

S. Kolathayar et al. (eds.), *Best Practices in Geotechnical and Pavement Engineering*, Lecture Notes in Civil Engineering 449, https://doi.org/10.1007/978-981-99-8505-0_23

analysis of a 5-m high road embankment composed of the locally available soil of the Roorkee region reinforced with geotextiles and geogrids which resulted in high factor of safety. Bandyopadhyay et al. [3] performed numerical analysis on Plaxis for highways composed of various fly ash compositions with varied densities and compactive effort. Extreme horizontal displacements were seen to decrease with increase in dry density of the samples, and lime- and cement-induced fly ash reduced the displacements up to 77.93%. Hong-wei et al. [4] conducted large-scale shaking table model tests to observe the performance of geosynthetic reinforced embankment. The peak seismic earth pressure was observed to vary nonlinearly with height of embankment with the peak attained around the middle of the embankment foundation slope. Edincliler and Toksoy [5] attempted to analyze the performance of geogrid reinforced lightweight embankment under the action of seismic activity using Plaxis. The results established successful stabilization with decreased earthquake-induced dynamic forces with the displacements getting reduced up to 29%. Bouabdallah et al. [6] attempted to evaluate the effectiveness of rigid reinforcement inclusions in the soft soils of Algeria in terms of settlement, shear stress and shear stain. The highest shear strain was found to be located in the middle of soft clay and maximum settlement at 0.9 m from the reinforcement inclusion while least at 0.2 m.

Prasad et al. [7] carried out standard penetration test (SPT) to collect data from places around Godavari River to perform liquefaction susceptibility by LiqIT software. The results were quantified in terms of shear stresses and settlements arising in response to earthquake. Liquefaction is defined as sudden loss of strength and stiffness in saturated and cohesionless soils and is known to be a major cause of damage of infrastructure during an earthquake event. It was concluded that SPT blow count (N) below 14 is the most susceptible to liquefaction, whereas soils with effective stresses above 95 kN/m^2 and SPT above 20 blows are characterized as safe. Ates et al. [8] employed N and shear wave velocity (V_s) to evaluate liquefaction potential of soils using empirical relationships. It was observed that areas comprising of thick soft soil deposits and shallow groundwater levels are highly prone to liquefaction. Hossain et al. [9] performed case study of Bangladesh to assess the liquefaction susceptibility from geotechnical parameters and groundwater depth and subsequently prepare hazard zonation map. The area exhibited 42–78% chances of liquefaction during an earthquake event of magnitude 8 and peak ground acceleration of 0.36 g. Sari et al. [10] examined the behavior of geotextile reinforced embankments on soft clay subgrade with varied depths, slopes and height to observe the effect on resistance moment and the required number of geotextile strips and accordingly develop graphs. The requirement of reinforcement layers was observed to increase with height of embankment and increased depth of soft soil. Chakraborty and Sawant [11] employed fragility curve, used to estimate the damage probability of a structure to be induced by liquefaction at different surcharge. The road embankment was exposed to 9 different earthquake intensities with varying parameters of relative density of foundation soil, embankment geometry and width of liquefiable layer among which relative density and thickness of problematic layer concluded to be prime factors contributing to failure of road embankment. Hasiholan et al. [12] performed stability analysis of a toll road embankment in Java Island composed of sandy volcanic ash, shallow groundwater levels and prone to seismic activity of magnitude greater than 5.0. The site was found to be stable in both static and dynamic case when prone to increased excess pore pressures with a liquefaction potential of 3.83 kN/m with factor of safety of 1.989 and 1.912 under static and pseudo-static analyses, respectively.

2 Methodology

In order to analyze the seismic stability of road embankment resting on soft soil under saturated conditions, a 8-m high road embankment was made to rest on subgrade composed of weak geotechnical properties up to 8 m depth and 50 m horizontal extent (refer Table 1). The considered geometry was subjected to horizontal earthquake acceleration with its maximum peak reaching to 0.38 g, taken from the past records available from United States Geological Survey (USGS) database which simulated the disturbance caused in seismic zone IV (refer Fig. 1). An approximate global mesh size of 1.5 m was used to undergo seismic analysis. Equivalent linear model was employed to perform the analysis, and the default functions for damping ratio, G reduction function and G_{max} were estimated with the characteristic of soft soil. The factor of safety was calculated using Morgenstern-Price failure criterion. The entire analysis was performed using Slope/W, Sigma/W and Quake/W on the commercially available GeoStudio software. The employed geotechnical properties in the simulation are given in Table 1.

The results of Slope/W analysis without reinforcement showed a factor of safety of 1.138, with the slip surface exhibiting base failure and the liquefaction zone lying underneath. Hence, this area shows presence of loosely packed, saturated soil material holding the potential of losing their inherent shear strength under the influence of external disturbances (refer Fig. 2). In this case, reinforcement was an essential requirement in order to prevent the collapse of the structure under the action of seismic forces. In order to design the reinforcement, two main history points, at the corner of crest of road embankment (history point 9) and at the mid of the soft soil subgrade (history point 15), were designed, and the maximum horizontal disturbance was observed. Under the influence of seismic force of 0.38 g, the peak

Table 1 Employed geotechnical properties

Geometry	Unit weight (kN/m^3)	Cohesion (kN/m^2)	Angle of internal friction (°)	Poisson's ratio
Foundation	16	0	30	0.15
Road embankment	18	20	34	0.4
Surficial crust	20	10	32	0.4
Granular drain	20	10	32	0.4

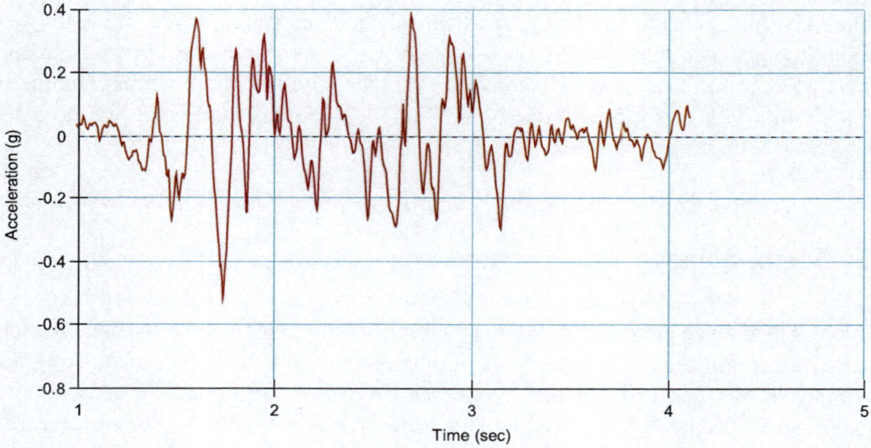

Fig. 1 Input seismic waves

displacement observed at the respective history points was approximately 0.7 and 0.4 g (refer Fig. 3). The increase in the horizontal acceleration in response to the seismic disturbance is attributed to the occurrence of liquefaction within the soft soil subgrade below the road embankment due to the action of increased pore-water pressure.

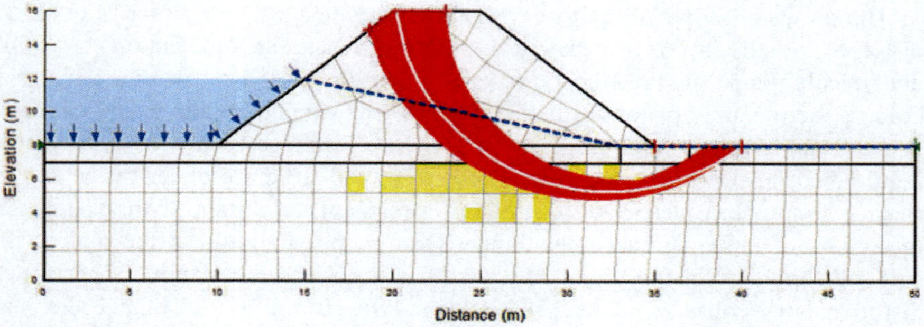

Fig. 2 Stability map without reinforcement

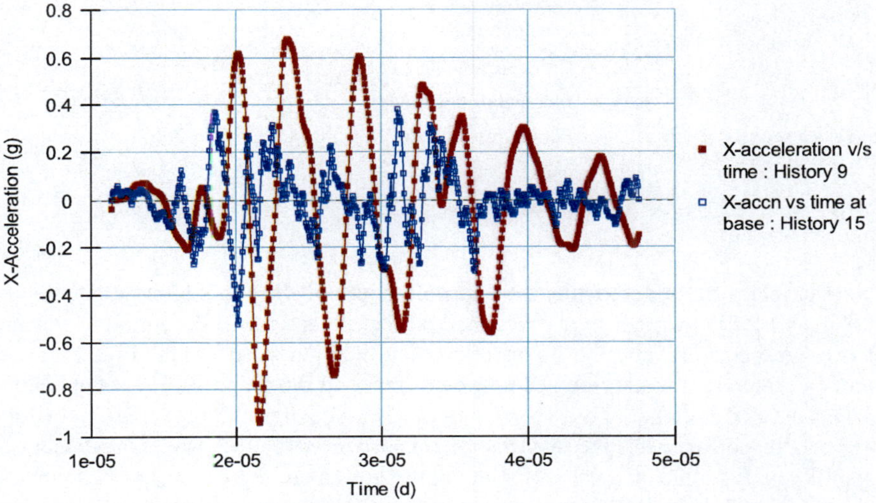

Fig. 3 *X*-acceleration versus time graph for the simulation without reinforcement

3 Results and Discussions

3.1 Effect of Geotextile Reinforcement on Stability of Road Embankment

Based on the results obtained from Slope/W and Quake/W for slip surface and location of liquefaction zone, geotextile reinforcement was designed to stabilize the structure. The mechanism involved in stabilization using geotextile is basically controlled by shear (friction) resistance on the top and bottom interface and bearing resistance offered by the supporting member. In GeoStudio, these mechanisms are defined by pullout resistance (P_{r}) of geotextile which is calculated as

$$P_{\mathrm{r}} = 2L_{\mathrm{e}}\sigma_{\mathrm{v}}'\alpha F \tag{1}$$

where L_{e} refers to embedded length of geotextile, σ'_v is the effective vertical stress at the interface of soft soil-reinforcement, α is the factor for scale effect correction, and the pullout resistance factor (F) is calculated using reinforcement bond coefficient f_{b} and angle of internal friction (ϕ) as $f_{\mathrm{b}} \tan\phi$. According to the studies conducted by Park and Hong [13], 0.5 L was concluded to be the adequate length of geotextile inclusion, and hence, two geotextile strips were included just below the road embankment and along the liquefaction zone with properties illustrated in Table 2. Keeping the bottom boundary fixed in X/Y plane and side boundary in Y direction, the seismic horizontal acceleration equivalent to 0.38 g was applied (refer Fig. 1).

Table 2 Geotextile reinforcement properties

E-modulus (*E*)	10,000 kPa
Cross-sectional area (*A*)	0.05 m^2
Moment of inertia (*I*)	2 m^4
Pullout resistance (P_r)	13.93 kPa
Tensile capacity (T_c)	23.09 kN/m
Reduction factor (α)	1

Post inclusion of geotextile, the factor of safety of the simulation augmented to 2.813 and 1.793 (Morgenstern-Price method) before and under the action of seismic forces, respectively, and the liquefaction zone was surpassed (refer Fig. 4). Also, the *X*-acceleration with reference to time dropped to 0.6 g and 0.21 g at the history points defined at corner of crest (history 9) and middle of bottom foundation (history 15) as a response to seismic disturbance caused (refer Fig. 5). The enhancement in stability and decline of acceleration along the horizontal direction can be attributed to the soil arching and pullout resistance associated with geosynthetic reinforcement. Due to the soil arching phenomenon, the surcharge and the disturbance caused by the seismic forces get transferred to the yielding portion amid the interface of soft soil and the geotextile reinforcement thereby creating stress redistribution and strengthening the soft subsoil. Post installation of geotextile in the simulation, the dilation gets inhibited in the inherent geomaterial and the non-dilating zone is shifted to the end anchorage so as to control soil dilatancy and increasing the pullout resistance value thereby declining the *X*-acceleration in comparison with the unreinforced section.

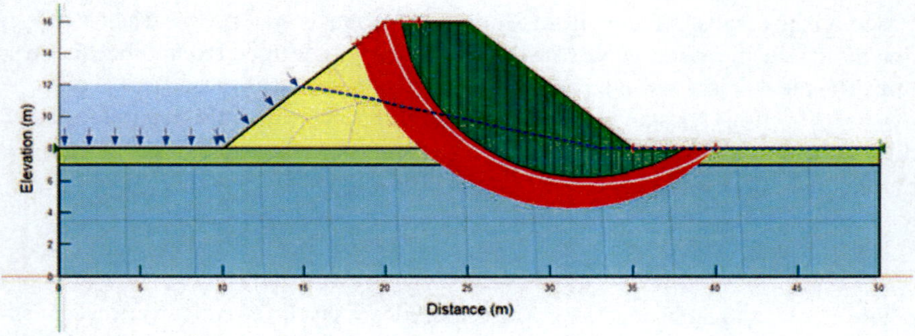

Fig. 4 Stability map after inclusion of geotextile reinforcement

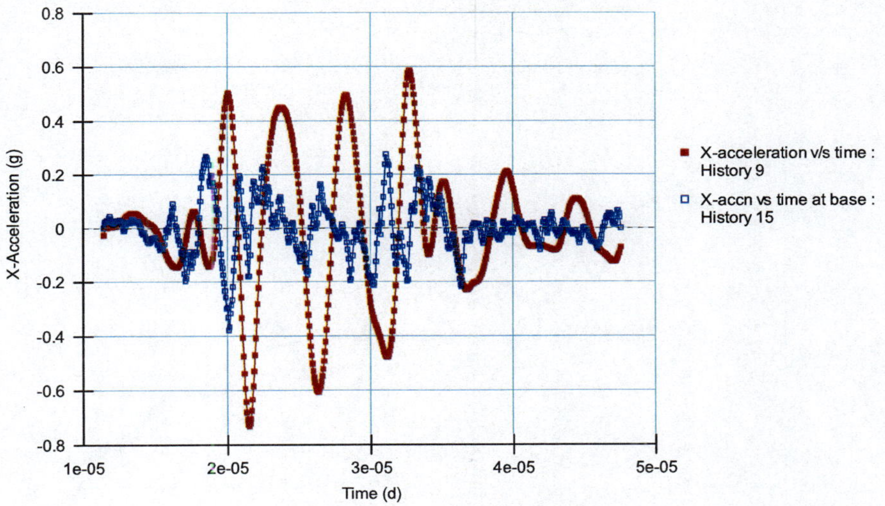

Fig. 5 X-acceleration versus time graph for the simulation with reinforcement

3.2 *Effect on Horizontal Displacement and Vertical Stress on Geotextile Inclusion*

On the basis of location of slip surface and prominent zone of liquefaction, the topmost layer of geotextile was placed in the toe of embankment. Figure 6 illustrates the variation of horizontal displacement with time with data recorded along the history points, and it can be clearly seen that initially a peak displacement of 0.01 m occurred at the bottom of soft soil foundation in response to the applied horizontal seismic acceleration which gradually dropped with time. This can be attributed to the enhancement of the strength of the foundation post geotextile installation, and past researches have concluded toe reinforcement to be effective in restricting the deformation at the toe and improving the seismic performance of the road embankment. This in turn assists in smooth embankment construction process without traffic suspension. A decline in the slip surface zone is also observed due to decrease in the substantial height of road embankment for sliding failure. Analytically, the horizontal body force was calculated using Eq. 2 employing density (ρ), volume of each mesh component (V_c), which resulted in least vertical stress of 115 kPa at the interface of soft soil and reinforcement (refer Fig. 7) and increased stresses getting redistributed over the depth of foundation quantified using the horizontal seismic coefficient and peak acceleration observed (α_H) (refer Eq. 3) (Oda et al. [14]). The same results were obtained from equivalent linear dynamic analysis in Quake/W.

$$f_H = \alpha_H \rho V_c \tag{2}$$

$$K_{H\,\max} = \frac{\alpha_H}{g} \tag{3}$$

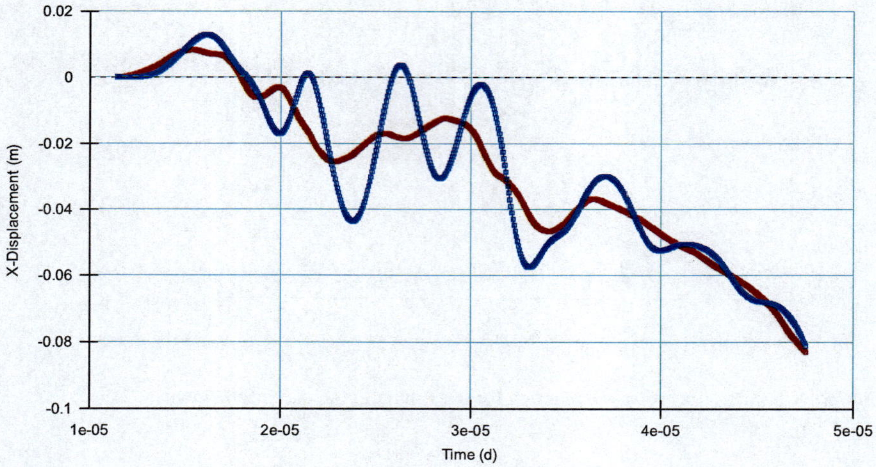

Fig. 6 Horizontal displacement versus time post geotextile installation

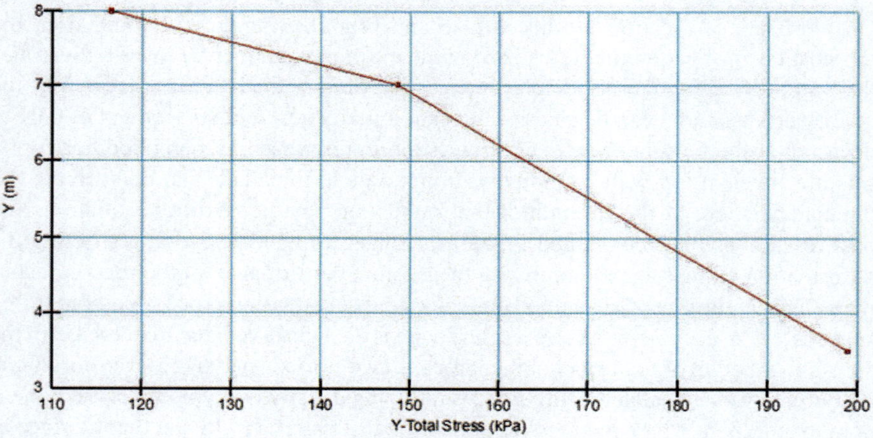

Fig. 7 Plot showing variation of vertical total stress with depth

3.3 *Distribution of Stress States in Soft Soil Subgrade Post Reinforcement*

Figure 8 depicts the distribution of q/p' ratio stress states, also known as Von Misses stresses, along the considered simulation. q and p' are critical state triaxial parameters quantified using the major, minor and deviatoric principal stresses as (refer Eqs. 4 and 5). The distribution of the shear stresses within the road embankment under the impact of pore-water pressure and seismic forces was seen to be limited up to 0.6 due to the inclusion of reinforcement layers.

$$q = \sigma_1 - \sigma_3 \tag{4}$$

$$p' = \frac{\sigma_1 + \sigma_2 + \sigma_3}{3} \tag{5}$$

Maximum concentration of the stresses of 1.8 and above was found to be around and below the interface of geotextile inclusion as the function of geotextile was to restrict the deformation within the soft soil and control the slip surface with generation of high stresses. The results were in relation with past findings where the inclusion of geotextile strips has resulted in maximum concentration of Von Misses stresses around the reinforcement binder course layer. Subsequently, high vertical displacements are seen to be located at the edges of geotextile layers where there is a drop in the q/p' ratio (Mahmoud and Hassan [15] and Sakai et al. [16]).

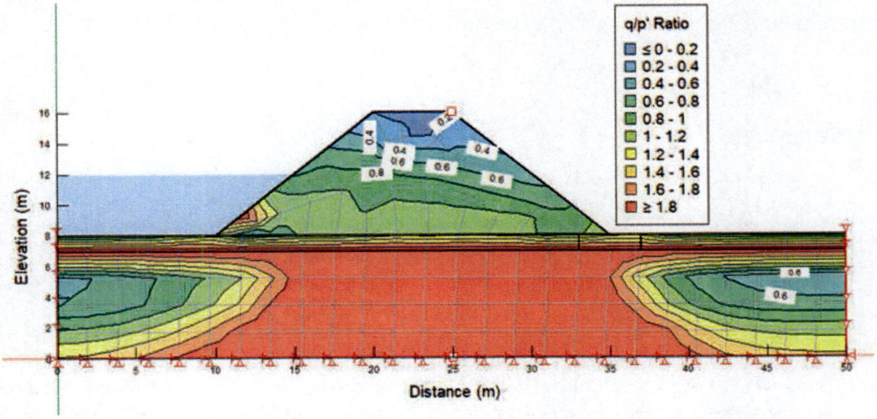

Fig. 8 Contours for distribution of q/p' ratio along reinforced road embankment

4 Conclusions

Improving infrastructure of an area has become an important need of hour. However, with increase in population and decrease in land availability, construction on problematic soils such as soft soils is encountered. Complete replacement of such soils is a tedious task due to high cost and labor requirement incurred, and therefore, stabilization techniques are used. This paper attempts to analyze the effect of geotextile reinforcement in soft soil underlying an 8-m high road embankment under the influence of seismic forces equivalent to peak ground acceleration of 0.38 g.

The geotextile reinforcement resulted in restricting liquefaction of soft soil in the presence of high pore-water pressure. The factor of safety of the road embankment before and post an earthquake event was seen to increase to 2.813–1.793, respectively, against 1.258 in the case of road embankment constructed on unreinforced soft soil. The horizontal accelerations at the crest and base of the subgrade material also dropped in comparison with the unreinforced simulations, and the decline was attributed to the soil arching mechanism and pullout resistance of the included geotextile strips. Geotextile reinforcement was also found to be effective in declining deformation in the subgrade material. Reinforcing the soft soil subgrade foundation at and below the toe of the road embankment with geotextiles resulted in offering enhanced resistance to seismic forces and subsequently declining the horizontal acceleration development within the embankment and its foundation. The reinforced zone also restricted the slip surface to the top surface composed of comparatively stronger geomaterial thereby maintaining the structure intact. Therefore, adequately designed geotextile reinforcement according to the geotechnical properties and external forces acting can be considered as an effective stabilization technique for increasing the strength and stabilizing soft soil thereby decreasing the involved cost and labor.

References

1. Koseki J (2012) Use of geosynthetics to improve seismic performance of earth structures. Geotext Geomembr 34:51–68. https://doi.org/10.1016/j.geotexmem.2012.03.001
2. Sharma P, Mouli B, Jakka RS, Sawant VA (2020) Economical design of reinforced slope using geosynthetics. J Geotech Geol Eng 38:1631–1637. https://doi.org/10.1007/s10706-019-01118-2
3. Bandyopadhyay K, Bhattacharjee S, Ghosh S (2011) Numerical approach for analysis of highway fly ash embankment. In: Proceedings of Indian geotechnical conference, Kochi
4. Hong-wei Z, Ling-kan Y, Guang-xing X (2020) Analysis of the stability and seismic behavior of geosynthetic reinforced embankments under earthquake. J Meteorol Sci 17(5):1269–1280. https://doi.org/10.1007/s11629-019-5519-3
5. Edincliler A, Toksoy YS (2022) Effects on geogrid reinforcement on the seismic performance of lightweight embankments. In: 7th EuroGeo conference, materials science and engineering. https://doi.org/10.1088/1757-899X/1260/1/012020
6. Bouabdallah F, Goudjil K, Salah M (2023) The effect of rigid inclusions on the dynamic response of highway embankment. Eng Technol Appl Sci Res 13(1):9843–9848. https://doi.org/10.48084/etasr.5400

7. Prasad CH, Chary MR, Thangamani K (2019) Analysing and preventing the problems of liquefaction in soils. Int J Eng Res Technol 8(12):748–753. IJERTV8IS120317
8. Ates A, Keskin I, Totic E, Yesil B (2014) Investigation of soil liquefaction potential around Efteni Lake in Duzce Turkey: using empirical relationship between shear wave velocity and SPT blow count (N). Advances in Materials Science and Engineering: 1–15. https://doi.org/10.1155/2014/290858
9. Hossain MS, Kamal AM, Rahman MZ, Farazi AH, Mondal DR, Mahmud T, Ferdous N (2020) Assessment of soil liquefaction potential, a case study for Moulvibazar town, Sylhet, Bangladesh. SN Appl Sci 2(777). https://doi.org/10.1007/s42452-020-2582-x
10. Sari PT, Lastiasih Y, Sugiarto S (2016) Proposed design graphs of geotextile reinforcement on soft clay under various field conditions. Civ Eng Dimension 18(2):109–116. https://doi.org/10.9744/CED.18.2.109-116
11. Chakraborty A, Sawant VA (2023) Fragility assessment of highway embankment resting on liquefaction-susceptible soil. Comput Geotech 161(105568). https://doi.org/10.1016/j.compgeo.2023.105568
12. Hasiholan F, Ismanti S, Rifa A (2023) Stability analysis of toll road embankment induced by liquefaction in Klaten Regency based on FEM numerical simulation. IOP Conf Ser Earth Environ Sci 1184(012003). https://doi.org/10.1088/1755-1315/1184/1/012003
13. Park J, Hong G (2021) Effective length prediction and pullout design of geosynthetic strips based on pullout resistance. Materials 14(6151):1–16. https://doi.org/10.3390/ma14206151
14. Oda K, Tokida K, Egawa Y, Nakahira A, Tanimura K (2008) Improvement technique of earthquake resistance of road embankment through numerical analyses. In: 14th world conference on earthquake engineering, China
15. Mahmoud H, Hassan EM (2019) Seismic performance of highway embankments. Mountain-Plains Consortium
16. Sakai T, Inukai S, Inagaki M, Nakano M (2023) Improvement in seismic resistance using replacement/counterweight fill method for existing high embankments on inclined ground constructed with various embankment materials. Soils Found 63(2). https://doi.org/10.1016/j.sandf.2023.101284

Numerical Analysis of Railway Formation with Geogrid-Reinforced Ballast and Blanket Layer

Veer Vikram Singh and Anil Kumar Sahu

1 Introduction

High-speed railways, in addition to providing a high level of mobility for people, are greatly advantageous as an environment-friendly means of transport [1]. The International Union of Railways (UIC) considers a commercial speed of 250 km/h to be the principal criterion to define high-speed rail. The current high-speed rail operating in India is Vande Bharat Express running at a speed of 180 km/h. By the year 2026, India is planning to operate its first bullet train which will run at a speed of 350 km/h. As the speed of railway transport is going to increase tremendously in upcoming years, the loads coming to railway formation are also going to increase simultaneously thus increasing the stresses on the existing formation.

High-speed railway ballast's dynamic behavior under moving loads was examined by Yetimoglu et al. [2]. Zhang et al. [3] determined the strength and deformation properties of the materials used in railway embankments. Leng et al. [4] performed high-frequency cyclic loading to study the deformation and degradation of railway ballast. For the existing subgrade soil, the increased stress can cause problems and sometimes can lead to failure.

For thousands of years, soils are mixed with different fibers, fabrics, and vegetation to improve quality and stability. Geosynthetics which are polymer products are used in civil engineering for decades. Leshchinsky and Ling [5] described geosynthetics as exciting material with a wide area of applications such as transportation, geotechnical, environmental, and hydraulics. Chen et al. [6] used geogrid reinforcement in the sand to increase its bearing capacity. A study was done on the function of geosynthetics in stabilizing rail track slabs [7]. Alabbasi and Hussein and Indraratna et al. [8, 9] studied the effects of geocell and microgrid confinement on the strength and deformation of geocell. Leshchinsky and Ling [10] studied the

V. V. Singh (✉) · A. K. Sahu
Department of Civil Engineering, Delhi Technological University, Delhi 110042, India
e-mail: vikramveervvs@gmail.com

S. Kolathayar et al. (eds.), *Best Practices in Geotechnical and Pavement Engineering*, Lecture Notes in Civil Engineering 449, https://doi.org/10.1007/978-981-99-8505-0_24

behavior of geogrid-reinforced railway embankments. Finite element modeling and lab tests were performed by Arulrajah et al. [11] to investigate the effects of geogrids in railway embankments. Esmaeili et al. [12] used a 3D model to study geocell-reinforced ballast. Nayyar and Sahu [13] modeled the railway embankment with geotextile, geofoam, and waste materials.

The primary goal of this study is the dynamic investigation of the behavior of displacements and stresses following the incorporation of the geogrid in the blanket layer. Many researchers tried to study the behavior of geogrid inclusion in railway formation, but more research is needed for dynamic analysis. For incorporating the dynamic analysis, 3D models were developed with the help of PLAXIS 3D software comprised of railway formation with and without geogrids. The models were analyzed for different numbers of geogrids.

2 Materials and Methods

This research paper's primary purpose is to introduce the geosynthetic between the blanket layer and ballast. Three-dimensional finite element software (PLAXIS) models and analyzes the railway formation.

2.1 *Railway Formation Geometry*

Figure 1 shows the railway segment that was taken into consideration for this study. The part has the following measurements: 35 m × 35 m × 9.5 m. The ballast layer is 350 mm thick, while the blanket layer is 600 mm thick. According to RDSO (2018) recommendations, a slope of 2H:1 V is used (Fig. 2). Sand is used to represent the subsurface and is modeled as the Mohr–Coulomb criteria. It is assumed that the soil has been drained. The Mohr–Coulomb criteria are also applied to the blanket and ballast, and their drainage type is assumed to be drained soil. Models of rails and sleepers use elastic materials.

Fig. 1 Dimensions of the numerical model

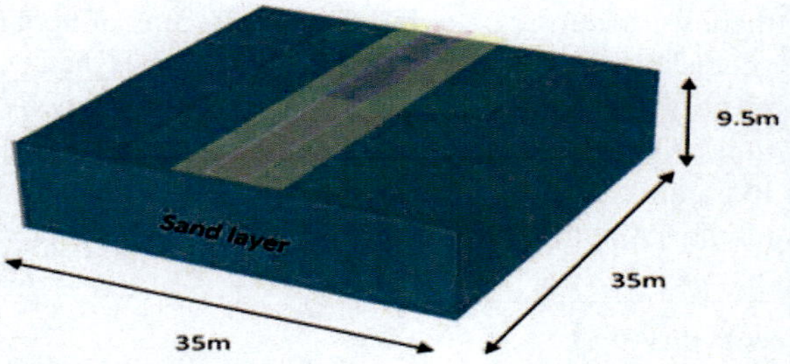

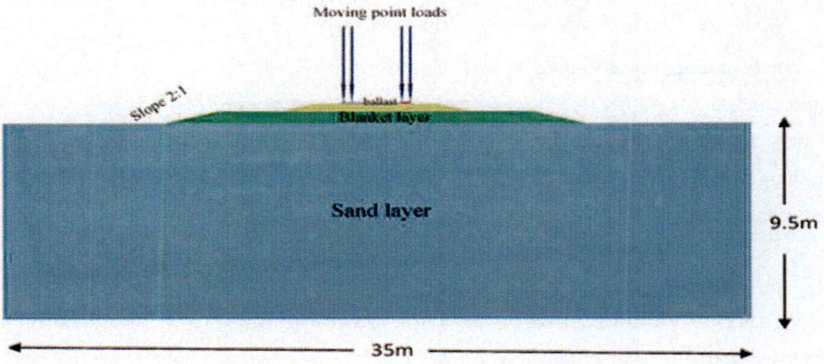

Fig. 2 Railway section for the modeling

2.2 *Material Properties and Modeling*

Tables 1 and 2 provide a summary of the mechanical characteristics of railway materials as well as the material characteristics of various materials employed in the simulation process.

The geogrids provide uniform distribution of loads over a larger area by increasing the stiffness of the base and resisting deformations. The geogrid used here is a planar geogrid, which is available in the PLAXIS 3D by default. Geogrids can be created in PLAXIS by choosing material as isotropic and by inputting the value of axial stiffness. The properties of geogrid are taken as per RDSO guidelines. The ultimate tensile strength is taken as 20 kN/m, and strain at ultimate tensile strength is taken

Table 1 Basic material properties of the soil layers for LE and MC models [14]

S. No.	Soil layers	γ_{sat} (kN/m^3)	γ_{unsat} (kN/m^3)	μ	ϕ °	*c* (kN/m^2)	ψ °	*E* (kN/m^2)
1	Ballast	21	19	0.3	65[a]	31[a]	5	110000[a]
2	Blanket	23	22	0.25	40	30	15	55,000
3	Sand	20	19	0.35	40	5	10	80,000

Remarks [a]Likitlersuang et al. [15]

Table 2 Input properties in PLAXIS 3D for rail and sleeper [14]

Parameter	Rails	Sleepers
Cross-sectional area (A) (m^2)	7.7×10^2	5.13×10^2
Unit weight (kN/m^3)	78	25
Young's modulus (E) (kN/m^2)	200×10^6	36×10^6
Moment of inertia around the second axis (I_3) (m^4)	3.055×10^{-5}	0.0253
Moment of inertia around the third axis (I_2) (m^4)	$5.13 \times \times 10^{-6}$	2.45×10^{-4}

Table 3 Model parameters for modeling the moving loads

Distance between the first and the last wagon axles (m)	21.7
Additional length for model (m)	6.5
Total additional length (right and left) (m)	13.0
Model length (m)	34.7
Sleeper distance (m)	0.6
Dynamic loads distance (m)	0.3

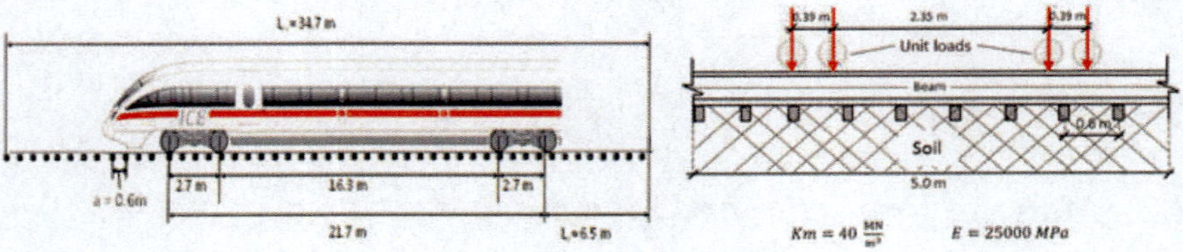

Fig. 3 Dimensions of an ICE train and calculated lengths for model [14]

as 10%. Two models are created for the simulation process. In the first model, the geogrid is installed at the interface between the blanket layer and the ballast. In the second model, the ballast layer is reinforced additionally with geogrids at intervals of 100 cm, and one layer is installed below the blanket layer for more reinforcement.

3 Finite Element Analysis

3.1 Finite Element Model

An ICE train model is taken for modeling, which is an intercity express generally running in Germany. The specifications used in the modeling are listed in Table 3. Dimensions of an ICE train and calculated lengths for the model are shown in Fig. 3.

The finite element model is illustrated in Fig. 4. The distance between the first and the last axle is 21.7 m, and the sleepers are laid at a spacing of 600 cm as per guidelines of RDSO.

3.2 Mesh and Boundary Conditions

10-noded element was selected to model the section. Two models are simulated for this purpose. Mesh sensitivity is performed to check the appropriate mesh size for the model. Owing the complexity and dynamic analysis, time taken by model for

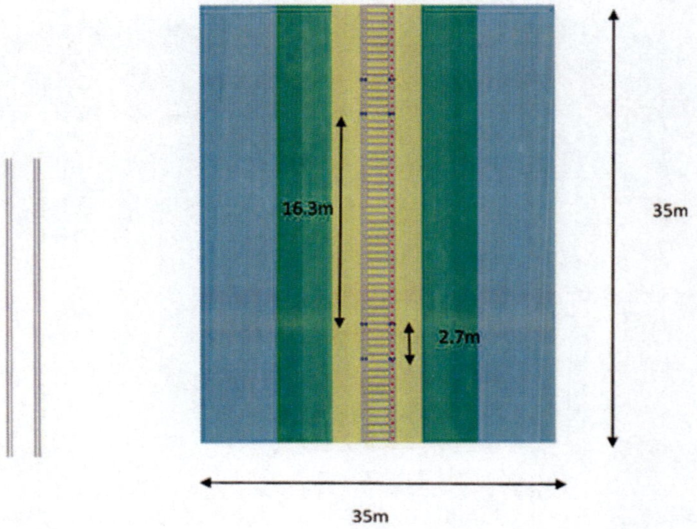

Fig. 4 Finite element model

calculation is much as the size of mesh is refined. Due to this the mesh size was selected as very coarse for both models. The number of soil elements in the first model was 2660, and the number of nodes was 4856. The average element size was 2.762 m, the maximum element size was 10.45 m, and the minimum element size was 2.00×10^{-3} m. The number of soil elements in the second model was 4617, and the number of nodes was 7983. The average element size was 2.416 m. The maximum element size was 10.06 m, and the minimum element size was 2.079×10^{-3} m. The boundary in both *X*- and *Y*-direction is normally fixed, but for *Z*-direction the boundary condition is taken as free. Elements and nodes are shown in Figs. 5 and 6, respectively.

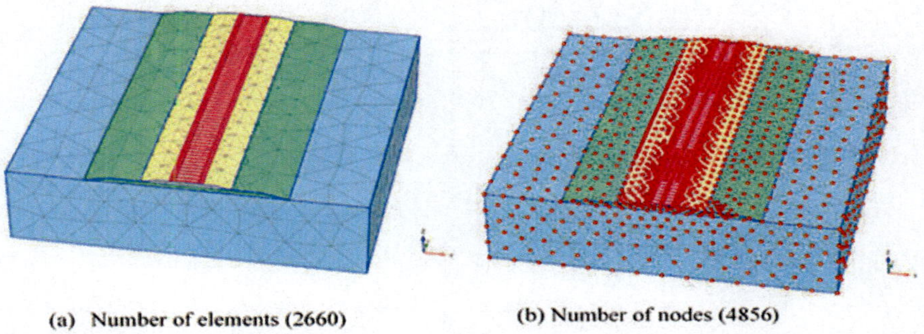

Fig. 5 Elements and nodes for model 1

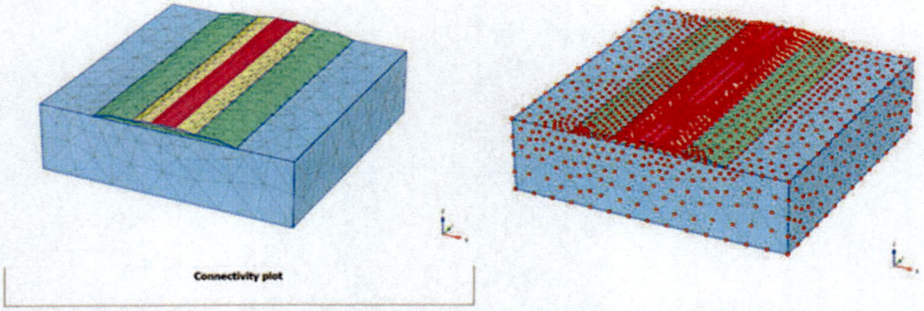

(a) Number of elements (4617) (b) Number of nodes (7983)

Fig. 6 Elements and nodes for model 2

3.3 Loading

Elastic materials were used in the design of the rails and sleepers. A moving point load was used to approximate the train loads at the rail. The portion traveled along the rail at a speed of 180 km/h while being subjected to an axle-applied wheel force of 90 kN. Figure 7 depicts how the axial load is organized. The total length of the track was taken as 35 m. At the speed of 180 km/h time taken by the train to pass this section was 0.7 s, but considering the total dimension of the train the time was taken as 1 s for dynamic analysis. The taken time was divided into 20 sub-steps to ease the calculation process.

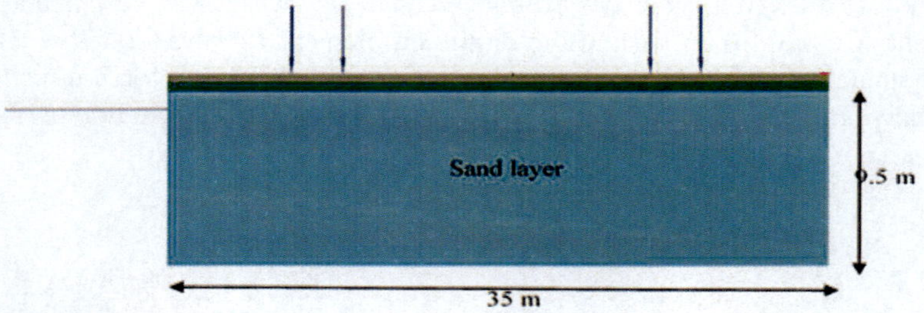

Fig. 7 Illustration of loading

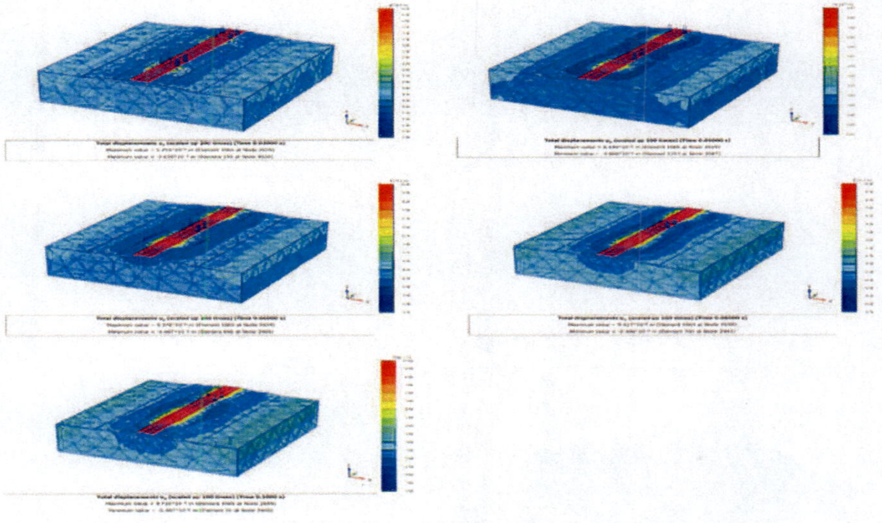

Fig. 8 Total displacements at different intervals without the geogrid

4 Results and Discussion

4.1 For Model 1

4.1.1 Vertical Deformations

The first model has two phases. Moving loads are applied during the first phase without the use of the geogrid reinforcement, and the geogrid is turned on during the second phase. Deformations, stresses, and shear velocities at various points and cross sections of the model are the outcomes of the simulation. Without the use of geogrid, Fig. 8 displays the total z-direction displacements at various time points. The total z-direction displacements with geogrid included are shown in Fig. 9 throughout a range of time intervals.

4.2 For Model 2

4.2.1 Vertical Deformations

Model 2 has two phases as well. Moving loads are applied during the first phase without the use of the geogrid reinforcement, and the geogrid is turned on during the second phase. Deformations, stresses, and shear velocities at various points and cross sections of the model are the outcomes of the simulation. The total displacements

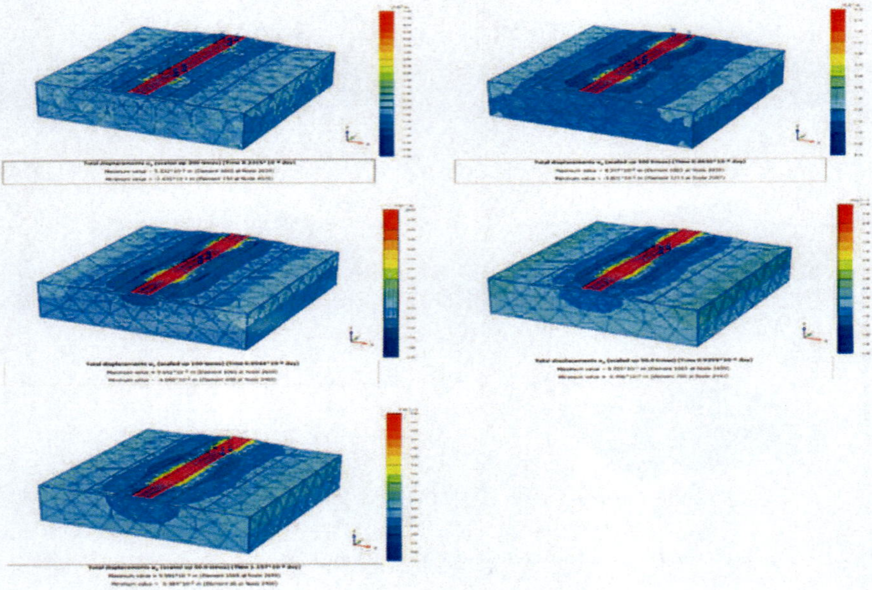

Fig. 9 Total displacements at different intervals with the geogrid

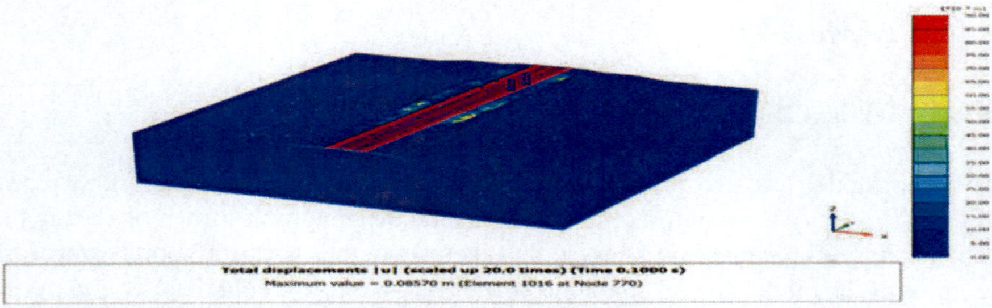

Fig. 10 Total displacement without the geogrid

in the z-direction at time interval are shown in Figs. 10 and 11 with and without the addition of the geogrid, respectively.

4.2.2 Stresses in the Rail Section

The axial forces along the length of the rail section are illustrated in Fig. 12. In Fig. 12a, b, it is clear that at some sections geogrid does improve the distribution of axial forces in the rail section and there is more uniform distribution of forces as compared to the model 1.

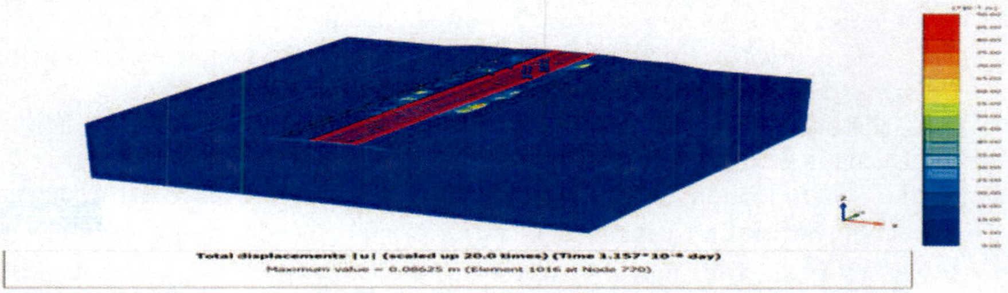

Fig. 11 Total displacements with the geogrid

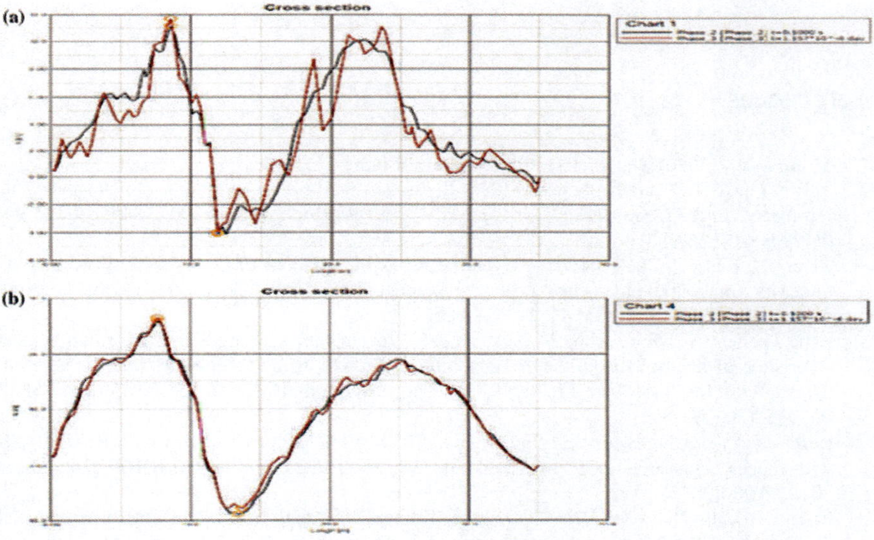

Fig. 12 **a** Comparisons of axial force and length of the train section with geogrid for model 2. **b** Comparisons of axial force and length of the train section without geogrid for model 2

5 Conclusion and Recommendations

A 90 kN moving train load with a speed of 180 km/h and time interval of 1 s was simulated using a three-dimensional finite element model. PLAXIS 3D software was used to simulate real-time conditions and to predict the deformations and stresses in the railway formation with and without the inclusion of geogrids at different depths. Dynamic analysis was used to calculate the actual results. Conclusions made from the analysis are:

1. After analyzing both models, it was inferred that using geogrid only at the interface of the blanket layer and the ballast did not restrain the deformation and the stresses significantly but there was a uniform distribution of loads in both the

Y-direction and z-direction. Model 2 showed a lot of improvements over Model 1. Using geogrids at various depths in the ballast layer did help in restraining the deformation and a reduction of about 30–40% is seen in the stresses.
2. Geogrids used as ground improvement for weak soil in railway formation helped in distributing the loads uniformly, reducing the values of deformation.
3. Values of deformations and stresses kept on changing across the length of the rail formation. Stresses are dependent of various factors like depth of geosynthetics, properties of geosynthetics, speed of the train, etc.

It is advised that more studies be done to ascertain the deformations and strains using lab experiments and tracking the outcomes to assess the actual behavior of the moving train.

References

1. Mundrey JS (2010) Tracking for high-speed rail. India RITES J
2. Yetimoglu T, Wu JTH, Saglamer A (1994) Bearing capacity of rectangular footings on geogrid-reinforced sand. J Geotech Eng 120(12):2083–2099. https://doi.org/10.1061/(ASCE)0733-9410(1994)120:12(2083)
3. Zhang X, Zhao C, Zhai W (2016) Dynamic behaviour analysis of high-speed railway ballast under moving vehicle loads using discrete element method. Int J Geomech 04016157. https://doi.org/10.1061/(ASCE)GM.1943-5622.0000871
4. Leng W, Xiao Y, Nie R, Zhou W, Liu W (2017) Investigating strength and deformation characteristics of heavy-haul railway embankment materials using large-scale undrained cyclic triaxial tests. Int J Geomech 17(9):04017074. https://doi.org/10.1061/(ASCE)GM.1943-5622.0000956(2017)
5. Leshchinsky B, Ling HI (2013) Numerical modeling of behavior of railway ballasted structure with geocell confinement. Geotext Geomembr 36:33–43. https://doi.org/10.1016/j.geotexmem.2012.10.006(2013)
6. Chen RP, Jiang P, Ye XW, Bian XC (2015) Probabilistic analytical model for settlement risk assessment of high-speed railway subgrade. J Perform Constr Facil 04015047. https://doi.org/10.1061/(ASCE)CF.1943-5509.0000789
7. Lenart S, Klompmaker J (2015) Geogrid reinforced railway embankment on soft soil—experiences from 5 years of field monitoring. https://www.researchgate.net/publication/281117609
8. Alabbasi Y, Hussein M (2021) Geomechanical modelling of railroad ballast: a review. Arch Comput Methods Eng 2021(28):815–839
9. Indraratna B, Khabbaz H, Salim W, Christie D (2006) Geotechnical properties of ballast and the role of geosynthetics in rail track stabilization. Ground Improv 10(3):91–101
10. Leshchinsky B, Ling H (2013) Effects of geocell confinement on strength and deformation behaviour of gravel. J Geotech Geoenviron Eng 139:340–352. https://doi.org/10.1061/(ASCE)GT.1943-5606.0000757(2013)
11. Arulrajah A, Abdullah A, Bo MW, Bouazza A (2009) Ground improvement techniques for railway embankments. Inst Civ Eng Ground Improv 162(GI1):3–14. https://doi.org/10.1680/grim.2009.162.1.3
12. Esmaeili M, Naderi B, Neyestanaki HK, Khodaverdian A (2018) Investigating the effect of geogrid on stabilization of high railway embankments. Soils Found. https://doi.org/10.1016/j.sandf.2018.02.005
13. Nayyar A, Sahu AK (2021) Numerical analysis of railway substructure with geocell-reinforced ballast geomechanics and geoengineering. https://doi.org/10.1080/17486025.2021.1928770

14. Shahraki M, Sadaghiani MRS, Witt KJ (2014) 3D modeling of train induced moving load on an embankment. www.plaxis.com/plaxis bulletin
15. Likitlersuang S, Pholkainuwatra P, Chompoorat T, Keawsawasvong S (2018) Numerical modelling of railway embankments for high-speed train constructed on soft soil. J Geo-Eng 13(3):149–159. https://doi.org/10.6310/jog.201809_13(3).6
16. Zhu XQ, Law SS, Huang L (2018) Identification of railway ballasted track systems from dynamic responses of in-service trains. J Aerosp Eng 31(5):04018060. https://doi.org/10.1061/(ASCE)AS.1943-5525.0000898
17. Ferdous W, Manalo A, Van Erp G, Aravinthan T, Ghabraie K (2018) Evaluation of an innovative composite railway sleeper for a narrow-gauge track under static load. J Compos Constr 22(2):04017050. https://doi.org/10.1061/(ASCE)CC.1943-5614.0000833
18. Sun Q, Indraratna B, Nimbalkar S (2015) Deformation and degradation mechanisms of railway ballast under high-frequency cyclic loading. J Geotech Geoenviron Eng 04015056. https://doi.org/10.1061/(ASCE)GT.19435606.0001375
19. Abadi T, Le Pen L, Zervos A, Powrie W (2019) Effect of sleeper interventions on railway track performance. J Geotech Geoenviron Eng 145(4):04019009. https://doi.org/10.1061/(ASCE)GT.1943-5606.0002022(2019)
20. Yu Z, Woodward PK, Laghrouche O, Connolly DP (2019) True triaxial testing of geogrid for high-speed railways. Transp Geotech 20. https://doi.org/10.1016/j.trgeo.2019.100247
21. Signes CH, Fernández PM, Roca JG, Franco RI (2016) Analysis of the bearing capacity of unbound granular mixtures with rubber particles from scrap tires when used as sub-ballast. Materiales de Construcción 66(324):e105
22. Leshchinsky B, Evans TM, Vesper J (2016) Microgrid inclusions to increase the strength and stiffness of sand. Geotext Geomembr 44:170–177. https://doi.org/10.1016/j.geotexmem.2015.08.003(2016)
23. El-kady MS, Azam A, Yosri AM, Nabil M (2023) Modeling of railway embankment stabilized with geotextile, geofoam and waste aggregates. https://doi.org/10.1016/j.cscm.2022.e01800
24. Koerner RB (2005) Designing with geosynthetics, 5th edn. Pearson, NJ
25. Indian railways code for the engineering department-revised edition (1982)
26. RDSO (1972) Civil Engineering Report no: C-127-report on a study of the characteristics of compacted and uncompacted expansive soils
27. Specification for railway formation RDSO/2018/GE: IRS-0004(D)
28. PLAXIS 3D tutorial manual 2020

Strength Analysis of Geotextile-Reinforced Subgrade

Pradeep Kumar, Kshitij Gaur, and Ashutosh Trivedi

1 Introduction

Geosynthetic reinforcement layers are extensively employed to enhance the performance and durability of pavement structures. In the context of unpaved roads, the California bearing ratio (CBR) is a crucial factor used to assess their performance. It serves as an indicator of subgrade soil strength for both paved and unpaved road applications. In current construction practices, geosynthetic layers are frequently utilized to improve weak or unsuitable soil subgrades for both paved and unpaved roads. The use of geosynthetics as reinforcements can effectively strengthen unpaved roads constructed on soft soil subgrades thereby increasing their lifespan [1]. The study investigates into the effects of replacing some of the aggregates with waste rubber tyre particles, replacing a portion of the bitumen with crumb waste rubber tyre and adding wax content to the bituminous samples [2]. The findings suggest that displacement behaviour, equivalent stress, and principle stress distribution more properly explain the basis of rutting in pavement subjected to moving load [3]. Using geotextile and jute fibre has improved the bearing capacity of granular subgrade soil [4]. The usage of natural geotextiles for improving unpaved roads and examining their combination with biaxial geogrid. Scanning electron microscope (SEM) photos showed that soil particles and the surface of the geogrid were bonded together

P. Kumar (✉) · K. Gaur · A. Trivedi
Delhi Technological University, Shahbad Daulatpur, Main Bawana Road, Delhi 110042, India
e-mail: pradeep.237cv006@nitk.edu.in

K. Gaur
e-mail: CEKshitijGaur@gmail.com

A. Trivedi
e-mail: atrivedi@dce.ac.in

P. Kumar
Department of Civil Engineering, National Institute of Technology Karnataka, Surathkal, Mangalore 575025, India

S. Kolathayar et al. (eds.), *Best Practices in Geotechnical and Pavement Engineering*, Lecture Notes in Civil Engineering 449, https://doi.org/10.1007/978-981-99-8505-0_25

during the soaked CBR and UCS testing [5–8]. In one such study, experimental and computational assessments were conducted to determine how geotextile reinforcement affected the bearing capacity of granular soils [9]. Another study described soft soil that had been strengthened with a non-woven geotextile functioning as a separator and covered in the sand was used for the CBR tests [10]. To gain a deeper understanding and analysis of the test results, FE analysis using PLAXIS software was conducted. Researchers have studied the impact of jute, geogrids, and geosynthetics on the performance of geomaterial [11–14]. Geosythetic-reinforced soil (GRS) technology is favoured for sustainable pavement construction, assessing subgrade strength via California bearing ratio (CBR) [15–18]. The study conducted lab tests on granular piles reinforced with geosynthetics (vertical encasement, horizontal strips, and combined reinforcement) in soft clay. By comparing lab results with PLAXIS 3D FEM software, they observed a notable increase in load capacity and stiffness, indicating the effectiveness of geosynthetics in enhancing pile performance [19]. This study uses flyash-based geopolymer with eggshell powder to stabilize clay at ambient temperature, providing eco-friendly green building solutions and improving soil strength [20]. The study investigated morphological and microstructural changes in FA, ESP, and S geopolymer, examining variable precursor concentrations (10–50%) in soft soil under fixed parameters, and geopolymerization effects were analysed using field emission gun-SEM, energy-dispersive X-ray spectrometer, and mapping techniques [21]. PLAXIS 3D finite element software was used to undertake a numerical analysis of reinforced floating granular piles put in soft clays. The reinforcement was implemented through the use of horizontal geogrid strips [22]. The subgrade refers to the layer of natural soil or engineered fill material that supports the pavement structure of a road. If the subgrade has poor strength, it can lead to several issues in road construction and maintenance. Weak subgrade can lead to cracking, rutting, settlement, and uneven road surfaces. Pre-construction strength assessment is vital, and techniques like soil stabilization, geotextile reinforcement, and subgrade replacement are used for improvement. This research compares jute and polypropylene geotextiles strength analysis and investigates ways to enhance subgrade strength for long-term road performance.

2 Experimental Set-Ups

2.1 Materials

One type of soil has been chosen for this study. Two types of geotextiles, i.e., woven jute textile (natural fibre) and woven polypropylene geotextile (synthetic fibre) are used in this work for analysing the effect on shear strength and CBR of soil.

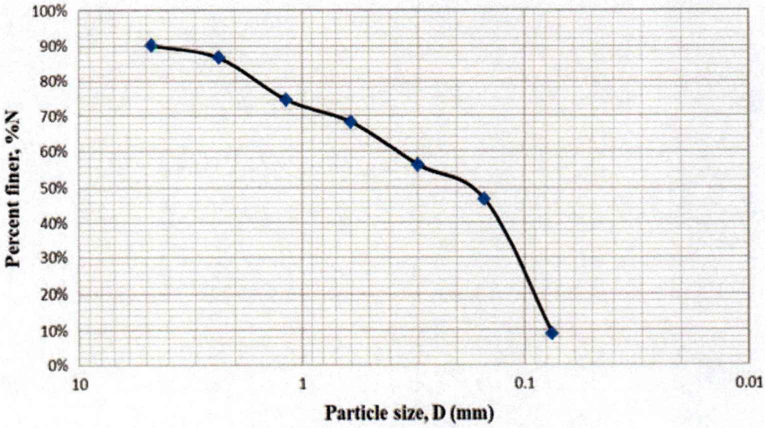

Fig. 1 Curve of soil's particle size distribution

2.2 *Soil Testing*

A soil sample from a nearby location was collected for the purpose of this study. To find the type of soil, the sieve analysis test is applied to a soil sample, and the results of this test yield a particle size distribution curve, as illustrated in Fig. 1. The soil can be categorized as poorly graded sand containing silt, abbreviated as SP-SM according to IS: 2720 Part-4 (1985). To ascertain the maximum dry density (MDD) and optimal moisture content (OMC) of soil, a standard proctor test also known as a light compaction test is performed on a soil sample that complies with IS: 2720 Part-7 (1980). Within this test, the soil sample is compacted in three equal layers in the mould of 1000 cc by applying 25 blows with a 2.6 kg rammer with a free fall height of 31 cm to each layer. The MDD and OMC achieved from the standard proctor test are 1.899 gm/cc and 12.94%, respectively, as shown in Fig. 2. A direct shear test (consolidated drained) is performed on remoulded soil to correctly estimate the shear characteristics of soil by preparing the soil sample on its maximum dry density by adding optimum moisture content. From this test, normal stress (σ) and shear stress (τ) curves are obtained. Moreover, laboratory tests are performed on soil geotechnical properties such as Atterberg limits and specific gravity, as shown in Table 1.

2.3 *Geotextile*

Geotextiles, which are permeable textiles, offer a wide range of benefits in various civil engineering and environmental applications and are used to improve soil characteristics. It has the ability to separate, drain, protect, reinforce, and filter when used

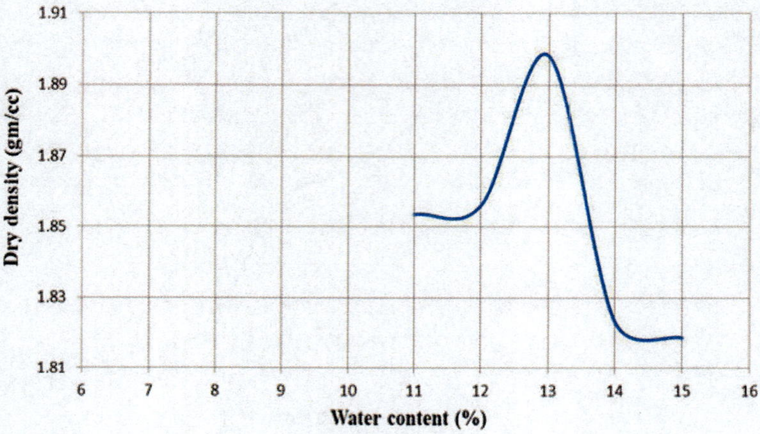

Fig. 2 Compaction curve of soil (based on standard proctor test)

Table 1 Soil's geotechnical characteristics

S. No	Characteristics	Values
1.	Gravel content (%)	9.7
2.	Sand content (%)	81.3
3.	Silt and clay content (%)	9
4.	Uniformity coefficient (*C*u)	4.9032
5.	Coefficient of curvature (*C*c)	0.39606
6.	Specific gravity (G)	2.63
7.	Limit for liquid (%)	30.5
8.	Plastic limit (%)	25
9.	Plasticity index (%)	5.5
10.	$\gamma_{dmax.}$ (gm/cc)	1.899
11.	OMC (%)	12.94

in contact with soil. Two types of woven geotextiles, i.e. jute textile (natural fibre) and polypropylene geotextile (synthetic fibre), are used here as depicted in Fig. 3a, b, respectively. Woven geotextiles are made by weaving two or more sets of threads, filaments, or other materials together. These threads are generally woven straight and parallel to each other. Table 2 displays various geotextile qualities as provided by the manufacturer.

Fig. 3 Geotextiles materials, a jute textile and b polypropylene (PP) geotextile

Table 2 Properties of geotextile

Properties	Jute textile	Polypropylene geotextile
Density (g/cm^3)	1.46	0.91
Tensile strength (MPa)	393–773	300–400
Elongation at break (%)	1.16–1.5	10–45
Young's modulus (GPa)	13–26.5	1.3

2.4 *Placement of Geotextiles*

Unreinforced soil samples with different geotextile layers are prepared and tested. Test results are compared to select the best reinforcing material for the California bearing ratio (CBR test) and direct shear test (DST).

Samples tested are as follows:

1. Samples without any geotextiles.
2. Direct shear test for soil sample by using jute as well as polypropylene geotextiles. In this test, a single layer of both geotextiles is mixed with an unreinforced soil sample.
3. CBR test for soil samples by using jute as well as polypropylene geotextiles. In this test, sample reinforced with single layer (*D*/2, *D*/3 and *D*/4), double layers (*D*/2, *D*/3 and *D*/3, *D*/4 and *D*/2, *D*/4), triple layers (*D*/2 and *D*/3 and *D*/4) as shown in Fig. 4.

3 Results and Discussion

IS light compaction, direct shear test, and CBR test are conducted according to IS: 2720 Part-13 (1986), IS: 2720 Part-16 (1987) respectively in the laboratory.

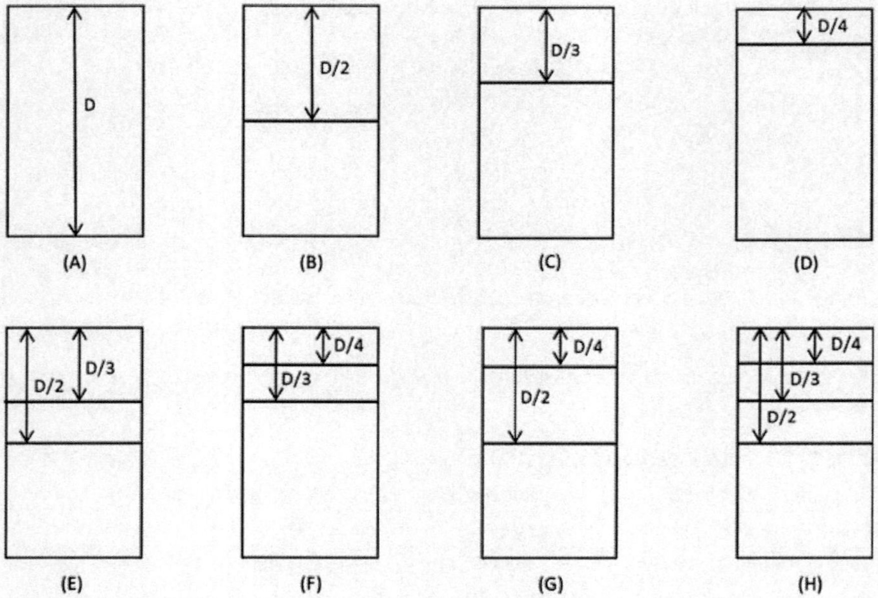

Fig. 4 Placement of geotextile in CBR mould

4 Direct Shear Test Results

The unreinforced soil sample is mixed with a single layer of jute as well as polypropylene geotextile to investigate how the addition of geotextiles impacts cohesion (c), angle of friction (ϕ), and overall shear strength of an unreinforced soil specimen on varied normal stresses (50, 100, and 150 kPa). These parameters are crucial in determining the soil's stability as well as bearing capability in addition to how they are affected by the addition of geotextiles can help improve soil reinforcement techniques and reduce the risk of soil failure. By mixing jute textile, it is found that the c and phi values increase which result in an increase in shear strength whereas by reinforcing polypropylene geotextile, the c and *phi* values decrease resulting in a decrease in shear strength with respect to unreinforced soil which is shown in Table 3. Reinforcing jute textile in soil significantly improves its cohesion value, which increases from 19.7 to 23.25 kPa, indicating a notable 18% increment. Additionally, the phi value, representing the internal friction angle, experiences a positive shift from 31.39 to 32.825°, indicating a 1.435° increment. As a result, the shear strength of the soil is enhanced due to these improvements. In contrast, when polypropylene geotextile is used for reinforcement, the cohesion value of the soil decreases from 19.7 to 13.96 kPa, illustrating a substantial 29% decrement. Similarly, the phi value decreases from 31.39

Table 3 Comparison table of shear strength parameters among unreinforced and reinforced soil

Parameters for shear strength	Unreinforced soil	Reinforced soil with jute textile	Reinforced soil with polypropylene geotextile
Cohesion (kPa)	19.7	23.25	13.96
Friction angle (°)	31.39	32.825	30.04

to 30.04°, showing a 1.35° decrement. Consequently, the shear strength of the soil is adversely affected, leading to a reduction in its overall stability. These findings emphasize the contrasting effects of utilizing jute textile and polypropylene geotextile as reinforcement materials. While jute textile exhibits positive reinforcement characteristics, enhancing the *cohesion* and *phi* values, polypropylene geotextile demonstrates a diminishing effect on these soil properties, resulting in reduced shear strength. Therefore, choosing the appropriate reinforcement material is crucial for optimizing the stability and strength of soil structures. Table 4 compares soil with and without reinforcement to show the increment and decrement in shear strength under varied normal stresses. In Fig. 5, a comparison between unreinforced and reinforced soil samples is shown on a graph that is plotted between normal stress and shear stress. The shear strength is calculated as:

$$S = c + \sigma_{\mathrm{n}} \tan(\phi) \tag{1}$$

4.1 CBR Test Results for Various Layers of Jute and Polypropylene (PP) Geotextiles

The CBR test involves mixing an unreinforced soil sample with single, double, and triple layers of geotextiles to determine the optimal depth of the layers that will yield a higher CBR value and increase the strength of the subgrade. The test also examines the impact of jute and polypropylene (PP) geotextiles on CBR values with the results compared between reinforced and unreinforced soil samples. Ultimately, this comparative analysis provides the effectiveness of geotextile reinforcement in enhancing soil strength and stability of the soil. A comparison between the outcomes of jute textile-reinforced soil samples and those reinforced with polypropylene geotextile can reveal important details about how well each material performs in terms of boosting the soil's stability and strength, aiding in selecting the most suitable reinforcing material for a particular application which are shown in Figs. 6 and 7 respectively. CBR values at 2.5 and 5 mm of penetration for jute and polypropylene geotextile are given in Tables 5 and 6. As the CBR (California bearing ratio)

Table 4 Comparison of shear strength among the geotextiles

	Shear stress (kPa) (unreinforced soil)	Shear strength (kPa) (unreinforced soil)	Shear stress (jute) (kPa)	Shear strength (jute) (kPa)	Increment in shear strength by reinforcing jute (%)	Shear stress (PP) (kPa)	Shear strength (PP) (kPa)	Decrement in shear strength by reinforcing polypropylene (%)
50	48.85	50.22	54.739	55.499	10.51	42.50	42.874	14.63
100	83.478	80.73	89.272	87.753	8.7	72.511	71.784	11.1
150	109.88	111.25	119.246	120.007	7.8	100.33	106.694	4.1

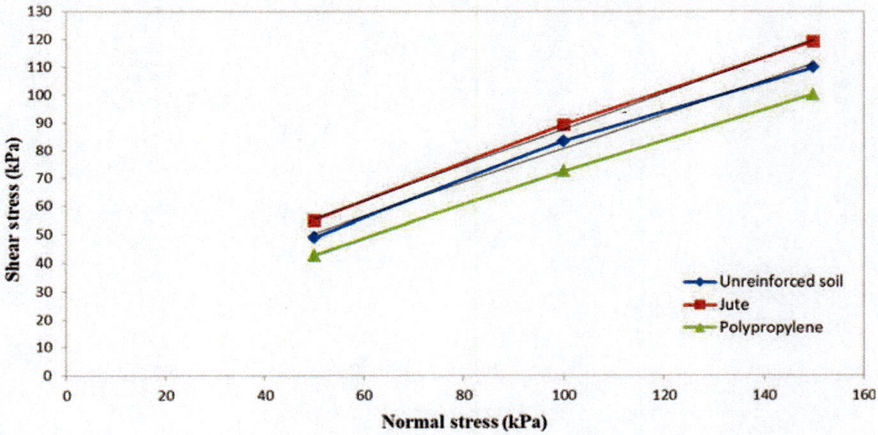

Fig. 5 Shear stress versus normal stress curve in soil with and without reinforcement

value at 5 mm penetration consistently shows a higher value compared to the 2.5 mm penetration, it is recommended to consider the CBR value at 5 mm penetration.

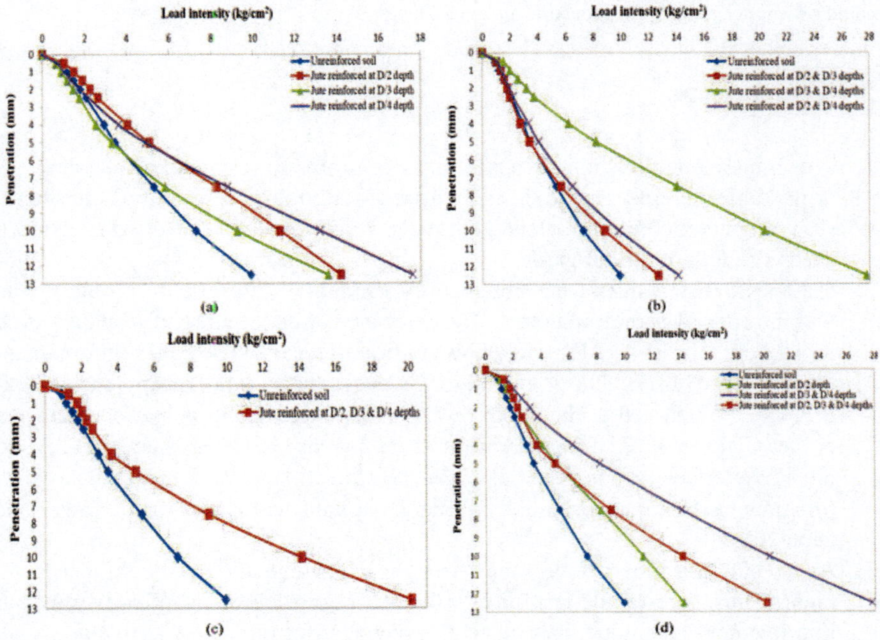

Fig. 6 Load intensity versus penetration plot for **a** soil reinforced with single, **b** double, **c** triple, and **d** optimum depths of jute textile among all layers

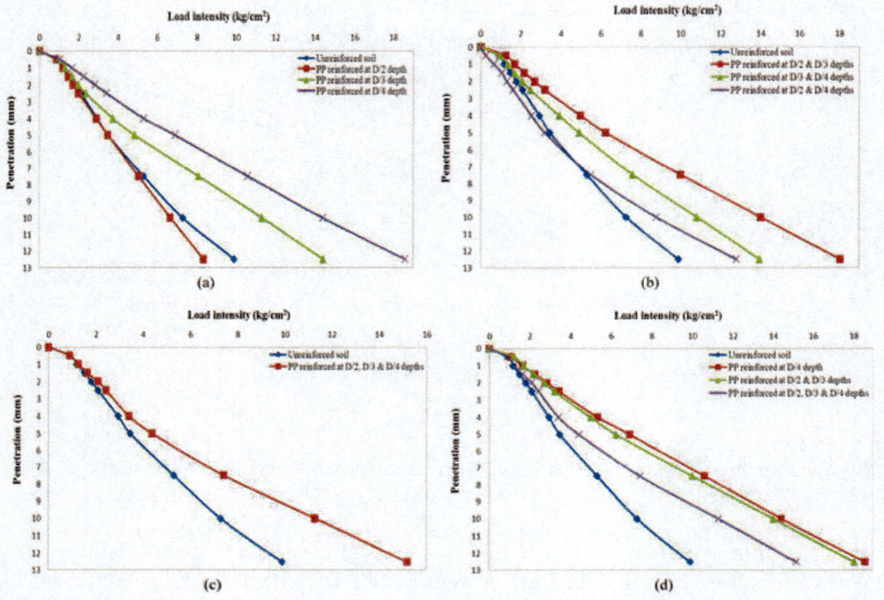

Fig. 7 Load intensity versus penetration plot for **a** soil reinforced with single, **b** double, **c** triple, and **d** optimum depths of PP geotextile among all layers

5 Conclusions

1. The implementation of jute textile as a reinforcement material has shown a remarkable increment in the shear strength of soil subgrade. In contrast, the inclusion of synthetic polypropylene geotextile as reinforcement resulted in reduced shear strength of the subgrade.
2. Enhanced CBR values were obtained experimentally when natural jute textile was introduced as reinforcement. The combined effect of jute textile at depths *D*/3 (4.23 cm) and *D*/4 (3.175 cm) shows optimum results. The dual reinforcement improved the CBR value from 3.32 to 7.82%. In the case polypropylene geotextile, when employed at the depth of *D*/4 (3.175 cm) from the surface shows a higher CBR value. The result represents an increase in CBR value from 3.32 to 6.6%. Nevertheless, it is essential to note that the utilization of jute textile as reinforcement exhibits a substantially higher CBR value compared to polypropylene geotextile.
3. As a result, the subgrade becomes stronger which can allow for a thinner pavement. Thus, jute textile-reinforced soil can be a viable and effective ground improvement technique, particularly in engineering projects that involve weak soil. Jute textile shows better results as reinforcement compared to polypropylene geotextile.

Table 5 Unsoaked CBR values of unreinforced and jute textile-reinforced soil at 2.5 and 5 mm penetration ($D = 12.7$ cm)

Unsoaked CBR	Unreinforced soil (%)	Jute reinforced in a single layer (%)			Jute reinforced in double layers (%)		Jute reinforced in triple layers (%)	
		D/2 (6.35 cm)	*D*/3 (4.23 cm)	*D*/4 (3.175 cm)	*D*/2 (6.35 cm) and *D*/3 (4.23 cm)	*D*/3 (4.23 cm) and *D*/4 (3.175 cm)	*D*/2 (6.35 cm) and *D*/4 (3.175 cm)	*D*/2 (6.35 cm) and *D*/3 (4.23 cm) and *D*/4 (3.175 cm)
$(CBR)_{2.5mm}$	3.004	3.8	2.53	3.2	2.9	5.53	3.2	3.75
$(CBR)_{5mm}$	3.32	4.85	3.2	4.7	3.25	7.82	3.9	4.75

Table 6 Unsoaked CBR values of unreinforced and polypropylene (PP) geotextile-reinforced soil at 2.5 and 5 mm penetration ($D = 12.7$ cm)

Unsoaked CBR	Unreinforced soil (%)	Polypropylene reinforced in a single layer (%)			Polypropylene reinforced in double layers (%)			Polypropylene reinforced in triple layers (%)
		D/2 (6.35 cm)	*D*/3 (4.23 cm)	*D*/4 (3.175 cm)	*D*/2 (6.35 cm) and *D*/3 (4.23 cm)	*D*/3 (4.23 cm) and *D*/4 (3.175 cm)	*D*/2 (6.35 cm) and *D*/4 (3.175 cm)	*D*/2 (6.35 cm) and *D*/3 (4.23 cm) and *D*/4 (3.175 cm)
$(CBR)_{2.5\ mm}$	3.004	2.82	3.4	4.9	4.6	3.6	2.35	3.5
$(CBR)_{5\ mm}$	3.32	3.32	4.63	6.6	5.94	4.7	3.1	4.2

References

1. Singh M, Trivedi A, Shukla SK (2019) Strength enhancement of the subgrade soil of unpaved road with geosynthetic reinforcement layers. Transp Geotech 19:54–60
2. Mohan R, Gupta A, Gaur K (2021) Utilization of bitumen, aggregate and wax with rubber tyre in a flexible pavement. In: International conference on advances in civil engineering, vol 1197, no 1, p 012017
3. Kumar Y, Trivedi A, Shukla SK (2023) Damage evaluation in pavement-geomaterial system using finite element-scaled accelerated pavement testing transp. Infrastruct Geotech. https://doi.org/10.1007/s40515-023-00309-y
4. Hossain AM, Adnan A, Alam AM (2015) Improvement of granular subgrade soil by using geotextile and jute fiber. Int J Sci Technol Soc 3(5):230–235
5. Sudarsanan N, Karpurapu R, Amirthalingam V (2015) The use of natural geotextiles in reinforcing the unpaved roads. Department of Civil Engineering, Indian Institute of Technology Madras, Chennai-600 036, India
6. Mittal A, Shukla S (2019) Effect of geosynthetic reinforcement on strength behaviour of weak subgrade soil. Mater Sci Forum 969:225–230. (Trans Tech Publications Ltd., Switzerland)
7. Viveka T, Kumar SN, Chamberlin SK (2021) Stabilization of slopes of sandy soils by using geosynthetics. In: International conference on advances in civil engineering, vol 1197, p 012081. Guntur, India
8. Shukla S, Mittal A (2022) Effect of jute fibre reinforcement on strength, thickness and cost of low-volume rural roads. Mater Today Proc 62(Part 12):6749–6754
9. Rashidian V, Naeini AS, Mirzakhanlari M (2018) Laboratory testing and numerical modelling on bearing capacity of geotextile-reinforced granular soils. Int J Geotech Eng 12(3):241–251
10. Bergado DT, Youwai S, Hai CN, Voottipruex P (2001) Interaction of nonwoven needle-punched geotextiles under axisymmetric loading conditions. Geotext Geomembr 19:299–328
11. Singh M, Trivedi A, Shukla SK (2020) Influence of geosynthetic reinforcement on unpaved roads based on CBR, and static and dynamic cone penetration tests. Int J Geosynthetics Ground Eng 6(2):13
12. Singh M, Trivedi A, Shukla SK (2020) Unpaved test sections reinforced with geotextile and geogrid. Mater Today Proc 32(Part 4):706–711
13. Singh M, Trivedi A, Shukla SK (2020) Fuzzy-based model for predicting strength of geogrid-reinforced subgrade soil with optimal depth of geogrid reinforcement. Transp Infrastruct Geotechnol 7:664–683
14. Giroud JP, Han J (2004) Design method for geogrid-reinforced unpaved roads. I. Development of design method. J Geotech Geoenviron Eng 130(8):775–786
15. Alobaidi I, Hoare DJ (1998) The role of geotextile reinforcement in the control of pumping at the subgrade-subbase interface of highway pavements. Geosythetic Int 5(6):619–636
16. Tingle JS, Jersey SR (2005) Cyclic plate load testing of geosythetic-reinforced unbounded aggregate roads. In: Transportation research record: journal of the transportation research board, No. 1936, transportation research board of the national academies. Washington, D.C, pp 60–69
17. Raja MNA, Shukla SK, Khan MUA (2021) An intelligent approach for predicting the strength of geosythetic-reinforced subgrade soil. Int J Pavement Eng 23(10)
18. Al-Refeai TO (2000) Behavior of geotextile reinforced sand on weak subgrade. J King Saud Univ Eng Sci 12(2):219–232
19. Hasan M, Samadhiya NK (2017) Performance of geosynthetic-reinforced granular piles in soft clays: model tests and numerical analysis. Comput Geotech 87:178–187

20. Shekhawat P, Sharma G, Singh RM (2023) Strength characteristics of hazardous wastes flyash and eggshell powder-based geopolymer-stabilized soft soil cured at ambient temperature. Arab J Geosci 16:127. https://doi.org/10.1007/s12517-023-11234-2
21. Shekhawat P, Sharma G, Singh RM (2023) Morphology and microstructure of waste material-based geopolymer with flyash, eggshell powder, and soft soil. Mater Lett 334:133621. https://doi.org/10.1016/j.matlet.2022.133621
22. Hasan M, Samadhiya NK (2016) 3D numerical analysis of granular piles with internal horizontal geogrid strips in layers. In: Proceedings of national conference on Indian geotechnical conference (IGC), IIT Madras, Chennai, India

Geotechnical Behaviour of Expansive Soil Reinforced with Fibre

Jatin and Kongan Aryan

1 Introduction

Expansive soil, also known as shrink-swell soil, is a type of soil that undergoes significant volume changes. These volume changes can lead to a variety of problems and challenges. One such soil is the loamy soil present in the Jodhpur district of Rajasthan state of India. Building on expansive soil can be challenging and expensive. Extra precautions and engineering techniques, such as deep foundations or soil stabilization methods, may be required to mitigate the effects of expansive soil, increasing construction costs and complexity. To address these problems, various strategies can be employed, including soil stabilization techniques around structures built on expansive soil.

Chemical compounds can be added to the soil to increase its strength. Past research in this field has demonstrated the potential use of marble dust, lime treatment and gypsum [1–8].

According to Ahirwar and Chore (2022), the strength of the soil combined with marble dust decreased as the amount of marble dust increased [1]. A study by Purohit and Wayal (2007) indicated that with addition of gypsum and dune sand led to decrease in swelling pressure in the soil further reducing cracking instance [2]. Dhemla and Bundela (2015) performed CBR test on six different Rajasthan locations with Barmer having weakest soil has shown improvement with jute fibre addition [3]. By using bentonite to change the behaviour of marble dust, Kumar and

Jatin (✉) · K. Aryan
Delhi Technological University, Shahbad Daulatpur, Main Bawana Road, Delhi 110042, India
e-mail: jatinsingh_2k21gte09@dtu.ac.in

K. Aryan
e-mail: arkongan@dce.ac.in

S. Kolathayar et al. (eds.), *Best Practices in Geotechnical and Pavement Engineering*,
Lecture Notes in Civil Engineering 449, https://doi.org/10.1007/978-981-99-8505-0_26

Jha (2021) were able to increase soil strength by reducing the MDD and OWC of sand and increasing them with the substitution of bentonite [4].

Lime treatment is a common method used to mitigate the expansive nature of soil by stabilizing it. However, the use of lime treatment or bentonite's bulk utilization for soil strengthening produces effects like increased soil alkalinity, soil compaction problems and long-term stability concerns. Therefore, the goal of this study is to increase soil strength utilizing natural fibres without further harming the soil ecosystem by changing the soil's fibre percentage and fibre length. Moreover, the attempt has been made to confirm the experimental work using Finite Element Analysis with ABAQUS software.

2 Material and Methodology Followed

2.1 Materials and Its Classification

The natural soil utilized in this research was gathered by the author at a construction site in Jodhpur, Rajasthan, India. Widely used term for it is expansive soil. The author filtered the soil through a 75 μm sieve using water before oven-drying it. It was found that more than 80% soil passes through the sieve and 20% is retained. On the selected area of soil, they performed elementary index tests. The hydrometer test findings as in Fig. 1 revealed that the soil sample had a 44% clay, 5% silt and 51% sand composition. The liquid limit and plasticity indices of the soil were discovered to be 52.9215 and 32.2615, respectively. Table 1 provides a list of the soil sample's characteristics.

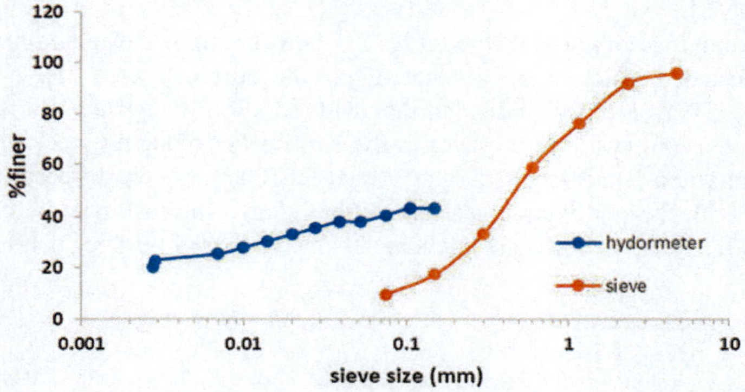

Fig. 1 Gradation curve of soil

Table 1 Synopsis of characteristics of applied soil

Properties	Values
Specific gravity (G)	2.5
Liquid limit (%)	52.92
Plastic limit (%)	20.66
Plasticity index (%)	32.26
Soil classification	CH
Maximum dry unit weight (g/cm^3)	14.927
Optimum moisture content (%)	15.3
Sand content (%)	51
Clay content (%)	44
Silt content (%)	5
D10	0.128
D30	0.269
D60	0.644

2.2 *Methodology Followed*

The unconfined compressive strength (UCS) test technique specified in IS 2720 (Part 10): 1991 was employed by the author. The sample size used for the experimentation adhered to the IS code provisions. The compacted soil specimens were cylindrical in shape with a diameter of 38 mm and a length 76 mm resulting in a length to diameter ratio of 2. Further, the length of sample was increased to 80 and 84 mm leading to length to diameter ratio of 2.1 and 2.3, respectively. Hand compaction was utilized for sample preparation. A two-part split mould with a diameter of 38 mm and grease as a lubricator for soil sample extraction was employed.

For the sample preparation, a suitable amount of oven-dried soil in a dry state was mixed to obtain a consistent mix. The sample was then combined with the appropriate amount of water to achieve the optimum water content value and corresponding dry unit weight of the soil. In the case of fibre, the fibres' weight and desired percentage were determined according to the weight of the soil used, and these fibres were gradually mixed by hand into the soil sample before adding water. The water was then slowly and thoroughly mixed with the soil and fibre mixture to ensure uniformity and prevent fibre accumulation. Hand mixing was employed to uniformly add water to the soil and fibre mixture. With and without fibre, the soil was compacted in five equal layers. The two parts of the mould were braced to prevent any movement during compaction, and each layer was uniformly distributed during the compaction process.

To achieve the compaction, a mould-sized plate was placed on one end with hammer falling from 30 cm height. 25 no of blows were applied with rod bearing weight of 2.5 kg. The prepared sample was then inserted into the experimentation equipment, and the rate of loading was 1.27 mm/minute. Displacement and load data

Fig. 2 Unconfined compressive strength test equipment

were recorded using a dial gauge and proving ring, respectively, at 30-s intervals. The entire process was recorded with a tripod-mounted camera, and pictures were taken at the final point. Equipment used for experimentation is shown in Fig. 2.

The author used front view images to analyse the crack pattern and their decrement in the soil sample. Table 2 represents the synopsis of overall test programme.

3 Examining Test Findings

Unconfined compressive strength is the maximum axial compressive stress that a soil sample can bear under zero confining stress. This stress is calculated by Eq. 1 (Fig. 3).

$$\sigma = \frac{P}{A_c}. \tag{1}$$

The strain was obtained from the observation of the experimentation. Average strain in this study is represented using ε_a. It is a measure of the overall strain distribution in the material under the applied load. For each test run for the current

Table 2 Synopsis of test programme

Test legend	Soil type	*f* (%)	Fibre category (mm)	*l* (mm)	Sample length (mm)
T1	Soil 1	–	–	–	76
T2	Soil 1	–	–	–	80
T3	Soil 1	–	–	–	84
T4	Soil 2	2.0	Coir	10	76
T5	Soil 2	2.5	Coir	10	80
T6	Soil 2	3.0	Coir	10	84
T7	Soil 3	2.0	Jute	10	76
T8	Soil 3	2.5	Jute	10	80
T9	Soil 3	3.0	Jute	10	84

Footnote f = Fibre content; l = Fibre length

Fig. 3 ABAQUS model

study, they plotted the stress–strain curve. Utilizing the peak value of the stress–strain plot, the soil's unconfined compressive strength (σ_{pc}) was calculated. The author calculated the modulus of resilience using graph plotting package of Origin Pro software by OriginLab computing the total area under stress–strain curve up to elastic limit in the graph as shown in Fig. 4 for Soil 1. The unit of modulus of resilience (U_r) reported in this analysis is J/m^3. The author discusses the findings in terms of a proportionate increase in average strain ε_a (%), peak compressive strength σ_{pc} and modulus of resilience U_r. Table 3 sums up the details of test results. The compressive strength test exhibits a noteworthy impact on strength of the soil.

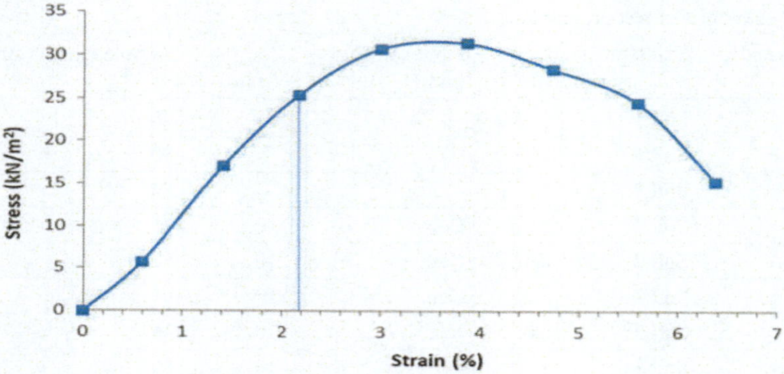

Fig. 4 Modulus of resilience calculation graph

Table 3 Synopsis of test results

Test legend	σ_{pc} (kN/m^2)	ε_a (%)	U_{rf} (J/m^3)	$\sigma_{pcf}/\sigma_{pcu}$	$\varepsilon_{af}/\varepsilon_{au}$	U_{rf}/U_{ru}
T1	18.4	4.58	–	–	–	–
T2	29.57	6.62	–	–	–	–
T3	31.36	3.87	–	–	–	–
T4	41.28	10.35	91.79	1.32	2.27	1.66
T5	44.75	11.93	138.67	1.43	2.61	2.50
T6	55.72	7.83	165.34	1.77	1.80	2.98
T7	57.04	9.93	135.31	1.82	2.17	2.44
T8	65.05	13.08	218.52	2.08	2.86	3.94
T9	69.78	10.21	296.62	2.23	2.23	5.35

σ_{pc} = peak compressive stress; ε_a = average strain; U_r = modulus of resilience

3.1 *Finite Element Analysis*

Finite Element Analysis (FEA) is a numerical method used to solve complex engineering problems by dividing a physical structure or system into smaller, simpler elements. For this study, the software used for construction and analysis of model is ABAQUS by Dassault Systems. A UCS model was created with specifications of 38 mm diameter and 76 mm length. Figure 3 depicts the image of the model built in the ABAQUS for analysis. An 8 node element was created. The static general step for analysis was used. Analysis was nonlinear in nature. The global seeding tool was used for meshing. Under loading, one side was restricted with ZSYMMETRIC loading, and other was loaded with uniform load. Post-analysis values of force and displacement at various nodal points were obtained.

3.2 Crack Propagation Comparison

The author used the tripod-mounted camera for recording the experimentation procedure. Adobe Photoshop was used to edit and clarify the pictures for comparison purpose. The pictures taken at the end of the experimentation was used for comparing the cracking behaviour of soil sample under load for reinforced and unreinforced conditions.

4 Results and Discussions

4.1 *Influence of* F

For study of influence of the fibre content the author used the Soil 2 and Soil 3 with proportions of 2, 2.5, 3% fibre with coir and jute fibre respectively with a constant length of 10 mm. Figure 5 depicts the comparison between unreinforced, coir and jute fibre at varying fibre content. In Fig. 5, it can be clearly observed that both coir and jute fibre when reinforced in 3% fibre content in the soil provides highest peak as compared to fibre content of 2 or 2.5%. The coir fibre shows an improvement of 10.75% from 2 to 2.5% fibre content and 19.28% from 2.5 to 3% fibre content in stress value. Furthermore, jute fibre shows 14.3% increase in peak stress value from 2 to 2.5% fibre content and 7.65% increase from 2.5 to 3% fibre content. Amongst coir and jute fibre, later has shown 25.24% improvement over the former in terms of stress values. In Fig. 6, comparison of peak compressive stress and average strain ratio for coir and jute reinforced soil was done. In stress comparison, jute shows improvement of 14.2% and 7.21% in 2% to 2.5% and 2.5% to 3%, respectively, and coir fibre shows improvement of 14.2% and 7.21% in 2% to 2.5% and 2.5% to 3%, respectively. In comparison with coir and jute fibre, at 3% fibre content jute has shown 25.98% increment in stress value as compared to coir fibre. In average strain comparison coir fibre depicts an increase of 20.27% and 4.5% for 2 to 2.5% and 2.5 to 3%, respectively, whereas in jute fibre for fibre content 2 to 2.5% and 2.5 to 3% depicts stress value increase of 20.32% and 17.04%, respectively. Figure 7 depicts the comparison of jute and coir fibre's modulus of resilience. For coir fibre with fibre content increase from 2 to 2.5% and 2.5 to 3%, the modulus of resilience improves by 50.06% and 19.2%, respectively. Jute fibre seems to have shown 60.14% and 35.78% for 2 to 2.5% and 2.5 to 3%, respectively. Amongst coir and jute fibre, later has shown increase in U_r by 79.82% against former. This indicates that jute fibre has more capacity to bear stress without permanent deformation than coir fibre.

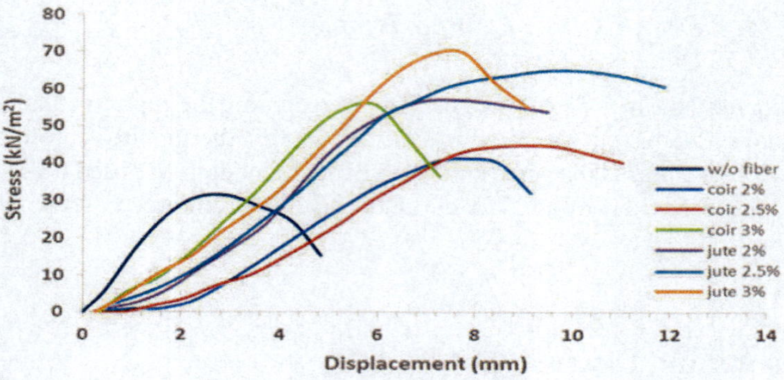

Fig. 5 Graph showing comparison of coir and jute fibre

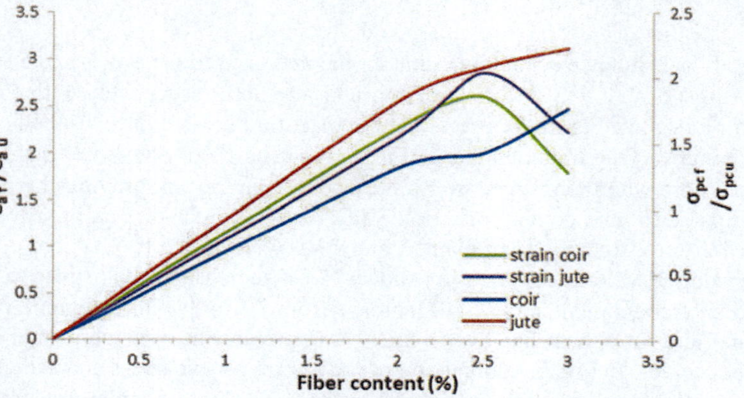

Fig. 6 Graph showing stress, strain comparison of coir and jute

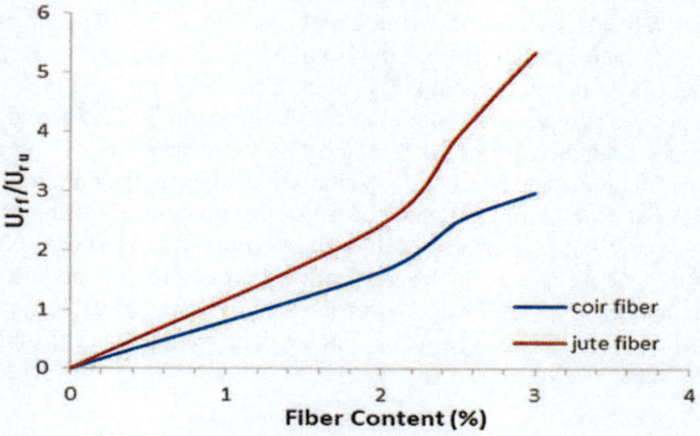

Fig. 7 Graph showing modulus of resilience of coir and jute

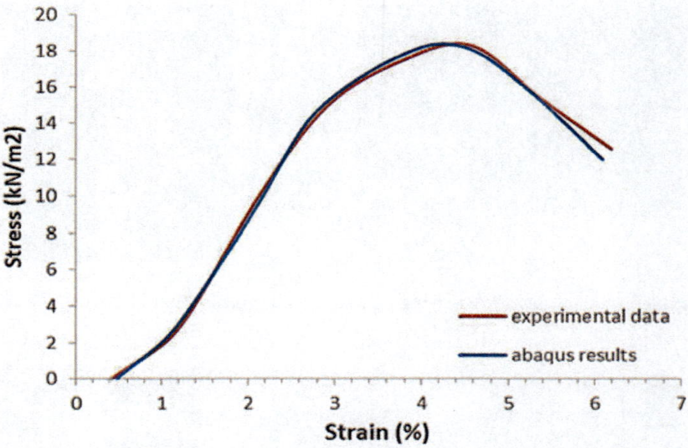

Fig. 8 Graph showing experimental and ABAQUS results

4.2 Finite Element Method Results

The values of displacement and force obtained from the experimentation were used to construct a stress–strain curve. Similarly, the results acquired from the model analysis were also processed to get hold of stress–strain curve. It was found that both the curves superimpose on each other. The graph of the results obtained from software analysis and experimentation is depicted in Fig. 8. The graph depicts that results obtained from the experimentation was counter verified by the software analysis.

4.3 Crack Propagation Results

The photographs were taken at the end of the experimentation to compare the crack propagation of the soil in unreinforced, reinforced with coir and jute fibres. Figures 9 and 10 provide comparison of coir and jute fibre respectively at varied fibre content. From Fig. 5, it can be observed with increase in fibre content the peak value becomes progressively higher which can be correlated to Fig. 9 such that with increasing fibre content the cracking delays and reduces as well, maximum delay in cracking and peak occurs in 3% fibre content. From Fig. 5, it can be witnessed that the jute fibre shows increase in peak value as the fibre content increases. This can be linked to Fig. 10 which shows delay and reduction in cracking as the fibre content increases. Amongst coir and jute fibre, it can be related from Figs. 5, 9 and 9 that there is least cracking and highest peak in the jute sample reinforced with 3% fibre content. It implies that reinforcement with jute at 3% fibre content not only provides the highest strength but also most cracking delay in the soil sample.

Fig. 9 From left to right, images showing comparison of coir fibre at 2, 2.5 and 3% fibre content

Fig. 10 From left to right, images showing comparison of jute fibre at 2, 2.5 and 3% fibre content

5 Natural Fibre Life Span and Preservation

The natural fibre used for this study are of 2 type's coir and jute fibre. Generally lasting 4 years, the coir fibre can be preserved for longer if treated properly. For coir fibre leaching is the first and foremost step to reduce the salt levels to control electrical conductivity. Coarser the fibre is more longevity it will provide. Optimum depth at which coir fibre is least exposed to environmental conditions can be other way to expand life of coir as long as 20 years. Jute being more susceptible to degradation can still last long if chosen properly. Jute fibre being more porous allows more space to circulation. Jute when treated properly and inserted at optimum depth can last more than 3 months. Geotextile membranes or plastic liners can be an alternative to prevent moisture contact to further increase the life span of natural fibres.

6 Conclusion

The following conclusion can be reached based on the analysis and interpretation of the results in this research. Reinforcing the vast soil of Rajasthan with natural fibres like coir and jute will increase its resilience. The peak strength σ_{pc}, average strain ε_a

and modulus of resilience U_r increase with fibre reinforcement. The three parameters mentioned depict the best strength capacity in 3% fibre content reinforcement.

For same fibre content jute is found to have provided better results than the coir fibre. It has been found that jute is able to bear 1.2 times the load than the coir fibre within the same fibre content categories. It implies that jute serves as a better fibre for reinforcement as compared to coir fibre provided with better preservation methods.

The verification of experimental results by ABAQUS model further concretes the fact that natural fibres can not only be an alternative to the soil reinforcement procedures used in Rajasthan's soil but also a improved way to conserve the already degrading soil bionetwork of the Rajasthan.

References

1. Ankush Kumar J, Arvind Kumar J, Shivanshi (2020) Geotechnical behaviour and micro-analyses of expansive soil amended with marble dust. Soils Found 60:737–751
2. Padma Nabh T, Vikram K (2017) Comparative study of soil physical characteristics of Jaipur district, Rajasthan. Afr J Environ Sci Technol 11(1)
3. Arvind K, Asheet B, Baljit Singh W (2007) Influence of fly ash, lime, and polyester fibres on compaction and strength properties of expansive soil. Am Soc Civ Eng 19(3)
4. Jitendra Singh Y, Suresh Kumar T (2016) Behaviour of cement stabilized treated coir fibre-reinforced clay-pond ash mixtures. J Build Eng 8:131–140
5. Ankush Kumar J, Ayush K, Arvind Kumar J (2021) Physical and swell behaviour of sand–bentonite and marble dust–bentonite mixes. In: Proceedings of the Indian geotechnical conference 2019, vol 134, pp 95–106
6. Arvind Kumar J, Ankush Kumar J, Shivanshi (2019) Potential of marble dust to improve the physical behavior of soil. Geotech Transp Infrastruct 28:189–201
7. Sivakumar Babu GL, Vasudevan AK (2008) Strength and stiffness response of coir fibre-reinforced tropical soil. J Mater Civ Eng Am Soc Civ Eng 20(9)
8. Pankaj D, Anup Kumar B, Arvind L (2015) Identification of weaker subgrade soil in Rajasthan and increment of CBR by jute fibre as additive, recent development in engineering, science and management, YMCA, Faridabad, May 2015

Effect of Coir Geocell on the Liquefaction Potential of Coastal Sand

Veena Jayakrishnan, K. S. Beena, and V. S. Vivek

1 Introduction

Soil liquefaction is a common feature of saturated loose sands and soils, where a solid behaves like a liquid in response to a dynamic loading. This may lead to dangerous real-world consequences. Liquefaction studies are emerging as prime importance in Kerala district due to the upcoming coastal highway project and semi high-speed railway corridor projects.

If a foundation soil is prone to liquefaction, no matter how many safety factors are taken into account while designing a superstructure, they won't be of any use during an earthquake. Hence, contractors and researchers take measures to improve the ground before construction due to the increasing infrastructural developments in the coastal area.

In recent years, engineers and scientists are constantly looking for new ground improvement techniques, which are cheaper and easily available. Exploring natural products is essential for making construction both economical and environmentally friendly due to the high cost of geosynthetics as well as growing environmental concerns about the excessive use of polymeric geosynthetics. As a result, the use of geocells is expanding quickly and is prevalent in many geotechnical engineering applications. Kolathayar [1] observed that there are few studies available on natural fibers under dynamic loading. In addition to HDPE geocells, geocells made out of natural materials have great scope for potential applications in ground engineering.

According to the research that have been published so far, coir products have not been used to their full potential as reinforcing materials. Coir, a biodegradable and environment-friendly material, is an excellent replacement for its synthetic counterparts. Natural fiber known as coir is taken from the husk that covers the coconut shell. Hence, the use of coir geocell as a replacement for synthetic geocells is emerging

V. Jayakrishnan (✉) · K. S. Beena · V. S. Vivek
Cochin University of Science and Technology, Kochi, Kerala, India
e-mail: veena.j.90@gmail.com

S. Kolathayar et al. (eds.), *Best Practices in Geotechnical and Pavement Engineering*, Lecture Notes in Civil Engineering 449, https://doi.org/10.1007/978-981-99-8505-0_27

nowadays [2–4]. Shaking table studies were conducted by many researchers to study the liquefaction phenomena [5–8]. Chauhan and Mittal, 2022, [4] have conducted experimental research on rigid box shake table tests to study the effectiveness of utilizing geocell to mitigate liquefaction and observed a marginal increase in soil liquefaction resistance. But the effect of boundary condition is not taken into account. Thus, a laminar box shake table studies are emerging nowadays.

Coir cannot be used for permanent reinforcing applications since it is a natural fiber and degrades over time due to biodegradation. Lekha [9] shown that interlocking coir mat between polyester and other robust materials can boost its durability. Cement and bitumen coatings can also make coir fibers more durable [10]. Researchers have suggested that coir's lifespan is relatively suitable to use as a reinforcing material based on the findings of durability experiments [11].

The purpose of the current research was to assess soil reinforcement using geocell as an anti-liquefaction measure. The liquefaction studies with geocell (synthetic/natural) as three-dimensional reinforcement in shake table are limited. Since there aren't any previous experiments employing a laminar box shake table apparatus, the work shows how effectively reinforcing fully saturated coastal sand deposits with coir geocell can reduce liquefaction potential.

2 Materials Used

2.1 Coastal Sand

The samples were collected from coastal area of Cherthala beach (76.292821 E, 9.656752 N) situated in Alappuzha district, Kerala, India. Due to their cohesionless, saturated and loose nature, these are vulnerable to liquefaction [12]. The properties of the sample are given in Table 1 and may be noted that it is poorly drained fine grained soil, which are most susceptible to liquefaction.

Figure 1 shows the light brownish color sand collected (Fig. 1a) and its microscopic image (500 × resolution) (Fig. 1b). The sample collected was in pure crystalline form and spherical structure as observed from Fig. 1b. Many scientists have concluded that natural sand used in buildings like retaining walls and slopes may be strengthened with natural geosynthetics, like coir geocells, to increase stability and sustainability since natural sand with regular shapes liquefies easily [13–16].

2.2 Coir Geocell

Coir geocell used for the study was procured from the National Coir Research and Management Institute (NCMRI), Trivandrum. The image of the coir geocell used and its properties are shown in Fig. 2 and Table 2, respectively. Coir geocell of

Table 1 Properties of coastal sand

Property	Value	Codal provision
Specific gravity	2.63	IS:2720 [Part 3/sec1-1980]
Permeability (mm/s)	0.0007	IS:2720 [Part 4]-1985
Effective size D_{10} (mm)	0.21	IS:2720 [Part 14]-1983
Mean grain size D_{50} (mm)	0.45	IS:2720 [Part 14]-1983
Minimum void ratio	0.539	IS:2720 [Part 14]-1983
Maximum void ratio	0.765	IS:2720 [Part 14]-1983
Coarse sand (%)	0	IS:2720 [Part 14]-1983
Medium sand (%)	51	IS:2720 [Part 14]-1983
Fine sand (%)	49	IS:2720 [Part 14]-1983
Fines content (%)	0	IS:2720 [Part 14]-1983
Soil classification	SP	IS:2720 [Part 4]-1985

(a) (b)

Fig. 1 **a** Coastal sand and **b** digital microscope image

various heights were utilized, i.e., 200 mm (full height geocell (FHGC)), 150 mm (intermediate height geocell (IHGC)), and 100 mm (half height geocell (HHGC)) to analyze the effectiveness of geocell height used to mitigate liquefaction potential of sand.

3 Experimental Investigations

Laminar box shake table tests were conducted to study the effectiveness of reinforcing coir geocell in coastal sand to increase liquefaction resistance. The test set up at Cochin University of Science and Technology, Kerala [12], was used, and the detailed design of the laminar box shake table was mentioned in [8, 12]. The laminar had

Fig. 2 Coir geocell used for the study

Table 2 Specifications of coir geocell

Specifications	Value
Pocket size	200 mm × 200 mm
Strip thickness	6.18 mm
Cell depth	200 mm
Diagonal length of single cell	280 mm

an overall internal dimension of 1000 mm × 600 mm × 800 mm. The major parts of the laminar box are the base plate with a supporting frame, laminar layers, and the flexible membrane. For the present study, the specifications of the experimental setup were as follows:

Frequency throughout the test = 1 Hz

The amplitude of acceleration of the shake(sinusoidal wave) = 0.24 g

Duration of shake = 20 s.

The samples were filled in a laminar box up to a height of 600 mm, and the pore pressure was measured at the central section at three depths (150, 300 and 450 mm from the base). Pore pressure sensors-TML strain gauge type transducer of type-KPE-200KPB were used for data acquisition. The dry sand bed was initially prepared by carefully using a funnel to fill each layer of the laminar box with the proper weight of sand for a relative density of 30%. The coir geocell was placed at the central portion (at a depth of 300 mm from base) as shown in Fig. 3.

The pore pressure sensors were installed at the proper locations to study the buildup and dissipation of excess pore pressure (EPP) developed during the shake. After the sensors were installed, sand was progressively moistened from the bottom in a regulated manner to prevent disturbing the layers and preserve the relative density.

Fig. 3 Placing of coir geocell at the central portion of the test setup

The tests were conducted on virgin coastal sand and sand reinforced with coir geocell of various heights. A total of four tests were conducted and the generation of EPP was recorded and analyzed to study the effectiveness of using these geocells to improve the liquefaction resistance of coastal sand.

4 Result and Discussions

The EPP values of four tests at the central section (top, middle and bottom) were recorded to study the effect of reinforcing coastal saturated sand with coir geocell. By dividing the maximum EPP by the initial effective vertical stress at the corresponding level, the maximum excess pore pressure ratio (EPPR) was calculated. A value of 1.0 for EPPR indicates liquefaction. The EPP versus time graph of four tests recorded at the top sensor is shown in Fig. 4, and it may be observed that a marginal decrease in the maximum pore pressure generation with the reinforcement of coir geocell irrespective of the geocell height. A similar trend was observed at the middle and bottom sensors also.

The maximum EPP and EPPR values at three positions and the final settlement observed were summarized in Table 3, and it may be noted that the maximum EPP value at all positions dropped from 5.9 kPa with the use of coir geocell irrespective of the height of geocell. The maximum drop was observed when reinforced with 200 mm height geocell (FHGC), i.e., liquefaction resistance increased up to 44.5%. Liquefaction resistance of sand while using IHGC and HHGC was increased to 37.6 and 33.72%, respectively. Similar findings were reported by Chauhan and Mittal [4], when soil was reinforced with geocell of 350 mm height and subjected to a 0.3 g acceleration shaking. The liquefaction resistance increased by 31.6% and 41% at relative densities of 35% and 65%, respectively. When geocell is used, it acts more like a composite material with a high internal friction angle and shear strength.

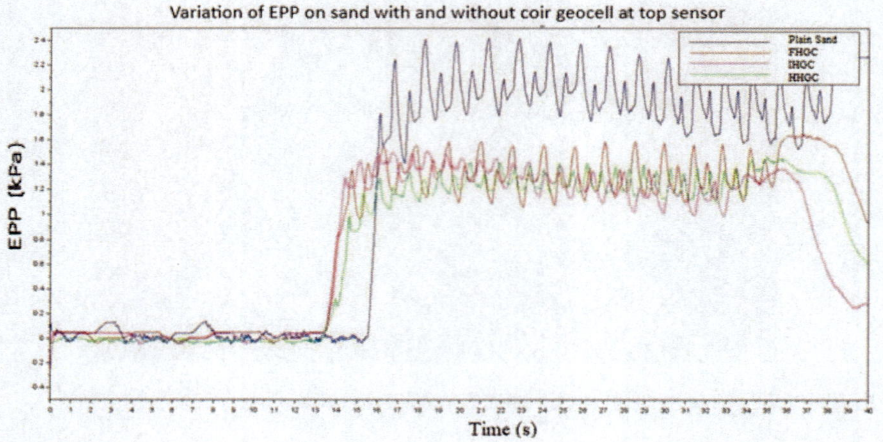

Fig. 4 Variation of EPP recorded in sand with and without reinforcement at top sensor

Table 3 Test results summarized

Test	Bottom sensor		Middle sensor		Top sensor		Final settlement (cm)
	Max. EPP (kPa)	Max. EPPR (kPa)	Max. EPP (kPa)	Max. EPPR (kPa)	Max. EPP (kPa)	Max. EPPR (kPa)	
Unreinforced sand	5.90	1.17	4.19	1.30	2.41	1.64	5.8
Sand with coir geocell of height 100 mm (HHCG)	3.91	0.78	3.45	0.97	1.65	0.98	5.4
Sand with coir geocell of height 150 mm (IHCG)	3.68	0.73	3.29	0.95	1.51	0.96	5.1
Sand with coir geocell of height 200 mm (FHCG)	3.27	0.65	3.16	0.93	1.45	0.94	4.0

From the results analyzed, it may be concluded that the confinement brought about by the presence of reinforcements like coir geocell may be the cause of the decrease in pore pressure. When comparing the final settlement, it may be observed that using a coir geocell with a maximum height result in a lower settlement than using the other two. This may be as a result of the extra confinement qualities the geocell displayed. Table 4 provides a thorough comparison of the present test findings from the literature with the similar investigations.

Table 4 Comparison of results of present study with similar study in shake table test

Reference	Soil	Type of geosynthetics	Test specifications			Liquefaction resistance increase in (%)
			Acceleration (g)	Relative density of soil (%)	Frequency (Hz)	
Chauhan and Mittal [4]	Poorly graded river sand	HDPE geogrid (4 layers)	0.1 g	35	4	23
			0.3 g	35		20
				50		12
				65		25
		HDPE geogrid (5 layers)	0.1 g	35		29
			0.3 g	35		24
				50		22
				65		35
		Geocell (350 mm height)	0.1 g	35		36
			0.3 g	35		31.6
				50		30
				60		43
Present work	Poorly graded coastal sand	Coir geocell of height	0.24 g	30	1	
		200 mm				44.5
		150 mm				37.6
		100 mm				33.72

From the table, it may be observed that the use of geogrids as well as geocells (HDPE/ coir) has improved the liquefaction resistance of the reinforced sand in the range of 10–45%. The deviation in the percentage increase value from the literature presented may be due to the difference in type of soil, relative density, and other test specifications. The table clearly shows that as the sand is reinforced with coir geocell, liquefaction potential of sand is reduced satisfying the general trend reported in literature.

Hence, from the results analyzed, it may be concluded that coir geocell reinforcements can lower the liquefaction potential of coastal sand. Accordingly, this study proposes that if the ground is reinforced with coir geocells prior to the building of new embankments (railways or roads) and subterranean infrastructure, such as reservoirs, the severity of damages might be greatly reduced.

5 Conclusions

To reduce the tendency for sand to liquefy, an experimental study was done to evaluate the efficiency of strengthening loose saturated coastal sand deposits with coir geocell. Literature survey in this area showed that such experimental studies are scanty. A laminar box shake table was used for the testing, with a constant shaking acceleration of 0.24 g and frequency of 1 Hz. The generation of excess pore pressure was recorded at different locations of the central section of the sand bed. To determine if the sand has liquefied, the maximum excess pore pressure ratio value is computed using the recorded EPP value. The following conclusions were observed:

1. The collected unreinforced coastal sand liquefied during the test conditions.
2. Using a coir geocell increased the sand's resilience to liquefaction, regardless of the geocell's height. The shear strength qualities of the reinforced sand may have greatly improved and it will behave more like a composite material.
3. The EPP decreased from 5.9 to 3.27 kPa while utilizing FHGC, indicating an enhanced liquefaction resistance of around 44.5%. The reported improved liquefaction resistance was 37.6% and 33.72%, respectively, after reinforcing with IHGC and HHGC. This increase in sand strength might have been caused by additional confinement properties of the geocell. The observed results corroborated well with similar literatures.
4. According to settlement studies, utilizing FHGC results in the least amount of settlement following a shake.

Accordingly, this study proposes that if the ground is reinforced with geocells prior to the building of new embankments and subterranean infrastructure, such as reservoirs, the severity of damages might be greatly reduced. The primary limitation of the current study is that the coir geocell utilized for the study was not treated before to use; thus, in the future, investigations on treated coir geocell may be undertaken to analyze the durability aspect as well. More research is vital to study the liquefaction phenomenon in Kerala coastal sand deposits using environmentally friendly alternatives. Thus, reinforcing liquefiable soils with biodegradable, low cost and easily available coir geocells are emerging nowadays by replacing the synthetic geocells. If adequately treated, this ground improvement approach may be used by more researchers in the future and might not just be restricted to experimental investigations; additional field study reports might be made available.

References

1. Kolathayar S (2019) Vibration isolation of foundation using HDPE and natural geocells-a review. In: International congress and exhibition sustainable civil infrastructures: innovative infrastructure geotechnology, pp 75–86. https://doi.org/10.1007/978-3-030-01920-4_7
2. Rahimi M, Tafreshi SNM, Leshchinsky B, Dawson AR (2018) Experimental and numerical investigation of the uplift capacity of plate anchors in geocell-reinforced sand. Geotext Geomembr 46:801–816. https://doi.org/10.1016/j.geotexmem.2018.07.010

3. Kolathayar S, Suja P, Nair V et al (2019) Performance evaluation of seashell and sand as infill materials in HDPE and coir geocells. Innov Infrastruct Solut 4:17. https://doi.org/10.1007/s41062-019-0203-6
4. Chauhan R, Mittal S (2022) Control of liquefaction potential by geosynthetic reinforcements—a study. In: Sitharam TG, Kolathayar S, Jakka R (eds) Earthquake geotechnics. Lecture notes in civil engineering, vol 187. Springer, Singapore. https://doi.org/10.1007/978-981-16-5669-9_1
5. Beena KS, Jayakrishnan V, Unni Kartha G, Shafnas PS (2021) Studies on the effect of fines content on liquefaction resistance using shake table tests. In: Patel S, Solanki CH, Reddy KR, Shukla SK (eds) Proceedings of the Indian geotechnical conference 2019. Lecture notes in civil engineering, vol 138. Springer, Singapore. https://doi.org/10.1007/978-981-33-6564-3_60
6. Varghese RM, Latha GM (2014) Shaking table tests to investigate the influence of various factors on the liquefaction resistance of sands. Nat Hazards 73:1337–1351. https://doi.org/10.1007/s11069-014-1142-3
7. Kartha GU, Beena KS, Thahir CPM (2019) Shake table studies on embankments on liquefiable soil. In: Soil dynamics and earthquake geotechnical engineering. Springer, pp 101–109. https://doi.org/10.1007/978-981-13-0562-7_12
8. Unni KG, Beena KS, Mahesh C (2018) Development of 1-D shake table testing facility for liquefaction studies. J Inst Eng Ser A 99:423–432. https://doi.org/10.1007/s40030-018-0299-2
9. Lekha KR (1997) Coir geotextiles for erosion control along degraded hill slopes. In: Proceedings of seminar on coir geotextiles, Coimbatore, India
10. Girish MS, Ramanatha Ayyar TS (2000) Improvement of durability of coir geotextiles. In: Proceedings for the Indian geotextiles conference, Bombay, India, pp 309–310
11. Lekha KR, Kavitha V (2006) Coir geotextile reinforced clay dykes for drainage of low-lying areas. Geotext Geomembr 24:38–51. https://doi.org/10.1016/j.geotexmem.2005.05.001
12. Beena KS, Jayakrishnan V, Alex A, Bindu CS (2021) Analysis of liquefaction potential of coastal sands using laminar box system. Indian Geotech J 1–16. https://doi.org/10.1007/s40098-021-00503-0
13. Balan K, Jency A (2015) Improvement of bearing capacity of soft soil using coir geocells. In: International conference on geotechnical engineering, Colombo, pp 415–418
14. Latha GM, Lakkimsetti B (2022) Morphological perspectives to quantify and mitigate liquefaction in sands. Indian Geotech J 52:1244–1252. https://doi.org/10.1007/s40098-022-00649-5
15. Lal D, Sankar N, Chandrakaran S (2018) Effect of reinforcement form on the behaviour of coir geotextile-reinforced sand through laboratory triaxial compression tests. Int J Geotech Eng 12:309–315. https://doi.org/10.1080/19386362.2016.1275428
16. Thallak SG, Saride S, Dash SK (2007) Performance of surface footing on geocell-reinforced soft clay beds. Geotech Geol Eng 25:509. https://doi.org/10.1007/s10706-007-9125-8

Experimental and Numerical Investigation on Soil Reinforcement using a Sustainable Material, Jute Geotextile

Abhijit Debnath, Jharna Debbarma, and Sujit Kumar Pal

1 Introduction

Construction of structures nowadays is being built on land consisting poor or soft soil. The primary purpose of the ground stabilization is to modify the soil properties at the site. Soil reinforcement is an effective and reliable method for enhancing the soils strength and stability. Numerous studies have analyzed the utilization of jute as a reinforcing material [1–6]. In the last two decades, the use of geotextiles as reinforcement has gained wide popularity in ground improvement work as they are easily available in the market at low cost, environmentally friendly, and easy to apply. Latha and Murthy [2] used triaxial compression test to determine the influence of reinforcement form on the strength improvement of geosynthetic-reinforced sand. A number of unconfined compressive strength (UCS) experiments on unreinforced fly ash and reinforced fly ash with jute geotextiles (JG) have been conducted. Results show that with the increase in the number of JG layers for reinforced fly ash specimens, the UCS values increase, and the maximum increment is about 525% with four layers of reinforcement [3]. Tuna and Altun [4] conducted interface direct shear tests to examine how geotextile inclusions affect the mechanical behavior of granular soils. Ranganathan [5] reported a 'status paper' on jute geotextiles' development and potential, mainly focusing on trends in world jute production and consumption. An experimental and numerical study on the utilization of sustainable material to modify the geotechnical properties of silty clay has been studied. Results show that adding JG reinforcement improves the shear strength (SS) of the reinforced samples, lowering compressibility and increasing the bearing capacity of reinforced soil below foundations [6]. To understand the strengths of clay that has been geopolymer-treated with flyash and eggshell powder, an experimental investigation has been performed

A. Debnath (✉) · J. Debbarma · S. K. Pal
National Institute of Technology Agartala, Tripura 799046, India
e-mail: abhijit.nita2020@gmail.com

S. Kolathayar et al. (eds.), *Best Practices in Geotechnical and Pavement Engineering*, Lecture Notes in Civil Engineering 449, https://doi.org/10.1007/978-981-99-8505-0_28

Table 1 Soil properties used in the present study

Property	Values
Specific gravity (G)	2.69
Liquid limit (%)	20.15
Plastic limit (%)	10.13
Maximum dry density (kN/m^3)	18
Optimum moisture content (%)	15
Sand (%)	48
Silt (%)	39
Clay (%)	11

Table 2 Properties of JG

Property	Values
Type of JG	Woven (Gray)
Weight	750 GSM
Tensile strength	22 kN/m
Pore size	250 micron

[7]. Despite the fact that much research has been conducted on JG reinforcement, a comprehensive investigation is still required.

Thus, this study attempts to perform the triaxial behavior of soil reinforced with JG to be used for various geotechnical projects. Unconsolidated undrained triaxial tests were implemented on a series of soil samples reinforced with JG sheets by altering the number of reinforcing layers of JG reinforcement. Along with the experimental study, a numerical analysis is also performed in finite element (FE) software PLAXIS 2D to understand better the results obtained from the triaxial compression test.

2 Materials

The present study choses locally available silty sand with clay soil and JG sheet collected from the local market has been used. Table 1 shows the soil properties. The thickness of the JG sheet is 2 mm. The properties of JG are given in Table 2.

3 Experimental Work

Triaxial tests were implemented on soil samples strengthened with JG sheets at confining stress of 100, 150, and 200 kPa [2]. The sample diameter is 38 mm, and the height is 76 mm. The soil samples are prepared conforming to the Indian standard

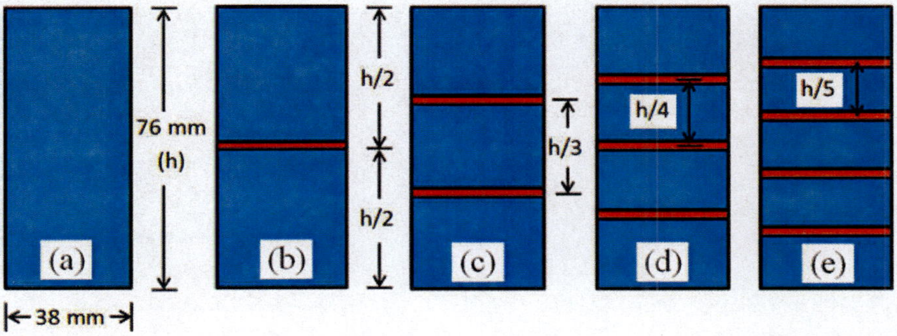

Fig. 1 Experimental triaxial test configurations: **a** without JG sheet, **b** single layer of JG sheet, **c** double layer of JG sheet, **d** triple layer of JG sheet, and **e** four layers of JG sheet

code [8]. All the soil samples for the experiment were made at constant dry density and moisture content. The samples are prepared at dry of optimum to acknowledge the actions of relatively dry soil, which can be used to construct geotechnical structures such as retaining walls, pavement constructions, etc. Different samples are reinforced by changing the number of JG layers and the spacing between the layers. Figure 1 shows the arrangement of the JG sheet in the soil specimen for the triaxial tests.

4 Numerical Study

In the current study, the two-dimensional FE analysis of the soil reinforcement using a JG sheet is carried out. FE-based software PLAXIS was used by a number of investigators for numerical analysis [9, 10], and thus was employed in the current study. The Mohr–Coulomb model has been successfully applied by several researchers to simulate the nonlinear stress–strain behavior of soil [11–14]. The FE program can be used in axis-symmetric modeling and plain strain modeling. An axis-symmetric analysis is implemented using the Mohr–Coulomb basis to model the nonlinear stress–strain behavior of soil and JG sheet. The FE analysis is done by 3-nodded triangular elements, and basic boundary conditions are used to model the soil reinforced with a JG sheet, as shown in Fig. 2. The vertical boundaries are consider free in the vertical direction and restricted in the horizontal direction, and the bottom boundary is restricted in both directions. Numerical studies are conducted on unreinforced and JG sheet-reinforced soil. Then, JG is applied at various positions of sub-grade soil. In all soil samples, material properties are assumed to be linearly elastic, and the magnitude of loading and the number of JG layers are varied. JG sheets are laid in four layers to evaluate the performance characteristics. The stress, strain, and displacement of the soil structure are considered as output, while the applied load at the top of the soil layer is input.

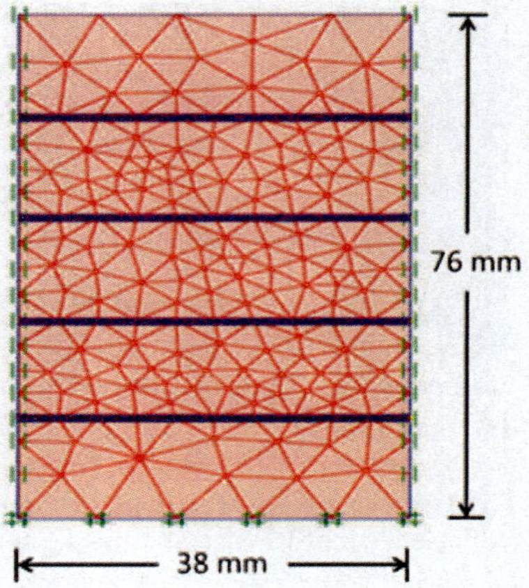

Fig. 2 Mesh pattern of soil sample reinforced with four layer of JG sheet

5 Results and Discussions

5.1 *Experimental Results*

Effect of JG Reinforcement on Deviator Stress. In the triaxial test, the pressures on the soil specimen are applied at a strain rate of 1.5% per minute up to the sample fails or up to a strain of 20%, whichever reaches earlier. Based on the experimental test results deviator stress versus strain curve for the soil specimens reinforced without and with the number of layers of JG sheets under different confining pressure determined as shown in Fig. 3. Figures depict that increasing the JG layer numbers increases the deviator stress significantly. Moreover, higher confining pressure results in a higher value of deviator stress.

Effect of JG Reinforcement on SS Parameters. The values of the SS parameter of soil reinforced with various arrangements of JG sheets are given in Table 3. Results show that there is a substantial effect of JG reinforcement on the SS parameters of soil. It is noticed that the maximum increase in the amount of c and ϕ values of soil is 76% and 46% respectively due to the insertion of four layers of JG sheets. It is to be stated here that the arrangement of reinforced soil samples (at constant dry density and moisture content) beyond four layers of JG sheet is not practicable; hence, the test results on soil reinforced with five or more layers of JG sheets are neglected. The rise in SS parameters of soil reinforced with JG sheet is because of the fact that the reinforcing elements interrelate with soil particles automatically by the surface friction and intermesh. The function of mesh or bond between the soil and geotextile is to transfer the pressure from the soil to the JG sheet by initiating the tensile strength of JG, which results in a decrease in strain and an increase in load-bearing capacity of soil specimen as indicated by the above figures. Therefore, it is noticed that the insertion of JG in the soil as a reinforcing element enhances the SS and load-bearing capacity of sub-grade soil.

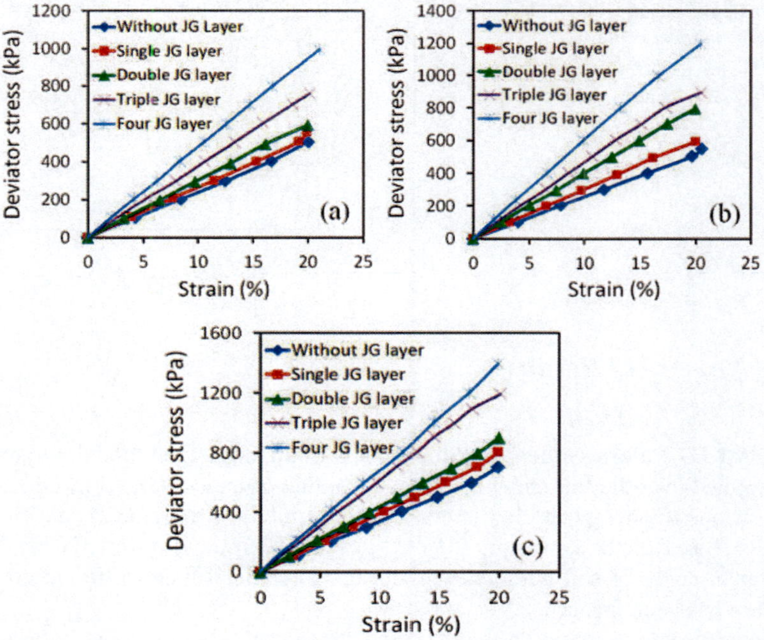

Fig. 3 Deviator stress versus strain curve for JG reinforced soil by varying number of layers at confining stress of **a** 100 kPa, **b** 150 kPa, and **c** 200 kPa

Table 3 SS parameters of soil reinforced with JG sheet

No. of layers	c (kN/m^2)	ϕ (Degree)	% increase (c)	% increase (ϕ)
0	54	26	–	–
1	65	28	20	8
2	68	30	26	15
3	70	35	30	35
4	95	38	76	46

Effect of JG Reinforcement on Strength Improvement. The strength improvement factor of reinforced soil specimens at a confining pressure of 200 kPa is shown in Fig. 4. The improvement factor represents the increase in strength. It was computed by dividing the reinforced soil sample's deviator stress by the unreinforced soil sample's deviator stress. Figure 4 depicts that the strength of the soil rises with the rise in the JG sheets. In four layers JG soil specimen, the soil's strength is two times that of soil without reinforcement.

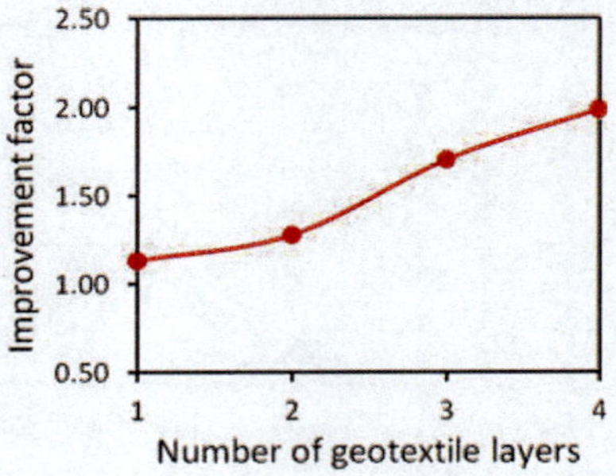

Fig. 4 Improvement factor versus the number of JG layers at confining pressure 200 kPa

5.2 *Numerical Results*

Effect of JG Reinforcement on Settlement. In the numerical model, the pressure was applied over the soil under the various confining stress (100, 150, and 200 kPa) with various JG sheets to realize better the results obtained from the triaxial compression test. Based on the data acquired from the triaxial compression test, the load versus settlement curve of soil with and without reinforcement under different confining pressure has been made.

Figure 5 indicates that the reduction in the amount of immediate settlement is found due to the reinforcement with the JG sheet. The settlement becomes lesser with the increased number of JG layers. In the present study, maximum reduction of settlement is achieved when four geotextile layers are applied. Figure 5c shows that the stress, strain, and displacement improvement due to JG is more significant if higher deviator stress is applied. The compressive strength and SS of soil is found to be increased with the increased number of JG sheets. The experimental and numerical investigation results clearly indicate that the sub-grade soil properties improved substantially due to the reinforcement with the JG sheet.

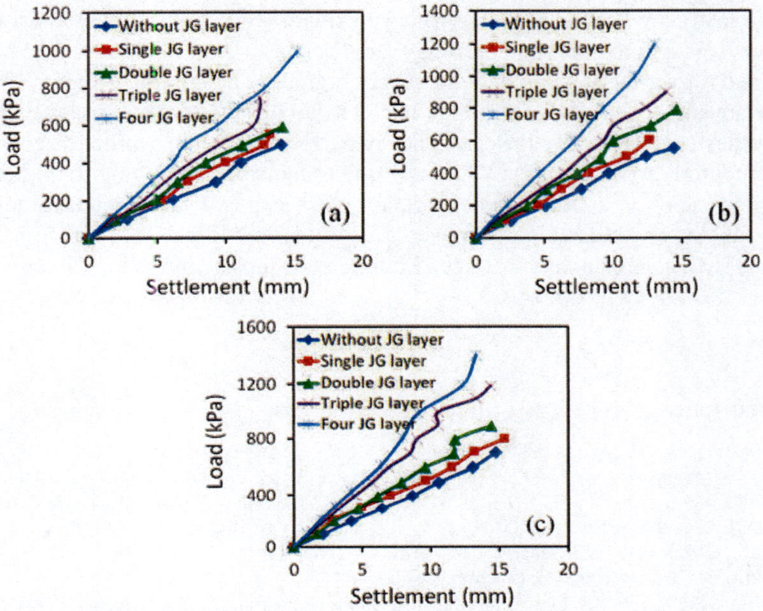

Fig. 5 Load versus settlement plot with the number of JG layers at confining stress of **a** 100 kPa, **b** 150 kPa, and **c** 200 kPa

6 Conclusion

The current study represents the results of soil reinforced with JG using a triaxial compression test and FE analysis. The following conclusions are made:

- After applying JG, the stress was distributed into a wider region of sub-grade soil. Hence, the sub-grade soil experienced lower stress, strain, and displacement.
- The SS parameters of sub-grade soil increase due to inserting a JG sheet in the soil. When the number of JG sheets is increased, then the strength parameters of the soil increase to a great extent. The highest value of an increase in c and ϕ values of soil are obtained for four layers of JG sheets.
- The strength of the soil increases with the increased number of JG layers. In four layers JG soil specimen, the soil's improvement factor is two times that of soil without reinforcement.
- From the current study, it is concluded that soil's bearing capacity increases and the quantity of instant settlement reduce when the soil is strengthened with JG sheets.
- Sub-grade soil reinforced with JG is an excellent geotechnical engineering material and can be applied in several geotechnical engineering implementations, such as reinforced earth retaining walls and reinforced embankments.

Jute fibers are used to control surface erosion of soil, embankment construction on weak soil, and road pavement strengthening and can be used to reduce the railway track and runway settlement built on poor sub-grade. Though jute geotextile has the above advantages, at the same time, it has some disadvantages also, such as jute geotextiles have relatively lower tensile strength compared to synthetic geotextiles, are susceptible to degradation when exposed to moisture, ultraviolet radiation, and biological activity, is prone to degradation when exposed to chemicals, including acids, alkalis, and some organic solvents, useful in low-impact or short-term applications, such as erosion control, landscaping, and slope stabilization in non-critical areas.

References

1. Saride S, Vedpathak S, Rayabharapu V (2014) Elasto-plastic behavior of jute-geocell reinforced sand subgrade. In Geo-congress 2014: geo-characterization and modeling for sustainability. ASCE, Georgia, pp 2911–2920
2. Latha GM, Murthy VS (2007) Effects of reinforcement form on the behavior of geosynthetic reinforced sand. Geotext Geomembr 25(1):23–32
3. Bera AK, Chandra SN, Ghosh A, Ghosh A (2009) Unconfined compressive strength of fly ash reinforced with jute geotextiles. Geotext Geomembr 27(5):391–398
4. Tuna SC, Altun S (2012) Mechanical behaviour of sand-geotextile interface. Scientia Iranica 19(4):1044–1051
5. Ranganathan SR (1994) Development and potential of jute geotextiles. Geotext Geomembr 13(6–7):421–433
6. Bahloul KM (2022) Utilization of sustainable material to improve the geotechnical properties of silty clay-experimental and numerical study. J Eng Res 6(5):56–63
7. Shekhawat P, Sharma G, Singh RM (2023) Strength characteristics of hazardous wastes flyash and eggshell powder-based geopolymer-stabilized soft soil cured at ambient temperature. Arab J Geosci 16(2):127
8. IS: 2720 (1993) Determination of the shear strength parameters of a specimen tested in unconsolidated undrained triaxial compression without the measurement of pore water pressure. Part 11, Bureau of Indian Standards
9. Debnath P, Debnath A, Pal SK (2022) A numerical study on laterally loaded single CFG pile embedded in layered soil. In: Proceedings of Indian geotechnical conference, Kochi
10. Koley S, Debnath A, Pal SK (2023) Impact of plastic waste on the 3D consolidation characteristics of sandy-silt with clay soil. In: Proceedings of International Conference on Interdisciplinary Approaches in Civil Engineering for Sustainable Development, IACESD-2023, Bengaluru, India
11. Debnath A, Pal SK (2023) Influence of surcharge strip loads on the behavior of cantilever sheet pile walls: a numerical study. J Eng Res 11(1):100029
12. Debnath A, Pal SK (2023) A numerical analysis on anchored sheet pile wall subjected to surcharge strip loading. J Eng Res: 100088
13. Debnath A, Pal SK (2022) Influence of surcharge strip loads on the behaviour of cantilever sheet pile walls: a numerical study by ABAQUS. In: Proceedings of Indian geotechnical conference, Kochi
14. Debnath A, Pal SK (2023) Behavior of cantilever concrete diaphragm wall under sequential excavation of front-fill soil: a numerical study. In: Earth retaining structures and stability analysis. Springer, Singapore, pp 395–407

Advanced Geophysics and Rock Mechanics

Application of Ground Penetrating Radar in Infrastructure Projects—A Simulation Approach

Sanjay Rana, Varun Narayan Mishra, and Praveen Kumar Rai

1 Introduction

Infrastructure projects require accurate information on subsurface conditions to ensure safe and efficient construction and maintenance [1]. Ground Penetrating Radar (GPR) has emerged as a reliable non-destructive geophysical method for subsurface investigation in these projects [2]. GPR provides high-resolution data on subsurface structures, voids, and utilities [3], aiding in the design, construction, and maintenance of roads, bridges, tunnels, and other infrastructure elements.

2 Basic Principles of Ground Penetrating Radar

The GPR method relies on the transmission of electromagnetic waves into the ground and the analysis of their reflections from subsurface features [1]. The two key components of GPR are electromagnetic theory and signal propagation [2]. This section discusses the basic principles of GPR, including electromagnetic theory [1], GPR signal propagation [3], and target reflection and detection [2].

S. Rana (✉) · V. N. Mishra
Amity Institute of Geoinformatics and Remote Sensing (AIGIRS), Amity University, Sector 125, Noida, India
e-mail: sanjay@parsan.biz

S. Rana
PARSAN Overseas Pvt. Ltd., New Delhi, India

P. K. Rai
K.M.C. Language University (U.P. State Govt. University), Lucknow, India

S. Kolathayar et al. (eds.), *Best Practices in Geotechnical and Pavement Engineering*, Lecture Notes in Civil Engineering 449, https://doi.org/10.1007/978-981-99-8505-0_29

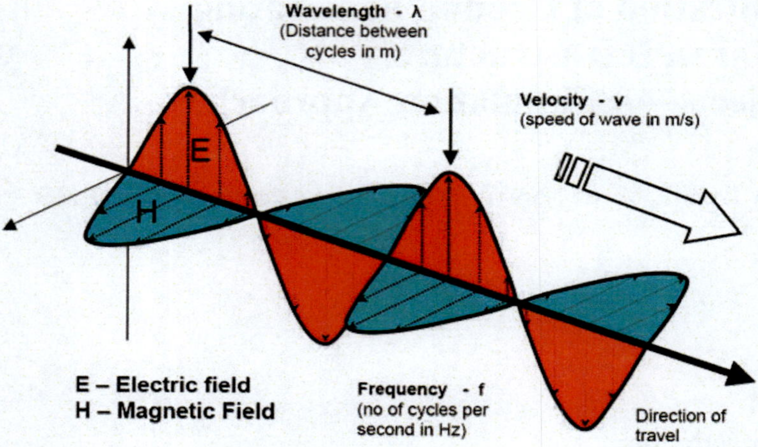

Fig. 1 Electromagnetic wave propagation in a material [4]

2.1 Electromagnetic Theory

GPR is a non-destructive geophysical technique using electromagnetic waves to explore subsurface structures [1]. Electromagnetic waves consist of oscillating electric and magnetic fields, with their propagation speed in materials dependent on dielectric properties [3]. At boundaries between materials with different dielectric properties, these waves partially transmit and reflect. Figure 1 represents the electromagnetic wave propagation in a material.

2.2 GPR System

A GPR system contains a transmitter, receiver, and antenna that emit, detect, and measure reflected signals, respectively.

2.3 Signal Propagation and Attenuation

Electromagnetic waves attenuate due to absorption, scattering, and dispersion [1]. Frequency affects attenuation and resolution. GPR signals can propagate through subsurface materials in several modes, i.e. Direct Wave, Reflected Wave, Refracted Wave, and Guided Wave.

2.4 Data Processing and Interpretation

Raw GPR data is processed and interpreted to infer subsurface features. Visualization techniques can be employed to better understand the spatial distribution and relationships between subsurface features, such as

- Profile view: Displaying the processed GPR data as a two-dimensional radargram, with distance along the survey line on the horizontal axis and depth or two-way travel time on the vertical axis.
- Time-slice or depth-slice view: Creating horizontal cross-sections through the GPR data volume at specific times or depths, providing a plan view of the subsurface features.
- 3D visualization: Combining multiple GPR profiles or time-slices to create a three-dimensional representation of the subsurface, which can be viewed and manipulated interactively.

2.5 GPR Target Reflection and Detection

Target reflection depends on dielectric contrast and geometry. Detection relies on identifying patterns in GPR data. Detection is influenced by frequency, depth, size, attenuation, and noise [5].

3 Applications of GPR in Infrastructure Projects

Ground Penetrating Radar (GPR) has proven to be a valuable tool in infrastructure projects due to its non-destructive nature, high-resolution imaging capabilities, and ability to detect subsurface features. The following sections discuss two key applications of GPR in infrastructure projects: road and bridge assessments.

3.1 Pavement Evaluation

GPR can be used to assess the thickness, layering, and overall condition of pavement structures, including asphalt, concrete, or composite pavements [6]. By detecting subsurface anomalies, such as voids, cracks, or moisture intrusion, GPR can provide valuable information for pavement maintenance, rehabilitation, and design decisions. Specific applications of GPR in pavement evaluation include

- Measurement of layer thickness: The use of GPR to measure thickness of various asphalt layers is already well established. Either air-coupled or ground-coupled antenna can be used for measurements. Air-coupled systems are non-contact, meaning they do not touch the surface during surveying, which makes them faster than ground-coupled systems. They are typically used for a quick scan of subsurface to indicate locations where in-depth testing with other systems, including ground-coupled GPR, is needed. Ground-coupled systems, in contrast, must remain in contact with the surface during surveying. GPR signals are reflected at layer interfaces where different dielectric constants exist within the asphalt pavements. The data collected are processed and analysed to measure the thickness of the asphalt pavement.

- Subgrade characterization: Over recent years, the application of GPR technology in evaluating pavement conditions has gained significant traction. Monitoring the state of the subgrade and tracking its changes over time offers crucial insights into the changes linked to the degradation of the pavement. This, in turn, aids in the proactive forecasting of upcoming road maintenance needs. By leveraging the electromagnetic theory, a model can be formulated to deduce the density of subgrade soils. This takes into account the dielectric and volumetric characteristics of the primary soil constituents: solid particles, water, and air. Such an approach enhances the prediction of not just surface wear and tear but also potential threats to the underlying pavement infrastructure.
- Void Detection: Cavities can form under pavements due to a variety of factors, including water line ruptures, sewer failures, settling of inadequately compacted materials, soil degradation into sinkholes, or erosion from rainwater. Such cavities can lead to pavement issues, from minor fissures to severe breakdowns. GPR offers an efficient and economical solution to detect these underlying cavities. The information gathered through GPR can be analysed as individual profiles or combined from several parallel scans to produce three-dimensional data sets. These data clusters can be inspected from various angles to visualize the horizontal spread and occasionally the depth of the anticipated cavities. Regular GPR assessments aid in pinpointing specific zones requiring intervention before the cavities expand, leading to increased repair costs.
- Moisture Detection: Evaluating the state of the subgrade and tracking its time-based fluctuations offers crucial insights into shifts linked to the degradation of the pavement. A technique has been formulated to deduce the density and moisture levels of subgrade soils utilizing a 2 GHz GPR scanning system combined with a straightforward exponential model. This technique employs the principles of electromagnetic mixing theory to reverse compute the density of subgrade soils. The findings indicate that using this non-invasive approach to predict soil density and rigidity aids in the proactive forecasting of both surface wear and potential threats to the underlying pavement layers. This technology proves highly efficient in pinpointing zones of moisture penetration, facilitating focused maintenance and remediation actions [7].
- Identification of buried utilities: Utility mapping presents an expanding avenue for many surveying enterprises. While tools like radio frequency (RF) pipe and cable detectors, along with electromagnetic locators, are prevalent, their limitation is in detecting only conductive (metallic) entities. There's a growing demand in utility surveys to identify, pinpoint, and trace subterranean non-metallic pipes and

cables. In such scenarios, GPR emerges as the preferred technology. As the GPR traverses the terrain, it records its location, and the integrated software structures the incoming signal data. This information is showcased in a radargram for the user's inspection and can be subsequently exported for in-depth analysis. Advances in GPR technology have made detection more accurate and more accessible. Details of this application are presented in Sect. 3.3.

3.2 Bridge Deck Condition Assessment

GPR is a valuable tool for assessing the condition of bridge decks and identifying potential issues that may impact their structural integrity [6]. GPR can be used to evaluate the condition of concrete, asphalt, or composite bridge decks, and detect the presence of reinforcement, prestressing tendons, or embedded utilities. Key applications of GPR in bridge deck condition assessment include

- Concrete deterioration assessment: GPR can identify areas of concrete deterioration, such as cracking, delamination, or spalling, by detecting variations in dielectric properties or changes in signal amplitude and continuity.
- Reinforcement and tendon mapping: GPR can locate and map the position of reinforcing steel, prestressing tendons, or embedded utilities within the bridge deck, providing essential information for structural evaluation, maintenance, and repair planning [1].
- Detection of moisture ingress and corrosion: GPR can detect areas of moisture ingress or corrosion in bridge decks by identifying changes in dielectric properties or signal attenuation, allowing for targeted maintenance and repair efforts.
- Evaluation of bridge deck overlays: GPR can assess the thickness and condition of bridge deck overlays, such as asphalt or protective coatings, providing information for maintenance planning and overlay design. Figure 2 represents the map showing bridge deck deterioration.

GPR plays a vital role in infrastructure projects, particularly in the assessment and evaluation of roadways and bridges. By providing detailed information about

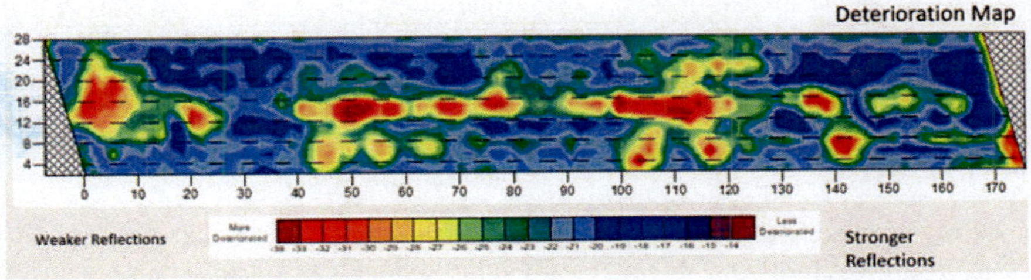

Fig. 2 Bridge deck deterioration map—red shows highest deterioration while blue shows least deterioration [8]

pavement structures, subgrade properties, and bridge deck conditions, GPR enables informed decision-making for maintenance, rehabilitation, and design, ultimately contributing to the safety and longevity of infrastructure assets.

3.3 Application of GPR in Utility Mapping

Ground Penetrating Radar (GPR) is a powerful tool for utility mapping and locating subsurface features, offering a non-destructive, high-resolution, and efficient method for detecting and mapping buried utilities and objects. The following sections discuss two key applications of GPR in utility mapping: locating underground utilities and mapping buried objects.

3.3.1 Locating Underground Utilities

GPR can be used to locate and map various underground utilities, such as water, gas, and sewer lines, electrical conduits, telecommunication cables, and other subsurface infrastructure elements. By detecting the presence of these utilities, GPR can help prevent accidental damage during excavation or construction activities, ensure compliance with safety regulations, and facilitate maintenance and repair planning. Specific applications of GPR in locating underground utilities include

- Detection of non-metallic utilities: GPR can detect non-metallic utilities, such as plastic or fibre-optic cables, which may not be detectable using traditional metal detectors or electromagnetic locators.
- Depth estimation: GPR can provide approximate depth information for buried utilities, assisting in excavation planning and avoiding potential damage to the utilities.
- Utility mapping and classification: GPR can help create comprehensive utility maps by identifying and classifying different types of buried utilities based on their geometry, size, and material properties.
- Leak detection: GPR can be used to detect subsurface water leaks in pipelines or sewer systems by identifying areas of high moisture content or changes in soil properties around the utilities. Figure 3 represents loss of signal due to increased conductivity caused by water saturation.

3.3.2 Mapping of Buried Objects

GPR can be used to map various buried objects, such as storage tanks, underground structures, or archaeological features, providing valuable information for site investigation, risk assessment, and land development planning [1]. Specific applications of GPR in mapping buried objects include

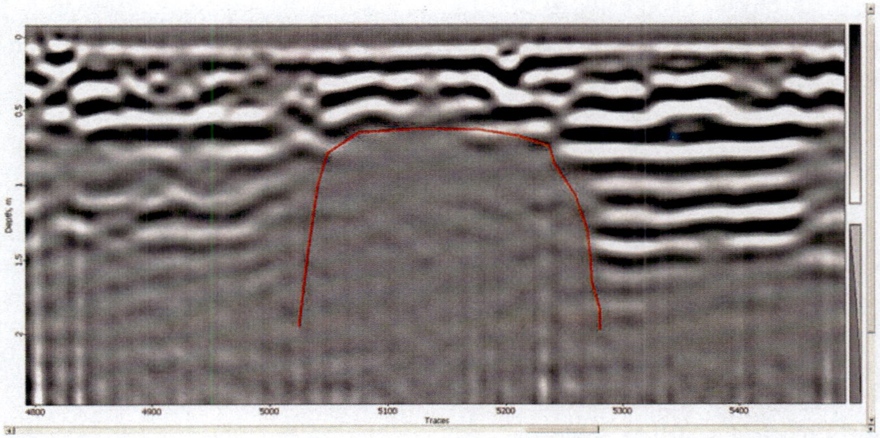

Fig. 3 Loss of signal due to increased conductivity caused by water saturation [9]

- Detection of underground storage tanks (USTs): GPR can locate buried USTs, such as fuel or chemical storage tanks, and assess their condition, helping to identify potential environmental hazards or compliance issues.
- Archaeological investigations: GPR can detect buried archaeological features, such as foundations, walls, or artefacts, providing non-invasive and efficient means for archaeological site investigation and preservation.
- Landfill and waste disposal site mapping: GPR can map the extent and composition of landfill or waste disposal sites, helping to assess their environmental impact and plan remediation efforts.
- Geotechnical investigations: GPR can identify buried geological features, such as bedrock, soil layers, or cavities, providing essential information for geotechnical site investigation and engineering design [5].

GPR offers significant advantages in utility mapping and the detection of buried objects, enabling the efficient and accurate location of underground utilities, mapping of subsurface infrastructure, and identification of buried objects for various applications. By providing detailed information about the subsurface environment, GPR allows for better planning, risk assessment, and decision-making in land development, construction, and maintenance projects.

3.4 Application of GPR in Tunnel Investigations

Ground Penetrating Radar (GPR) has become an invaluable tool for tunnel investigations, offering a non-destructive, high-resolution, and efficient method for detecting and mapping subsurface features related to tunnel construction and maintenance.

The following sections discuss two key applications of GPR in tunnel investigations: pre-construction site investigations and tunnel lining and support structure inspection.

3.4.1 Pre-construction Site Investigations

GPR can be used to conduct pre-construction site investigations for tunnels, providing essential information about subsurface geological and hydrogeological conditions, as well as the presence of any utilities or buried structures. By obtaining detailed information about the subsurface environment, GPR can help minimize construction risks, optimize tunnel design, and reduce costs. Specific applications of GPR in pre-construction site investigations include

- Geological mapping: GPR can identify subsurface geological features, such as bedrock, soil layers, or fractures, providing crucial information for tunnel alignment, design, and construction planning.
- Hydrogeological assessments: GPR can detect the presence of groundwater, as well as the geometry and properties of aquifers, aiding in the assessment of potential tunnel dewatering requirements and groundwater management strategies [10].
- Detection of buried utilities and structures: GPR can locate, and map buried utilities or structures, such as pipelines, cables, or foundations, helping to avoid potential conflicts or damage during tunnel construction.
- Archaeological and environmental assessments: GPR can identify buried archaeological features or environmentally sensitive areas, facilitating the preservation of cultural heritage and environmental protection during tunnel construction [2].

3.4.2 Tunnel Lining and Support Structure Inspection

GPR can be employed to inspect the condition and integrity of tunnel lining and support structures, such as shotcrete, concrete, or steel reinforcements. By detecting subsurface anomalies or defects, GPR can provide valuable information for maintenance, repair, and safety assessments of tunnels [11]. Specific applications of GPR in tunnel lining and support structure inspection include

- Concrete thickness measurement: GPR can accurately measure the thickness of tunnel lining materials, such as shotcrete or concrete, providing essential data for quality control and structural evaluation.
- Reinforcement and tendon mapping: GPR can locate and map the position of reinforcing steel, prestressing tendons, or other embedded elements within tunnel linings, aiding in structural assessment and maintenance planning.
- Detection of delamination, cracking, or voids: GPR can identify areas of tunnel lining deterioration, such as delamination, cracking, or voids, by detecting changes

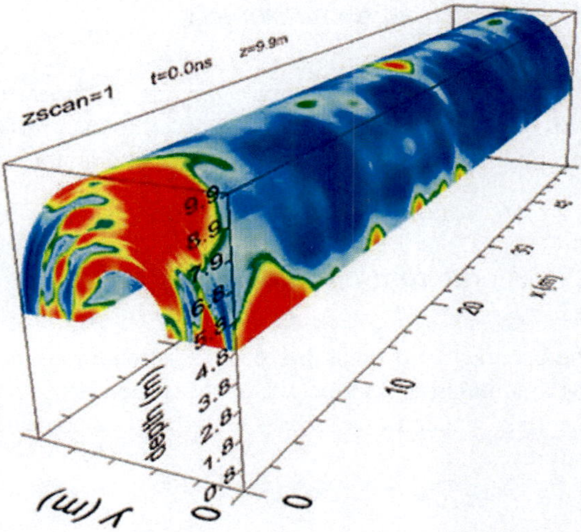

Fig. 4 3D GPR survey of tunnel to detect weak/ anomalous zones behind the lining [12]

in dielectric properties or signal amplitude and continuity, allowing for targeted maintenance and repair efforts.

- Assessment of ground support systems: GPR can evaluate the condition and effectiveness of ground support systems, such as rock bolts or grouting, by identifying variations in material properties or signal reflections from support elements. Figure 4 represents 3D GPR survey of tunnel to detect weak/ anomalous zones behind the lining.

GPR plays a significant role in tunnel investigations, contributing to pre-construction site assessments and the inspection of tunnel lining and support structures. By providing detailed information about subsurface geological and hydrogeological conditions, as well as the integrity and condition of tunnel infrastructure, GPR enables informed decision-making for tunnel design, construction, maintenance, and safety management.

4 Simulation Approach in GPR

Simulation is a valuable approach in Ground Penetrating Radar (GPR) research and applications, as it can help researchers and practitioners understand the complex electromagnetic interactions between GPR signals and subsurface features, optimize survey designs, and develop new data processing and interpretation techniques [13]. In the context of GPR, simulation involves the use of mathematical models and numerical algorithms to replicate the propagation, reflection, and scattering of electromagnetic waves in the subsurface environment [13]. The following sections provide an overview of the key aspects of the simulation approach in GPR:

4.1 Electromagnetic Models

The foundation of GPR simulation lies in the formulation of accurate and efficient electromagnetic models that describe the propagation of GPR signals through different subsurface materials and their interaction with subsurface features [1].

4.2 Simulation Parameters

The accuracy and reliability of GPR simulations depend on the correct definition of the simulation parameters, such as the GPR system properties, the subsurface materials' dielectric properties, and the geometry and dimensions of the subsurface features.

4.3 GPR Data Synthesis

Using the electromagnetic models and simulation parameters, GPR data can be synthesized by solving the governing equations and calculating the GPR signals' propagation and reflection in the subsurface environment. The synthesized GPR data can be used to investigate the effects of various factors on GPR signal behaviour, such as the antenna frequency, the subsurface material properties, or the geometry and orientation of the subsurface features [1].

4.4 Simulation Applications

GPR simulations can be used for various purposes, such as

- Survey design optimization: GPR simulations can help practitioners optimize survey designs by evaluating the effects of different GPR system parameters, survey geometries, or subsurface conditions on the GPR data quality and interpretability [10].
- Data processing and interpretation development: GPR simulations can aid in the development and testing of new data processing and interpretation techniques by providing controlled and synthetic datasets with known subsurface features and properties [10].
- Educational and training purposes: GPR simulations can serve as valuable educational and training tools, helping students and professionals gain a better understanding of GPR principles, system behaviour, and data interpretation [10].

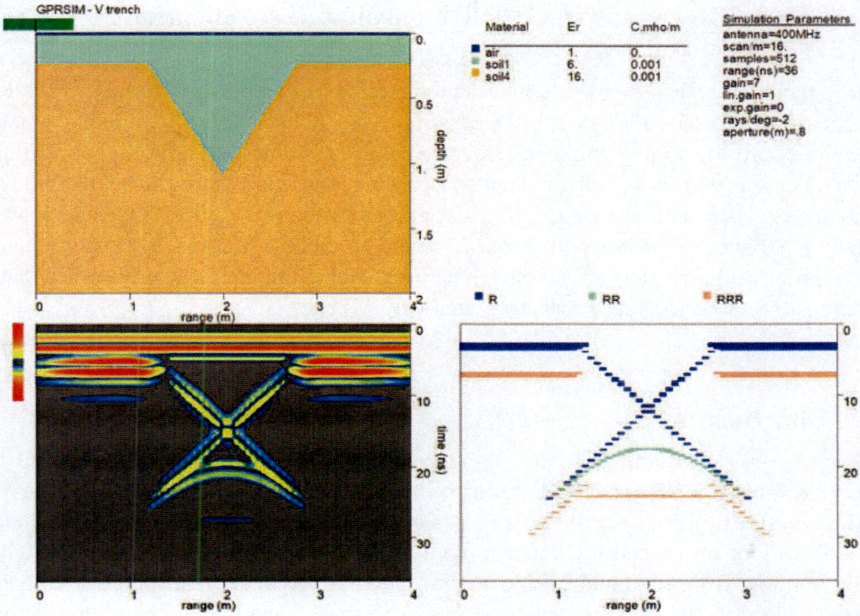

Fig. 5 Example showing importance of simulation [14]

The simulation approach in GPR plays a crucial role in advancing our understanding of GPR signal behaviour, optimizing survey designs, and developing new data processing and interpretation techniques. Figure 5 shows an example of the importance of simulation. A V trench gives a hyperbolic response which can be misinterpreted as a buried pipe in the absence of this understanding through simulation [10]

5 Challenges in GPR Applications

GPR applications face challenges such as signal attenuation, complex subsurface conditions, and data processing and interpretation. Signal attenuation depends on material conductivity, antenna frequency, and scattering from subsurface heterogeneities. Complex conditions arise from overlapping reflections, anisotropy, inhomogeneity, and interfering signals. Data processing challenges include selecting appropriate techniques, handling large data volumes, and accurate interpretation. Addressing these challenges optimizes GPR effectiveness and reliability.

6 Future Directions in GPR Research and Applications

GPR technology has evolved significantly and has potential for further advancements in advanced GPR systems, integration with other geophysical methods, and machine learning and AI applications. Future GPR systems may offer higher resolution, deeper penetration, real-time data processing, and multi-frequency capabilities. Integrating GPR with other geophysical methods can improve subsurface characterization, enhance detection, and increase survey efficiency. ML and AI can automate data processing, extract, and classify features, and integrate GPR with other data sources for more accurate subsurface analysis.

7 Conclusion

Ground Penetrating Radar (GPR) has proven to be an indispensable tool in the field of infrastructure projects, offering non-destructive, high-resolution, and efficient means for subsurface investigations. The applications of GPR in infrastructure projects are vast, ranging from road and bridge assessments, utility mapping, tunnel investigations, and more. Its ability to detect and map subsurface features, such as buried utilities, geological layers, and structural elements, has enabled better decision-making, risk assessment, and project planning in various civil engineering and construction projects. Despite the challenges associated with signal attenuation, complex subsurface conditions, and data processing and interpretation, GPR technology continues to evolve, with ongoing research and development paving the way for advanced GPR systems, integration with other geophysical methods, and the application of machine learning and artificial intelligence techniques. In conclusion, the application of Ground Penetrating Radar in infrastructure projects holds immense potential for improving the quality, safety, and efficiency of civil engineering and construction works. By leveraging the advancements in GPR technology and overcoming the associated challenges, practitioners can continue to harness the power of GPR for a wide range of applications, ultimately contributing to the sustainable development of infrastructure and the betterment of society.

References

1. Daniels DJ (2004) Ground penetrating radar, 2nd edn. The Institution of Electrical Engineers
2. Conyers LB (2012) Interpreting ground-penetrating radar for archaeology. Routledge, London
3. Jol HM (ed) (2009) Ground penetrating radar theory and applications. Elsevier, Amsterdam
4. Rana, Sanjay. Ground penetrating radar (under water) for condition assessment of stilling basins. https://www.parsan.biz/
5. Neal A (2004) Ground-penetrating radar and its use in sedimentology: principles, problems and progress. Earth Sci Rev 66(3–4):261–330

6. Alani AM, Banks K (2013) Applications of ground penetrating radar (GPR) in bridge deck monitoring and assessment. J Appl Geophys 97:45–54
7. Abdelmawla A, Kim SS (2020) Application of ground penetrating radar to estimate subgrade soil density. Infrastructures 5:12
8. Website https://insitutek.com
9. Rana, Sanjay. Ground penetrating radar survey for water line leakage detection in Bangalore, India, Project Report
10. Gómez-Ortiz D, Martínez-Moreno FJ, Grandjean G, Martín-Crespo T (2015) Ground penetrating radar for tunnel detection in challenging environments. J Appl Geophys 123:1–14
11. Annan AP (2005) GPR methods for hydrogeologic studies. In: ground penetrating radar: theory and applications. Elsevier, pp 185–218
12. Rana, Sanjay. Ground penetrating radar survey for tunnel lining defect investigation, Project Report
13. Giannopoulos A (2005) Modelling ground penetrating radar by GprMax. Constr Build Mater 19(10):755–762
14. Goodman D (2003) GPR-SIM: a simulation tool for ground-penetrating radar. In: Proceedings of the 11th international conference on ground penetrating radar (GPR'2003), pp 1–5

Applications of Geophysics in Structural and Geotechnical Engineering

Sanjay Rana, Varun Narayan Mishra, and Praveen Kumar Rai

1 Introduction

1.1 Background and Motivation

Geophysical methods have long been used in the exploration and characterization of Earth's subsurface, primarily in the fields of petroleum, mineral, and water resources [1]. In recent years, however, there has been a growing interest in applying geophysical techniques to the domains of structural and geotechnical engineering. This increased interest is driven by the need for non-destructive, cost-effective, and efficient tools for subsurface investigation, structural health monitoring, and site characterization. The application of geophysics in these fields has the potential to provide valuable information about the subsurface conditions, material properties, and structural integrity, leading to better design, construction, and maintenance of civil infrastructure [2].

S. Rana (✉) · V. N. Mishra
Amity Institute of Geoinformatics and Remote Sensing (AIGIRS), Amity University, Sector 125, Noida, India
e-mail: sanjay@parsan.biz

S. Rana
PARSAN Overseas Pvt. Ltd., New Delhi, India

P. K. Rai
K.M.C. Language University (U.P. State Govt. University), Lucknow, India

S. Kolathayar et al. (eds.), *Best Practices in Geotechnical and Pavement Engineering*, Lecture Notes in Civil Engineering 449, https://doi.org/10.1007/978-981-99-8505-0_30

1.2 Scope of the Paper

This technical paper aims to provide an overview of the applications of geophysics in structural and geotechnical engineering. The focus will be on the most promising geophysical methods and techniques, their principles, and how they can be used to address specific engineering challenges. The paper will also discuss the limitations and challenges associated with these methods and offer insights into future research directions and potential advancements in the field.

2 Overview of Geophysical Methods in Structural and Geotechnical Engineering

Geophysical methods have been widely used in various applications, including structural and geotechnical engineering, to investigate subsurface properties and conditions. This section provides an overview of the most common geophysical methods employed in these fields.

2.1 Seismic Methods

Seismic methods rely on the propagation of elastic waves through the subsurface, and the measurement of the travel time, amplitude, and frequency of these waves to infer subsurface properties [3].

Refraction and reflection. Seismic refraction and reflection methods involve the generation of seismic waves using a source (e.g. a hammer or an explosive charge) and the recording of the waves after they have travelled through the subsurface using an array of receivers (e.g. geophones or accelerometers). In refraction surveys, the travel time of the direct and refracted waves is used to estimate the velocities of subsurface layers and the depth to subsurface interfaces. In reflection surveys, the travel time and amplitude of reflected waves are used to map subsurface interfaces and estimate the properties of subsurface materials. Figures 1 and 2 represent the schematic representation of seismic refraction and seismic reflection method, respectively.

Surface wave analysis. Surface wave analysis techniques, such as Multichannel Analysis of Surface Waves (MASW) and Spectral Analysis of Surface Waves (SASW), are used to measure the dispersion of Rayleigh or Love waves, which propagate along the ground surface [5]. Figure 3 represents the typical MASW set-up. The dispersion data are used to estimate the shear-wave velocity profile of the subsurface, providing valuable information about the stiffness and dynamic properties of soils and rocks.

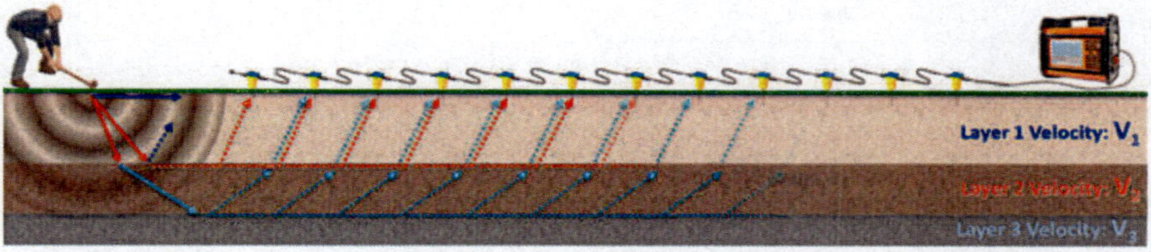

Fig. 1 Refraction seismic [4]

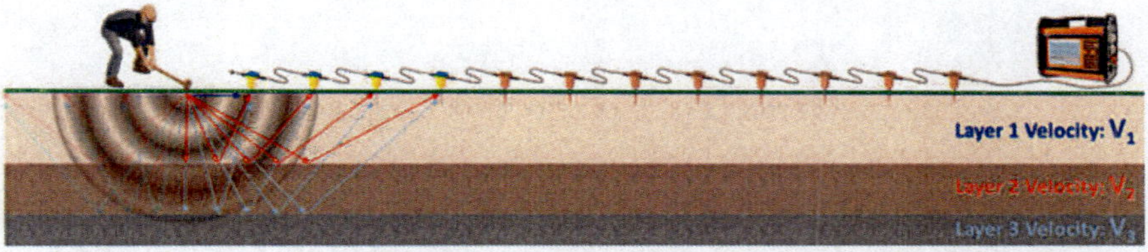

Fig. 2 Reflection seismic [4]

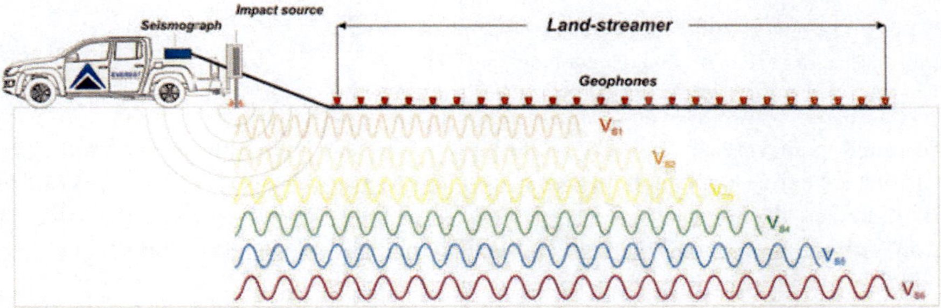

Fig. 3 MASW set-up and principle [6]

2.2 *Electrical Methods*

Electrical methods are based on the measurement of the subsurface's electrical properties, such as resistivity and conductivity, which can be related to subsurface material composition, porosity, and fluid content [1].

Electrical resistivity tomography. Electrical resistivity tomography (ERT) involves injecting a current into the ground using a pair of electrodes and measuring the resulting potential differences at the surface using another pair of electrodes. The acquired data are then used to generate a 2D or 3D image of the subsurface resistivity distribution, which can be related to the geotechnical properties of soils and rocks [7]. Figure 4 provides basic schematic of ERT set-up.

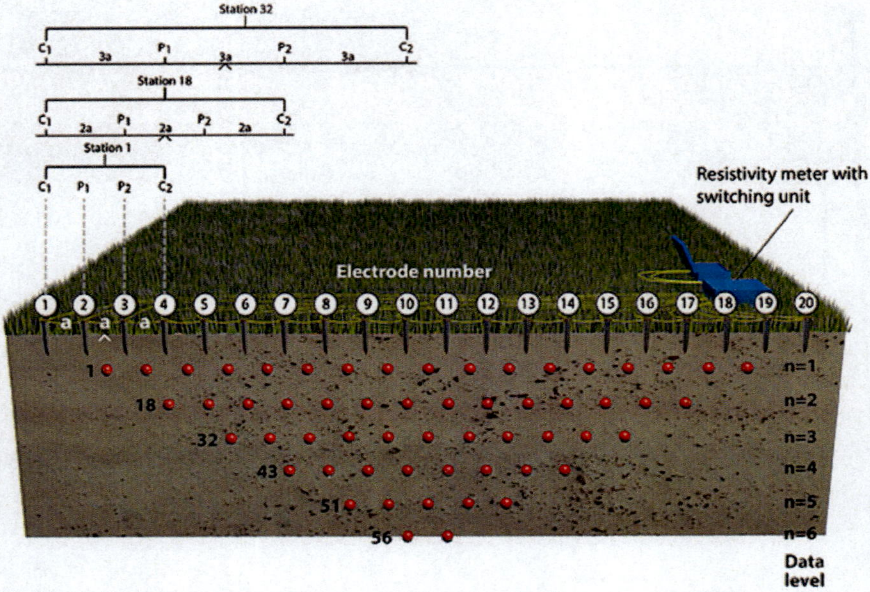

Fig. 4 Electrical resistivity imaging/tomography [7]

Induced polarization. Induced polarization (IP) is a technique that measures the subsurface's time-dependent response to an applied electric field [1]. The IP response is related to the presence of chargeable minerals, clay content, and fluid saturation, providing information about subsurface material composition and groundwater conditions.

2.3 *Electromagnetic Methods*

Electromagnetic methods use the interaction between electromagnetic fields and subsurface materials to estimate the subsurface's electrical and magnetic properties [1].

Ground-penetrating radar. Ground-penetrating radar (GPR) involves the transmission of high-frequency electromagnetic waves into the ground and the recording of the reflected signals [8]. GPR can provide high-resolution images of the subsurface and is particularly useful for detecting and mapping shallow subsurface features, such as utilities, tunnels, and voids.

Frequency-domain electromagnetics. Frequency-domain electromagnetic (FDEM) methods involve the generation of a primary electromagnetic field using a transmitter coil and the measurement of the secondary electromagnetic field induced in the subsurface using a receiver coil [1]. The amplitude and phase of the secondary field are related to the subsurface's conductivity and magnetic susceptibility, providing information about subsurface material properties and groundwater conditions.

2.4 *Gravity and Magnetic Methods*

Gravity and magnetic methods are passive geophysical techniques that measure the Earth's gravitational and magnetic fields, respectively [1]. These methods can be used to detect subsurface features associated with variations in density (gravity) or magnetic properties (magnetic). Applications in structural and geotechnical engineering include mapping bedrock topography, fault zones, and subsurface voids, as well as estimating overburden thickness, which can be useful for site characterization and foundation design.

3 Applications in Structural Engineering

Advancements in geophysical methods have led to applications in structural engineering. This section elaborates on some of these innovative uses, which contribute to improved assessment, monitoring, and understanding of the behaviour of various structures.

3.1 *Assessing the Integrity of Concrete Structures*

Geophysical methods, such as ground-penetrating radar (GPR), electrical resistivity tomography (ERT), and ultrasonic techniques, are being used to assess the integrity of concrete structures. These non-destructive methods can help identify issues such as corrosion of reinforcing steel, internal voids, and delamination. By detecting these issues early, engineers can perform targeted maintenance and repair, prolonging the service life of the structures. Figure 5 shows GPR profile of deteriorated and corroded sections of reinforcement in a concrete slab.

3.2 *Monitoring the Health of Bridges and Dams*

Geophysical methods can be used to monitor the health of critical structures like bridges and dams. For instance, seismic methods can be employed to assess the dynamic response of these structures and detect any changes in their natural frequencies, which may indicate damage or degradation. In addition, continuous monitoring

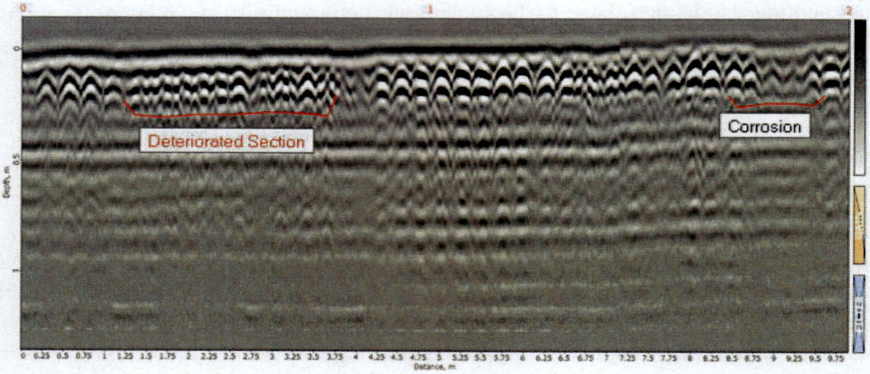

Fig. 5 GPR profile showing deteriorated and corroded sections of reinforcement in a concrete slab [9]

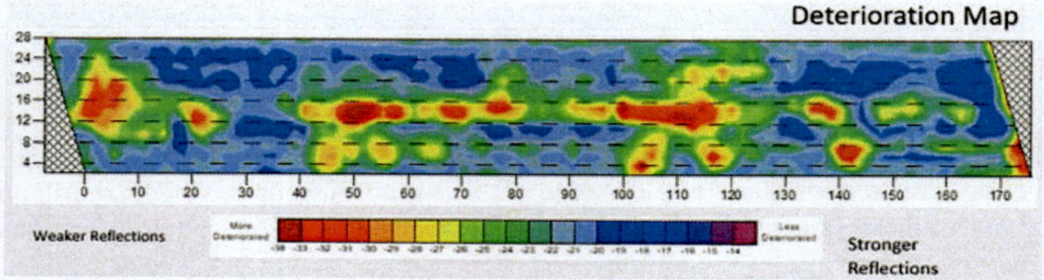

Fig. 6 Bridge deck deterioration map—red shows highest deterioration while blue shows least deterioration [10]

systems using fibre-optic sensors, strain gauges, or other sensing technologies can be integrated with geophysical methods to provide real-time data on the structural health of these vital assets. Figure 6 shows the map of bridge deck deterioration.

3.3 Detecting and Characterizing Cracks and Defects

Detecting and characterizing cracks and defects in structural elements is critical for maintaining their safety and performance. Geophysical techniques, such as acoustic emission, ultrasonic testing, and GPR, can be employed to identify the presence, location, and extent of cracks and defects in various materials, including concrete, steel, and masonry. This information can be used to guide maintenance and repair efforts, ensuring the continued safety and functionality of structures.

3.4 Investigating the Dynamic Behaviour of Structures

Understanding the dynamic behaviour of structures is essential for their design, analysis, and maintenance. Geophysical methods, such as seismic and microtremor techniques, can be used to determine the natural frequencies, mode shapes, and damping ratios of structures. This information can be used to calibrate numerical models, assess structural performance under dynamic loads (e.g. earthquakes and wind), and guide the development of mitigation strategies.

3.5 Evaluating the Impact of Natural Hazards on Structures

Natural hazards, such as earthquakes, landslides, and flooding, can have severe consequences for the built environment. Geophysical methods can be employed to evaluate the potential impact of these hazards on structures. For example, seismic site characterization using surface wave analysis can help determine the site's amplification potential during earthquakes, which can influence the design of structures in seismic-prone areas [5]. Similarly, ERT and GPR can be used to identify areas susceptible to landslides or flooding, allowing for the implementation of appropriate mitigation measures during the design and construction of structures.

These applications of geophysical methods in structural engineering demonstrate the potential for improved assessment, monitoring, and understanding of the behaviour of various structures, leading to safer, more resilient, and efficient built environments.

4 Applications in Geotechnical Engineering

Geophysical methods have found numerous applications in geotechnical engineering, contributing to a better understanding of subsurface conditions and properties. This section elaborates on some of these innovative uses, which enhance site characterization, subsurface feature mapping, and the evaluation of various geotechnical parameters.

4.1 Site Characterization and Ground Model Development

Geophysical methods, such as seismic refraction, ERT, and GPR, can be employed in site characterization and ground model development. These non-destructive techniques provide valuable data on the subsurface stratigraphy, depth to bedrock, groundwater table, and geotechnical properties of soils and rocks. Combining

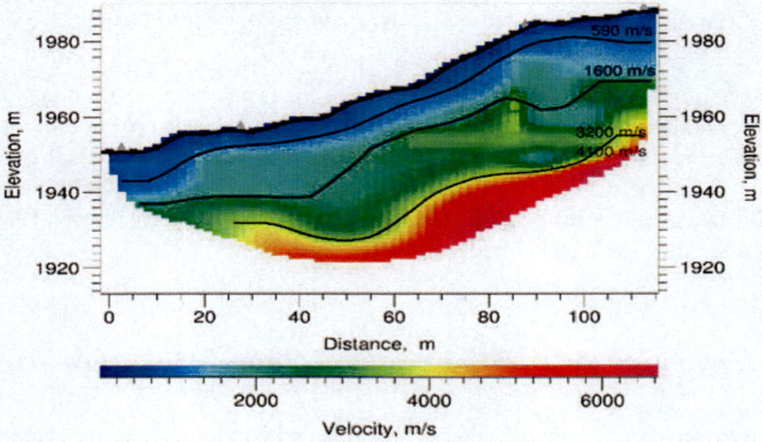

Fig. 7 Seismic refraction tomography showing various subsurface layers [11]

these data with traditional geotechnical investigation methods, such as drilling and sampling, allows for the development of more accurate and detailed ground models. Figure 7 shows the seismic refraction tomography.

4.2 *Identification and Mapping of Subsurface Features*

Geophysical techniques can be used to identify and map various subsurface features, such as voids, faults, sinkholes, and underground utilities. GPR, ERT, and magnetic methods are particularly useful in this regard, as they can provide high-resolution images of the subsurface, helping engineers and geotechnical professionals plan and design infrastructure projects more effectively. Figure 8 represents the 2D electrical resistivity imaging/tomography section.

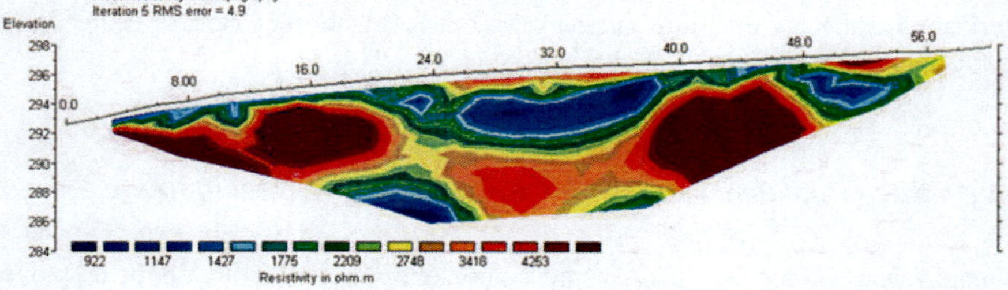

Fig. 8 2D Electrical resistivity imaging/tomography section showing a cavity (blue colour low resistivity zone between 24 and 40 m) at proposed bridge pier location [12]

4.3 *Landslide and Slope Stability Assessment*

Geophysical methods, such as ERT, seismic refraction, and surface wave analysis, can be employed to assess landslide hazards and slope stability. These techniques provide information on the subsurface structure, soil and rock properties, and groundwater conditions, all of which are critical factors in slope stability analyses [13]. By identifying potential areas of instability, geotechnical professionals can develop appropriate mitigation measures to reduce the risk of landslides.

4.4 *Liquefaction Potential Analysis*

Seismic methods, such as shear-wave velocity measurements, can be used to evaluate the liquefaction potential of soils during earthquakes. Liquefaction, a phenomenon in which saturated soils lose their strength and stiffness due to pore pressure increase during seismic events, can cause significant damage to structures and infrastructure. By determining the in situ shear-wave velocity of soils, engineers can assess the liquefaction susceptibility and develop appropriate design and construction strategies to mitigate potential risks.

4.5 *Evaluating the Geotechnical Properties of Soils and Rocks*

Geophysical methods, such as seismic refraction, surface wave analysis, and ERT, can provide valuable insights into the geotechnical properties of soils and rocks, including their stiffness, strength, and permeability. This information can be used to optimize foundation design, assess the suitability of materials for construction purposes, and predict the behaviour of soils and rocks under various loading conditions [14].

These applications of geophysical methods in geotechnical engineering demonstrate their potential for improving our understanding of subsurface conditions and properties, leading to better design, construction, and management of infrastructure projects.

5 Challenges and Limitations

While geophysical methods offer numerous advantages and applications in structural and geotechnical engineering, they also face certain challenges and limitations. This section discusses some of these issues, including data acquisition and quality, integration of methods, site-specific limitations, and data interpretation and uncertainty.

5.1 Data Acquisition and Quality

Acquiring high-quality geophysical data can be challenging, particularly in areas with complex subsurface conditions or in urban environments. Factors such as background noise, signal attenuation, and interference from nearby structures can affect the quality of acquired data. Additionally, the choice of appropriate survey parameters, such as electrode spacing, transmitter–receiver separation, and sampling frequency, is crucial to ensuring data quality and resolving the desired subsurface features.

5.2 Integration of Geophysical Methods

Each geophysical method provides information on different subsurface properties, and their integration can enhance the overall understanding of subsurface conditions. However, the integration of geophysical methods can be challenging due to differences in data formats, resolutions, and scales. Developing standardized workflows and data processing techniques to combine and interpret multi-method geophysical datasets is crucial for effective integration.

5.3 Site-Specific Limitations

Geophysical methods may face site-specific limitations that can affect their applicability and effectiveness. For example, certain methods may be less effective in areas with high subsurface heterogeneity, low contrast between target and surrounding materials, or where access is restricted. In such cases, a combination of geophysical methods or the use of alternative investigation techniques may be necessary to overcome these limitations.

5.4 Data Interpretation and Uncertainty

Interpreting geophysical data can be challenging due to the non-uniqueness of the inverse problem, which means that multiple subsurface models can produce similar geophysical responses. This non-uniqueness can lead to uncertainties in the interpretation of subsurface properties and features. Incorporating additional constraints, such as geotechnical data from boreholes or laboratory tests, can help reduce these uncertainties and improve the reliability of geophysical interpretations.

Despite these challenges and limitations, geophysical methods continue to play a vital role in structural and geotechnical engineering. Ongoing research and development efforts aim to address these issues and further enhance the applicability and effectiveness of geophysical methods in these fields.

6 Conclusions

In conclusion, geophysical methods have shown great potential for applications in structural and geotechnical engineering, providing valuable insights into subsurface conditions and properties. The integration of various geophysical techniques, such as seismic, electrical, electromagnetic, gravity, and magnetic methods, can enhance the overall understanding of subsurface features and contribute to more informed decision-making in infrastructure projects.

Despite the challenges and limitations associated with data acquisition, quality, integration, and interpretation, ongoing research and development efforts are addressing these issues and further advancing the applicability and effectiveness of geophysical methods. The integration of geophysical methods with remote sensing and geospatial technologies, as well as the incorporation of machine learning and artificial intelligence techniques, are promising avenues for future research and development.

As geophysical methods continue to evolve, they are expected to play an increasingly important role in structural and geotechnical engineering. The advancements in these fields will contribute to better design, construction, and management of infrastructure projects, ultimately leading to safer and more resilient built environments.

References

1. Telford WM, Geldart LP, Sheriff RE (1990) Applied geophysics, 2nd edn. Cambridge University Press
2. Pajewski L et al (2013) Applications of ground penetrating radar in civil engineering—COST action TU1208. In: 2013 7th international workshop on advanced ground penetrating radar, Nantes, France, pp 1–6. https://doi.org/10.1109/IWAGPR.2013.6601528
3. Green R (1974) The seismic refraction method—a review. Geoexploration 12(4): 259–284. ISSN 0016-7142
4. Website www.guidelinegeo.com
5. Park CB, Miller RD, Xia J (1999) Multichannel analysis of surface waves. Geophysics 64(3):800–808
6. Website https://everestgeophysics.com
7. Loke MH (2010) Electrical imaging surveys for environmental and engineering studies. A practical guide to 2-D and 3-D surveys. Electronic document, 60

8. Daniels DJ (2004) Ground penetrating radar, 2nd edn. Institution of Engineering and Technology
9. Rana, Sanjay. Ground penetrating radar survey for concrete wall inspection at Hidd power plant, Bahrain, Project Report
10. Website https://insitutek.com
11. Rana, Sanjay. Seismic refraction tomography at upper Karnali HEP Project, Nepal, Project Report
12. Rana, Sanjay. Electrical resistivity imaging & crosshole seismic test for cavity detection, Kota, India, Project Report
13. Zakaria MT, Mohd Muztaza N, Zabidi H et al (2022) Integrated analysis of geophysical approaches for slope failure characterisation. Environ Earth Sci 81:299
14. Owen W (2010) Geophysical methods for determining the geotechnical engineering properties of earth materials, California. Dept. of Transportation. Office of Geotechnical Support, 2010

Applications of Diaphragm Walls for Underground and Marine Structures—A Review

Shradha Shirodkar and Purnanand P. Savoikar

1 Introduction

Diaphragm walls (D-walls) are constructed to support large-scale construction projects like dams, tunnel designs, deep basements, enclosures, etc., to act as a foundation, retaining wall, support for the underground construction, or a way to prevent access to enable deep excavation. To ensure structural stability and water-tight construction, D-walls are constructed in the form of interlocking panels. D-walls generally have thicknesses between 60 and 150 cm and widths between 2 and 3.5 m. The depth at which diaphragm walls can be constructed is 60 m. Slurry walls and slurry trenches are other names for diaphragm walls, which are used to construct cast in situ reinforced concrete retaining walls. These walls provide permanent retaining walls with fewer construction joints that are rigid and affordable. In crowded regions, adjacent to existing structures, in locations with little headroom, or when an excavation is being done to a greater depth to produce stable slopes, D-walls are sometimes used. The D-walls has fewer joints as compared to a secant wall and can be advantageously used in areas where there is a need to reduce the disturbance to adjacent structures and for supporting very deep excavations. Disadvantages of a diaphragm wall are that they are not suitable in hard soils. Due to the need for excavating and pumping equipment, diaphragm wall construction is more expensive than other kinds of excavations. Diaphragm walls are suitable in urban areas with vast and existing infrastructure, where a highly rigid earth retention system is required. Figure 1 shows typical diaphragm wall under construction.

S. Shirodkar · P. P. Savoikar (✉)
Civil Engineering Department, Goa College of Engineering, Ponda, Goa 403401, India
e-mail: psavoikar@gmail.com

S. Kolathayar et al. (eds.), *Best Practices in Geotechnical and Pavement Engineering*,
Lecture Notes in Civil Engineering 449, https://doi.org/10.1007/978-981-99-8505-0_31

Fig. 1 Diaphragm wall. (https://www.designingbuildings.co.uk/w/images/4/40/Diaphragmwall1.jpg) [1]

2 Literature Survey

Krishnan et al. [2] analyzed the lateral response of a pile and a D-wall under seismic pressure on dredged soil using finite element method. Parameters such as soil characteristics, the structural properties, and the order of the dredging operations influencing the soil movements were studied. Yajnheswaran et al. [3] studied the open berth structures in marine soils which are supported by D-walls and anchor rods. In order to strengthen the structure, resist lateral stresses, and significantly reduce deflection, anchor rods may be added. Using the finite element software PLAXIS 3D, the deep draft berth of the new Mangalore port equipped with a diaphragm wall and anchor rods is analyzed in this work. For diaphragm walls with and without anchors, the displacements, shear forces, and bending moments were calculated. Comparisons were made for variation of bending moment, shear force, and deflection with depth of the diaphragm wall. Shah et al. [4] presented a tool for analysis and design of diaphragm walls using Visual Basic, which accounts for soil variation and provide speedy, optimized results. D-wall parametric analysis was carried out with variable wall width, concrete grade, soil property below dredge line, and diaphragm wall with secondary wall. It was concluded that the thickness of the wall has a greater impact on ultimate material costs than the quality of the concrete.

Jameson et al. [5] reported that the shoring was needed in exceptionally difficult circumstances for supporting of excavation systems for Alaskan Way Viaduct Replacement and the Port of Miami Tunnel. For these projects, integrated lateral support and groundwater control were provided by combining cutting-edge ground improvement techniques with conventional shoring techniques on a large scale. These projects were made practicable and cost-effective by the innovative assistance of excavation methods made viable by the most contemporary tools and equipment.

Liu et al. [6] examined the demolition of the deep excavation support system installed. A numerical modeling approach was employed to assess how the metro line deformed while the deep excavation support system was being destroyed. The results of the numerical analysis were found to be trustworthy when compared to the monitoring data and were used to assess the probable deformation of the nearby metro line and give instructions on building sequences. Kumar and Mumbai [7] illustrated the constructional and design components of three retaining systems that are frequently used in Indian cities, viz., soldier piles with wooden laggings, diaphragm walls, and continuous piles. Study described the experiences and elements that enable the selection of an appropriate retaining system, estimation of lateral earth and hydrostatic pressure distribution, constructional features, water-related issues, and execution delays. Karmegam [8] studied the usefulness of underground metro system which is very helpful in crowded places when the government is concerned about acquiring land and property. The construction process, structural analysis, and precise design of a diaphragm wall were discussed. For the case study, the sandy soil conditions in Chennai city are taken into account with actual geotechnical data to understand the structural behavior. Putchakayala et al. [9] explored the behavior of D-wall in various soil environments throughout Chennai. To understand the behavior of D-wall from a broader perspective, a parametric study has also been carried out by altering various parameters. Several evaluations have shown that the influence of different soil conditions on the design of D-wall is comparatively greater than that of any other characteristics.

Nguyen et al. [10] examined the deformation of the ground surface and the D-wall during the excavation of the city's first metro line in Ho Chi Minh City. It was found that the ground settlement increases while the lateral displacement of the D-wall decreases when the drop in water table level during the dry season is taken into account.

Robinson and Bell [11] presented the most important design considerations, the typical construction details that can help produce a finished product of high quality, common uses for D-wall, and their method of construction. Due to the limited amount of land available for development, the high land prices that go along with it, the growth of deep urban infrastructure, and their increasing use, they are becoming more common in densely populated urban areas. Diaphragm walls can be economically used in permanent works.

Ilies et al. [12] reported effect of calculation models which influence the design and result of embedment depths, moments, shear force, and structural displacements for different wall heights. This study shows how this method is helpful for cost-effective designing with limited time and cheapest tools. It is helpful in places which are crowded and it also improves site stability.

Raman et al. [13] compared the outcomes of the numerical analysis of diaphragm wall with those of the in situ monitoring. The study focussed on how the diaphragm wall interacts with the soil as well as the overall ground deformation regimes in the area around the location. It was reported that the measured soil movements were

20% lower than the predicted. Inclinometers are useful equipment for measuring and analyzing the lateral soil deformation brought on by earthmoving. Installing inclinometers is advised for similarly significant projects to create feedback analysis for performing modified design standards or more precise analytical models of soil characteristics for the lateral displacement of D-wall.

El-Nahas et al. [14] analyzed the diaphragm walls' geotechnical in situ behavior using data from a monitoring instrumentation program. On the size and shape of recorded soil deformations, the implications of bracing systems, wall stiffness, and soil profiles are explored. It was concluded that variations in bracing methods, wall stiffness, and soil profiles all contribute to variations in the magnitude and shape of lateral soil movement near these walls.

Sharma et al. [15] presented field monitoring data and numerical studies on the impact of a large excavation next to MRT tunnels. The displacement of the tunnel linings was tracked using a cutting-edge tunnel monitoring system with 1 mm accuracy. The system performance was determined to be quite satisfactory. In general, the estimated displacements of the tunnel lining were greater than the values that were measured. The tunnel lining stiffness has significantly affected the tunnel displacements. Although a stiff lining is more likely to experience larger bending moments, it is also more likely to experience less distortion. It could be advantageous to choose a flexible lining system if minor distortions have no impact on the tunnel functionality.

James et al. [16] reported that in comparison to Plaxis results, there was very less variation in results indicating the technique's suitability for use on diaphragm walls. From the fundamental ideas of pile foundations, formulas for calculating vertical axial forces are created by changing the equations to incorporate the unique requirements of diaphragm wall.

3 Design Considerations

According to the design procedure of D-wall, it is seen that the overall soil structure model for design is not used. Average characteristics of soil using fundamental soil factors discovered through basic experiments were determined. Before building the permanent underground structure, a constant, uniform design earth pressure loading to the braced diaphragm wall is applied. When applicable, the hydrostatic and surcharge loads are added. Plastic analysis and design are used for the D-wall proportioning. The D-wall is built in a slurry-constructed trench that contains a layer of extra slurry using precast concrete panels and similar materials. A permanent underground structure is designed to accommodate the extra backfill and underlying surcharge loads. Asymmetrical and symmetrical instances, groundwater fluctuations, and surcharges should all be taken into account. To simulate the desired bearing pressure, the structure supported by individual springs is assumed. D-wall's structural design is more sophisticated than other ERSS since it serves as a temporary and permanent structural system in cut-and-cover structures. Therefore, two different types of studies are needed for the structural analysis of D-wall in cut-and-cover structures: analysis during the construction stage and analysis at the permanent stage. The D-wall reinforcement in the soil face is governed by investigations into the construction stage. However, the D-wall reinforcement in the excavation face is governed by the permanent stage analysis. The construction stage analysis is critical for D-wall design since it shows a large increase in bending moment and shear force with relation to D-wall thickness when compared to permanent stage analysis. The concrete is to be positioned beneath the slurry when building the D-wall. In these situations, structural concrete's design compressive strength and shear strength must be lower than the selected concrete grade. Therefore, for D-wall design, the characteristic compressive and shear stress is taken into account as being 80% of the adopted concrete grade.

4 Case Study

Case 1

In this case, undrained analysis of a deep excavation in clayey soil using software hardening soil model of Taipei National Enterprise Center (TNEC) is carried out [17]. The TNEC is an 18-storey building with five basement levels constructed. The excavation for its site is completed using a top-down method across seven stages as shown in Fig. 2.

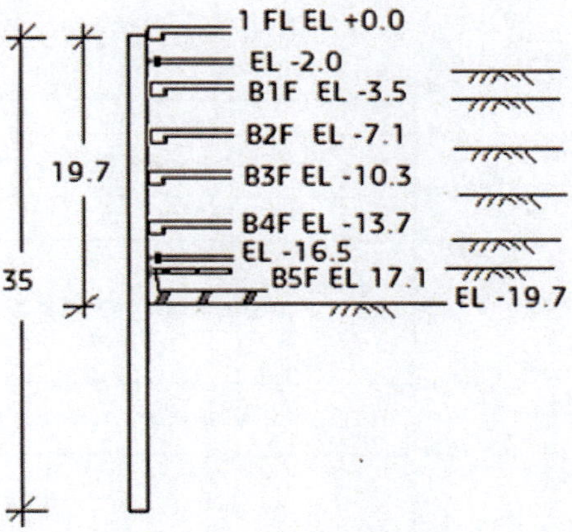

Fig. 2 TNEC excavation cross section [17]

The following stages are:

First stage: Excavation was done at 2.8 m with ground level at 0.
Second stage: Temporary struts (H400 × 400 × 13 × 21) at 2 m followed by excavation at depth of 4.9 m.
Third stage: Struts are demolished and slab is casted at 3.5 m depth. Further ground is excavated at 8.6 m.
Fourth stage: Slab was cast at 0 and 7.1 m followed by excavation at 11.8 m.
Fifth stage: Slab was cast at 10.3 m and excavation is done at 15.2 m.
Sixth stage: Slab was cast at 13.7 m followed by excavation which is done at 17.3 m. Temporary struts installed (H400 × 400 × 13 × 21).
Seventh stage: Excavation done at 19.7 m and struts of same dimension are installed (H400 × 400 × 13 × 21).

Struts are provided to prevent wall deformation before installing concrete slabs.

Case 2

Cross-section of the excavation support system and soil profile at Syed ALwi Project [18] is shown in Fig. 3. The framework of the excavation support system was formed by a 20 m diaphragm wall braced twice. About 1.0 m of groundwater is existed at the original level. The strut spacing is 4.5 m. The structural characteristics of the wall and struts as well as the HSS soil parameters used in the analysis are compiled. For the soft clay, the maximum anticipated wall deflections utilizing HSS ($E_{50} = 150_{\mathrm{Cu}}$), HS ($E_{50} = 150_{\mathrm{Cu}}$), and HSS ($E_{50} = 100_{\mathrm{Cu}}$) are 45, 50, and 57 mm, respectively. The largest wall deflection that could be measured was 49 mm. The HSS variant produces somewhat smaller wall deflections than the HS type for the same ($E_{50}/C_{\mathrm{u}} = 150_{\mathrm{Cu}}$).

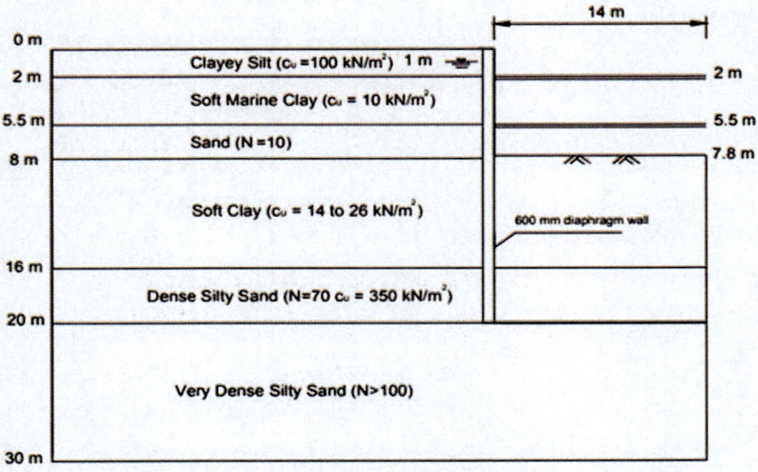

Fig. 3 Details of the excavation support system for Syed ALwi project [18]

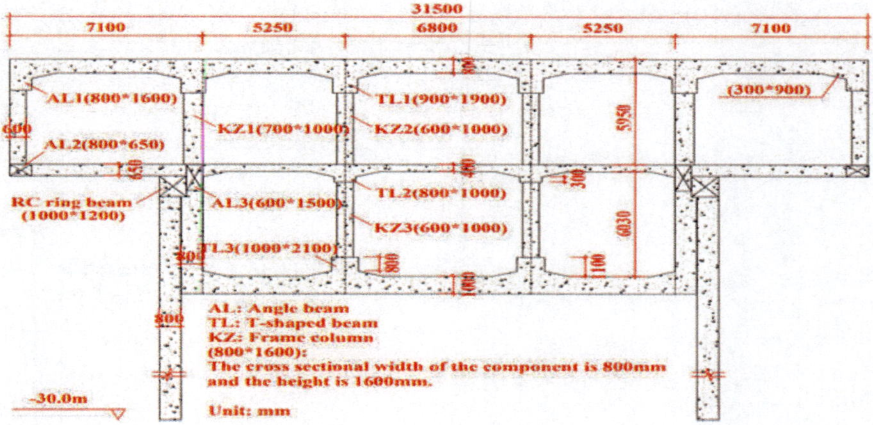

Fig. 4 Details of subway station [19]

Case 3

One of the biggest risks to the subsurface structure during and after powerful earthquakes is ground liquefaction. However, there are not many studies on the seismic response of large-scale subway underground structures on liquefiable sites for subway stations. Wang et al. [19] reported that the uneven uplift across the spans is a characteristic of the uplift of the unequal-span subway station. For the cantilever span, the top layer suffers the least damage when the diaphragm wall is erected, while the degree of the stress damage to the bottom plate of the tube station is increased. These new findings provide insight into the seismic design approach and performance for the unequal-span tube stations at a liquefaction site. At the cross-sections where the diaphragm wall intersects the bottom slab, the diaphragm wall is susceptible to significant tensile damage. Figure 4 shows the details of diaphragm wall for subway station.

Case 4

The process for the analysis and design of the D-wall [8] is detailed in this paper with the help of the subsequent real-world example. In this case study, a typical cut-and-cover structure with two levels is taken into account. With genuine geotechnical parameters discovered through borehole experiments, Chennai's sandy soil conditions are taken into account. The thicknesses of base slab, roof, and concourse slabs were 1.3 m, 1.2 m, and 1 m respectively, while the D-wall was 1 m thick and the clear width of station was 16.5 m. Behavior of the D-wall was studied by varying the thickness of D-wall. It is found that increase in thickness of D-wall increases the bending moment and shear force and decreases the horizontal deflection. When 1 m wall thickness was used, horizontal deflection was reduced to 18%, and by using 1.2 m wall thickness, horizontal deflection was reduced to 43% in comparison to

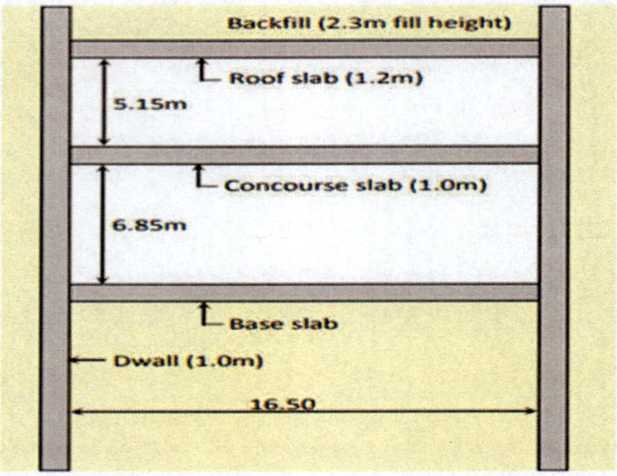

Fig. 5 Details of typical two-level cut-and-cover structure [8]

0.8 m thick wall. Figure 5 shows the details of a typical two-level cut-and-cover structure adopted for D-wall and concourse and roof slab.

Case 5

Five phases were used to conduct the literature review: design parameter analysis, building proximity analysis, earth and water pressure analysis, seismic analysis, and Indian standards analysis. Through the study of design parameters, we are able to understand the most suitable variables, optimum dimensions, concrete grade, soil conditions, and anchor rod location all may be used in the design. The effective bending moment, displacement, and shear force are the topics of the investigation. In the second section, it is discovered that diaphragm walls have a stronger impact on nearby buildings. Both water pressure and ground pressure are crucial in diaphragm wall design [20]. Figure 6 shows typical details of anchored diaphragm wall.

Fig. 6 Anchored diaphragm wall [20]

5 Applications

The type of D-walls used depends upon the site conditions and functions they need to perform. Diaphragm wall is used to retain soil for boundary walls of excavation for basements, underpasses, subways, etc. Load-bearing D-walls serve as good option for drilled piers in tall building foundations, bridge piers, etc. They are used as impermeable cutoffs to control the seepage below earthen dams. Benefits from using diaphragm walls lie in the places where there are lot of structures and construction activities going on. The type of analysis, either drained or undrained, must be carefully considered depending on the building period and soil permeability. Additionally, if clay is only found within a small portion of the location, it might not be best to construct the entire D-wall with clay as an overburden. Therefore, rather than planning the entire site and taking into account clay strata, additional costs can be required for a thorough examination to validate the extent of clay.

For soils with high permeabilities, such as sand or gravel, or for soils that are subjected to prolonged loading, a drained analysis is usually utilized. On the other hand, when simulating the short-term loading of a specific clay model, undrained soil properties must be taken into account since this analyzes the sample before it has been drained.

6 Conclusions

In the present study, a review of application of diaphragm walls in underground and marine conditions is discussed. It is observed that many parameters, including the soils characteristics, the structural properties and the order of the dredging operations, influence the behavior of diaphragm walls. The thickness of the wall has a greater impact on ultimate material costs than the quality of the concrete. The ground water table affects the behavior of wall since it was observed that the settlement increases while the lateral displacement of the D-wall decreases when the drop in water table level during the dry season is taken into account. Several evaluations have shown that the influence of different soil conditions on the design of D-wall is comparatively greater than that of any other characteristics.

References

1. https://www.designingbuildings.co.uk/w/images/4/40/Diaphragmwall1.jpg
2. Gokul Krishnan M, Sathyanarayanan D, Subha IP, Muthukkumaran K (2009) Behavior of an open type berthing structure under earthquake condition—a numerical approach. In: IGC 2009, pp 553–557
3. Yajnheswaran, Ranjan HS, Rao S (2015) Analysis of the effect of anchor rod on the behavior of diaphragm wall using plaxis 3D. Aquatic Proc 4:240–247

4. Shah PD, Vyas BA (2015) Parametric study of diaphragm wall. Int J Sci Tech Eng (IJSTE) 1(11):141–150
5. Jameson R, Starcevich J, Hanke R (2015) North American trends and developments of complex support of excavation systems. In: IFCEE 2015, GSP 256
6. Liu B, Xiu W, Reddy KR (2015) Effects of demolishing the deep excavation support system used for tall building construction on adjacent metro line, modeling and field comparison. In: IFCEE 2015 GSP 256, pp 1107–1121
7. Kumar M, Mumbai K (2018) Deep support system using diaphragm walls and contiguous piles
8. Karmegam R (2020) Analysis and design of diaphragm wall in sandy soil conditions. Int Res J Eng Tech (IRJET) 07(11):61–70
9. Putchakayala P, Karmegam R (2021) Effect of soil behavior on diaphragm wall in soils of Chennai City. Int Res J Eng Tech (IRJET) 8(2):1998–2003
10. Nguyen BP, Ngo CP, Tran TD, Bui XC, Doan NP (2022) Finite element analysis of deformation behavior of deep excavation retained by diaphragm wall in Ho Chi Minh City. Indian Geotech J 52(4):989–999
11. Robinson C, Bell A (2018) Principles of design and construction. Temp Works 175–192
12. Ilies NM, Farcas VS, Marius P (2014) Design optimisation of diaphragm wall. In: 8th International conference interdisciplinarity in engineering. INTER-ENG 2014, 9–10 October 2014, Tirgu Mures Romania
13. Rahman A, Taha M (2005) Geotechnical performance of embedded cast-in situ diaphragm walls for deep excavation. Slovak J Civil Eng
14. El-Nahas F, Eisenstein F, Shalaby A (1989) Behavior of diaphragm wall during construction of Cairo metro. Int Soc Soil Mech Geotech Eng
15. Sharma JS, Hefny AM, Zhao J, Chan CW (2001) Effect of large excavation on adjacent MRT tunnels. Tech Tunnelling Undergr Space 16(2001):93–98
16. James A, Kurian B (2021) Diaphragm wall retaining system—a simplified model for design loads. Aust J Civil Eng
17. Ou CY, Shiau BY, Wang IW (2000) Three-dimensional deformation behavior of the Taipei National Enterprise Center (TNEC) excavation case history. Can Geotech J 37(2):438–448
18. Lim KW, Wong KS, Orihara K, Ng PB (2003) Comparison of results of excavation analysis using WALLAP. SAGE CRISP, and EXCAV97, pp 83–94. Underground Singapore 2003, Singapore
19. Wang J, Guowei M, Zoung H, Duo Y, Fu J (2019) Influence of diaphragm wall on seismic responses of large unequal-span subway station in liquefiable soils. Tunneling Undergr Space Tech 91
20. Mini MP, James A, Kurian B (2018) A theoretical study on analysis of diaphragm wall. Int Res J Eng Tech

Finite Element Analysis of Twin Tunnels in Granitic Rock

Koshi Roshan James and Purnanand P. Savoikar

1 Introduction

Tunnels are essential since they made transportation simpler. Due to the lack of available land in emerging nations like India due to rapid urbanization, development, industrialization, and civilization, tunnels are crucial. It may be built underground in the event of a metro tunnel. Additionally, tunnels can be used for economic activity, water conductor systems, shelters, relaxation, and defense in developing nations [1]. During the construction of a tunnel, the rock's stability is crucial. Selecting the right excavation techniques in accordance with the grade of the surrounding rock and water seepage ensures the stability of the excavation in the surrounding rock. The chosen excavation methods should be studied to ensure the stability of the tunnel excavation after choosing the construction techniques based on engineering expertise during large-span tunnel construction [2]. To reproduce the effect of the tunneling process in the analytical model and to conduct realistic analysis of the tunnel behavior with lining interaction, an extensive knowledge of the process is necessary. Different components of the process, such as ground movements brought by digging, lining installation, and rock-liner behavior, must be idealized in the analytical model [3]. Gao and Davies [4] illustrated how the BEM produces more accurate stress evaluation results when compared to other domain approaches like the Finite Element Method (FEM) at similar levels of discretization [4].

Elsayed [3] studied the rock lining interaction for circular tunnel using finite element analysis using PLAXIS software with various dimension of tunnel such as 3 m, 4.5 m, and 6 m in diameter with surcharge depth of 10 m, 20 m, and 30 m, respectively. The various properties of rock like poor rock, moderate rock, and hard rock were analyzed to study radial displacement in rock mass, average radial displacement in lining system, and final displacement [3].

K. R. James · P. P. Savoikar (✉)
Civil Engineering Department, Goa Engineering College, Farmagudi, Goa 403401, India
e-mail: psavoikar@gmail.com

S. Kolathayar et al. (eds.), *Best Practices in Geotechnical and Pavement Engineering*, Lecture Notes in Civil Engineering 449, https://doi.org/10.1007/978-981-99-8505-0_32

Zhu et al. [5] studied the behavior of railway tunnel with large section constructed using different construction methods. The FEA software ANSYS was used to stimulate and analyze the stress and deformation in 2D and 3D. The results indicate that the bench cutting vault give more settlement then the two side walls. The simulation of the construction sequences in ANSYS was done to analyze the stresses in construction stages [5]. Mandal and Singh [1] analyzed the tunnel lining, a support system in the tunnel structure by using Midas GTS NX and USBR software. The results indicated that Midas GTX NX software gives lower values than the USBR software in all cases [1]. Sarkar et al. [6] examined the influence of over load pressure and soil properties on tunnel settlement. It was seen that when the soil layer changed from silty sand to clayey silt, the ground settlement decreased dramatically by 14–25% even though the tunnel depth remained the same [6]. Huo et al. [2] performed the FEM analysis using three step excavations on extra-large cross section of tunnel. Model was generated using FEM software Midas GTX NX with all the support system like rock bolt, middle pipe, steel arch, and shotcrete prescribed as per the geological and geotechnical condition and analyzed using nonlinear construction analysis. The results showed that the maximum vertical deformation of the tunnel vault and the middle of the invert was about 34 mm [2].

2 Project Case Study

2.1 Project Profile

National Highway Authority of India (NHAI) has come up with an up-gradation of the old highway NH17–NH 66 for Karwar-Kundapura highway, to reduce the time of travel by constructing the twin tunnels. The tunnels were constructed in Karwar which are 10.9 m in width and in 300 m long (West tube) and 400 m long (East tube) provided with two lanes. The site investigation was carried out and the rock samples were tested, which indicated the presence of granitic rock which comes under igneous type of rock with surcharge load of lateritic soil of depth 10 m above the granitic rock. The rock samples were tested and rock was classified as per the rock mass rating (RMR) values as good rock mass (61–80) and poor rock mass (21–40). The various support systems required of the stability of that rock in the section are decided based on the rock quality. Figure 1 shows the twin tunnels under construction.

2.2 Components of Tunnel

Portal. The actual doorway or main entrance of tunnel is known as portals. A portal indicates the intersection point between the underground opening and the ground

Fig. 1 Twin tunnel Karwar

surface. The steel rib of ISMB 250 at 500 mm c/c is embedded in 250-mm-thick shotcrete with 300-mm-thick RCC lining on it.

Steel ribsx. Steel members that have been bent into an arch shape are known as H-shaped steel ribs. Despite the availability of parts in a variety of sizes, only a handful is actually used. Traditional H-shaped steel ribs' geometrical and material characteristics for NATM tunneling ISMB 250 and 300 of 8-mm-thick web were used.

Pipe umbrella. The arch construction in the roof and spring line regions, as well as the stability of the face and in front of the face right after the excavation, is to be supplemented by pipe umbrellas. At least 30% of the pipe umbrella should extend past the following excavation's face. The umbrella reinforcement at the site is of perforated steel pipe of 88 mm diameter fully grouted about 12 m long and 300 mm c/c with 25% of overlapping along the excavation direction.

Rock bolts. They take up the load of the surrounding rock mass when rock begin to deform; this specially happens in fractured rock and the rock with multiples of crack formation. The bolt holds the rock in its position avoiding the deformation in the tunnel. The length of the rock bolt depends upon the failure zone of the rock. As the tunnel contains of 4-m-long rock bolt which is yield at 50% of its yield strength the number of rock bolts were decided depending upon the rock mass rating (RMR) values, 20 numbers of rock bolt for poor rock mass section and 8 numbers of rock bolt for good rock mass in staggered position were provided.

Shotcrete. Is a type of concrete that is sprayed on the wall surface after the reinforcing bars and steel mesh is provided. The hose-pipe nozzle with high pressure and velocity is sprayed and the concrete get placed on wall surface without any form work, and it is efficient that needs only a small industry to get manufactured. It creates strong chemical bonds between many materials. The shotcrete ingredients of M20 grade

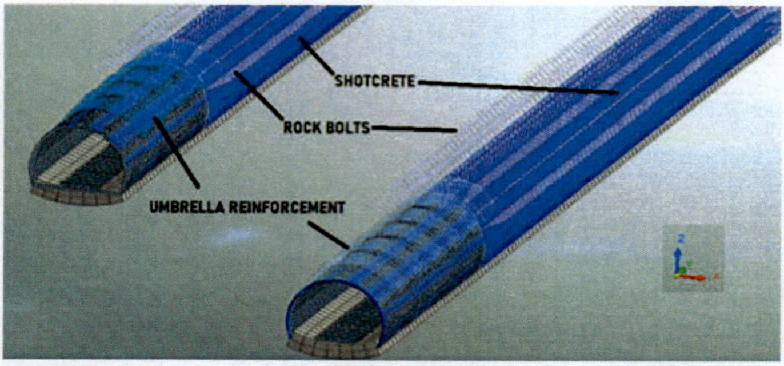

Fig. 2 Components of tunnel

with admixture of Normet and fiber reinforcement conforming to IS: 9012–1978. Figure 2 shows the components of the tunnel.

2.3 *Various Model Generated Using Midas GTS NX*

Figures 3 and 4 show some of the typical models generated using Midas GTS NX. Figure 3 shows the model generated for Cityringen Copenhagen Metro in Denmark while the model shown in Fig. 4 represents the tunnel system for Trans-Hudson Express in USA.

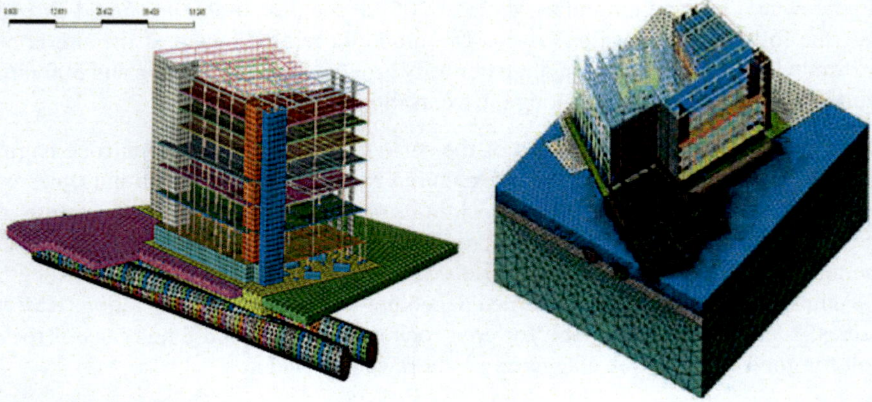

Fig. 3 Cityringen Copenhagen metro, Denmark

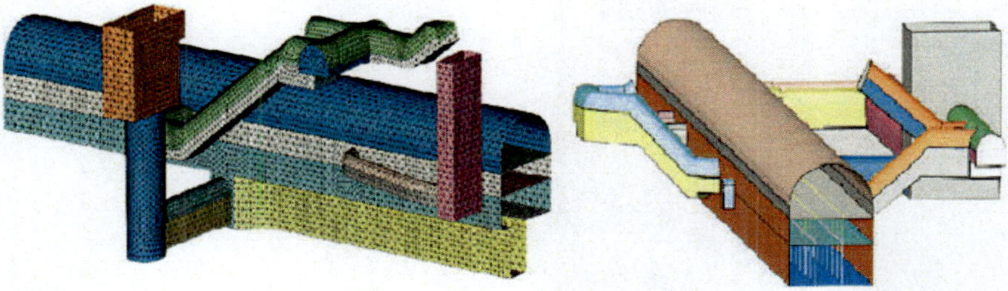

Fig. 4 Trans-Hudson express, USA

2.4 Modeling in Software

The 3D model of twin tunnel was developed using Midas GTS NX software, starting with the profile of the terrain of the tunnel where in the contour map of the site was plotted and exported in to the software. Profile of the tunnel developed is shown in Fig. 5.

Followed by the modeling, the interior components of tunnel like shotcrete, rock bolt, umbrella reinforcement, ISMB section at portal, are modeled as the actual tunnel. Finer meshing was done to increase the accuracy of the result and the boundary conditions, loads were applied, and analysis was performed. Figure 6 shows the interior components of tunnel.

The full tunnel section is classified as per the section such as north portal, south portal, west tunnel, and east tunnel also it classified as per various rock masses like good rock and poor based on then rock mass rating (RMR) value of the each rock which were encountered during the site investigation. Figure 7 shows the classification of various rock masses.

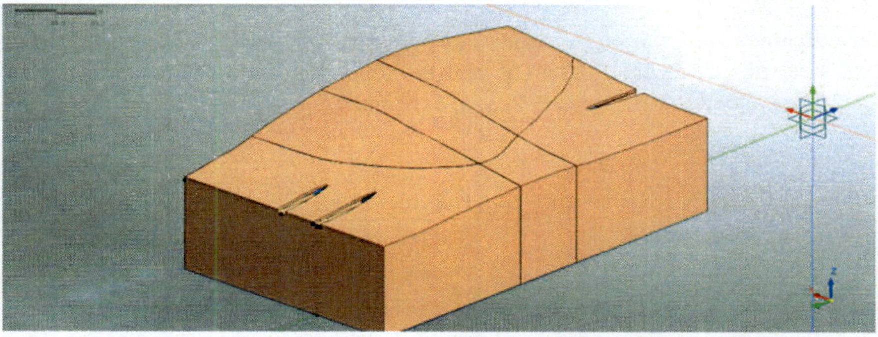

Fig. 5 Profile of the terrain of twin tunnel

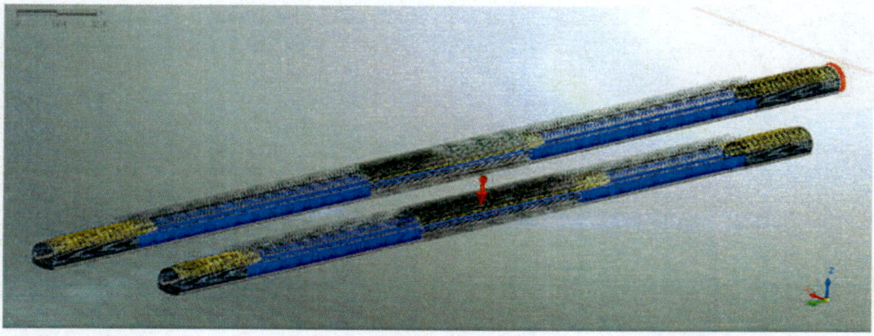

Fig. 6 Interior components of tunnel

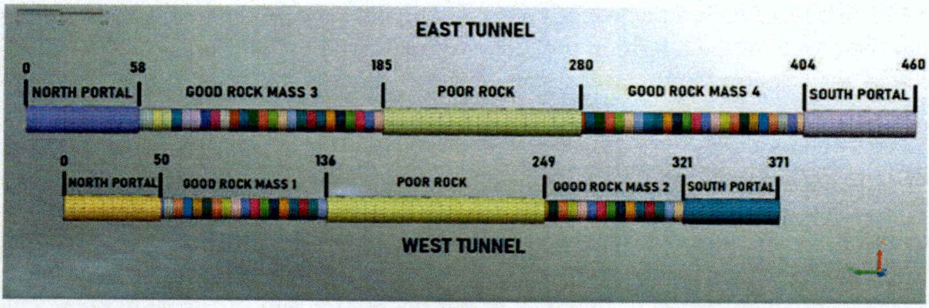

Fig. 7 Classification of various rock masses and sections

2.5 Material Properties of the Tunnel Model Used in Midas GTS NX Software

The properties of the various materials like poor rock, good rock, lateritic soil, structural steel ribs, rock bolt, shotcrete, and the umbrella reinforcement used in the analysis are tabulated in Tables 1, 2, 3, 4, 5, 6, and 7.

Table 1 Properties of the poor rock mass

Sl. no	Properties	Values
1	Elastic modulus (E)	10,000,000 kN/m^2
2	Poisson ratio	0.22
3	Unit weight	20 kN/m^3
4	Damping ratio	0.05

Table 2 Properties of the good rock mass

Sl. no	Properties	Values
1	Elastic modulus (E)	15,000,000 kN/m^2
2	Poisson ratio	0.3
3	Unit weight	28 kN/m^3
4	Damping ratio	0.05

Table 3 Properties of the laterite soil

Sl. no	Properties	Values
1	Elastic modulus (E)	5000 kN/m^2
2	Poisson ratio	0.3
3	Unit weight	16.2 kN/m^3
4	Damping ratio	0.05
5	Cohesion (c)	20 kN/m^2
6	Friction angle ϕ	36°

Table 4 Properties of the ISMB 300 steel ribs

Sl. no	Properties	Values
1	Elastic modulus (E)	210,000,000 kN/m^2
2	Poisson's ratio	0.2
3	Weight	0.46 kN/m

Table 5 Properties of the rock bolt

Sl. no	Properties	Values
1	Elastic modulus (E)	210,000,000 kN/m^2
2	Poisson ratio	0.3
3	Weight	0.0384 kN/m

Table 6 Properties of the shotcrete

Sl. No	Properties	Values
1	Elastic modulus (E)	5000 kN/m^2
2	Poisson ratio	0.3
3	Unit weight	25 kN/m^3

Table 7 Properties of the umbrella reinforcement

Sl. No	Properties	Values
1	Elastic modulus (E)	210,000,000 kN/m^2
2	Poisson ratio	0.3
3	Unit weight	20 kN/m^2

3 Result and Discussion

The mechanism involving the settlement and shear stress in the tunnel ultimately depends up on the strength of the rock when it is tested in the laboratory and surcharge load coming on the tunnel. The settlement and the shear stresses in the tunnel are measured on the tunnel surface. The rock mass rating (RMR) value gives the idea of the strength of rock containing the features like unconfined compressive strength of rock sample, spacing between the joints in rock sample The rock quality designation (RQD) and the conditions of the discontinuity and the shear stresses depend on the amount of the surcharge load on the cross sectional area of the tunnel per square meter. The results are shown for the analysis carried out for the 3D model of the tunnel shown in Fig. 7, using Midas GTS NX, along the length of the tunnel. Figure 8 exhibits the settlement at the crown for the west tunnel. The maximum settlement is found at the poor rock at chainage 110–210 m at mid-section whose value varies between 34 and 42 mm. The good rock mass 1 and 2 exhibits less settlement as compared to the poor rock mass, which varies between 31 and 37 mm. For the both end sections, also known as portal, the settlement varies between 9 and 12 mm.

Figure 9 shows the shear stress at the crown of the west tunnel due to overburden. It shows the maximum shear stress at the crown (between chainage 110 and 210 m) where poor rock mass exists. This stress value ranges between 110 and 120 kN/m^2. The shear value at the crown dips from 115 to 100 kN/m^2 for the good rock mass 1 (between chainage 50 and 136 m) and good rock mass 2 (between chainage 249 and 321 m). The shear stress values for the both north and south portal ranges between 20 and 70 kN/m^2.

Figure 10 shows the settlement at crown for the east tunnel. The maximum value is observed for poor rock mass at chainage 130–230 m where settlement of 42–39 mm is observed. The settlement for the good rock mass 3 (at chainage 58–185 m) and good rock mass 4 (at chainage 280–404 m) varies between 33 and 36 mm followed by the settlement for the portal at north and south which ranges from 9 to 11 mm, respectively.

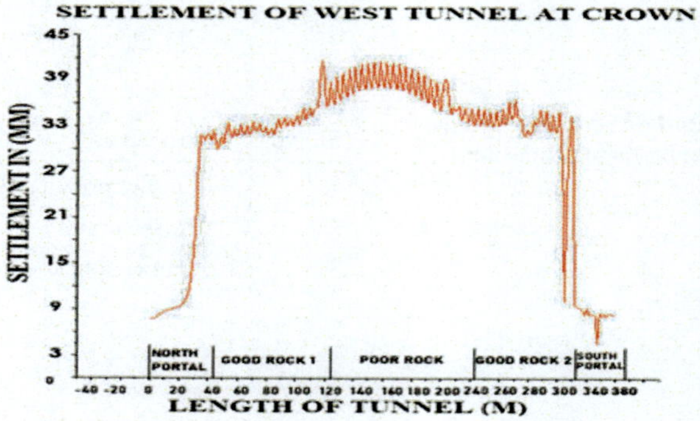

Fig. 8 Settlement of west tunnel at crown

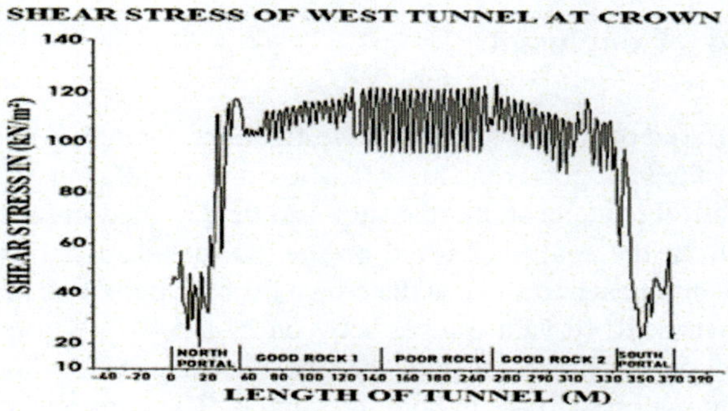

Fig. 9 Shear of west tunnel at crown

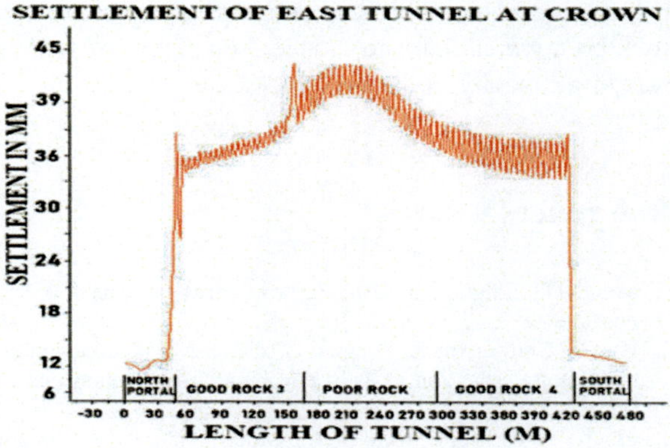

Fig. 10 Settlement of east tunnel at crown

Figure 11 shows the shear stress for the east tunnel at the crown. Shear stress varies between 105 and 120 kN/m^2 for the good rock mass 3 (between chainage 58 and 185 m) and rock mass 4 (between chainage 280 and 404 m) and also for the poor rock mass (between chainage 185 and 280 m).

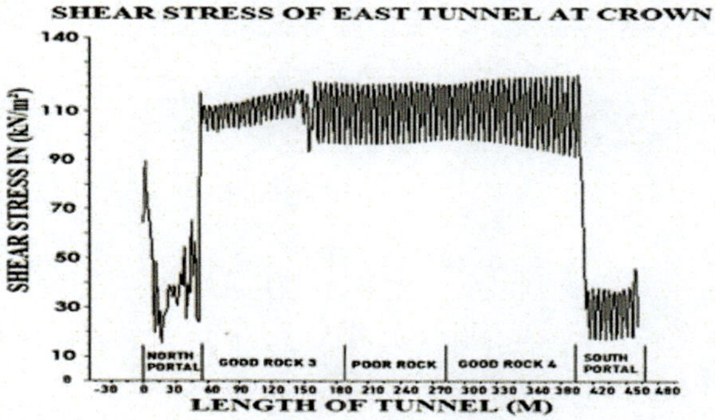

Fig. 11 Shear of east tunnel at crown

4 Conclusion

Based on the geotechnical investigations conducted and the rock mass rating (RMR) values for good rock mass (RMR value 61–80) and poor rock mass (RMR value 21–40) the tunnel support system can be designed in further stages. The results shown from the analysis carried on the 3D tunnel model exhibit the maximum value of settlement and shear at the crown over the poor rock mass section for both the tunnel whose RMR value ranges between 21 and 40. It is also concluded that the settlement and shear stresses in good rock mass and in portal section are less than those compared to poor rock mass in both the tunnels. The poor rock can be strengthened in further stages by providing steel ribs bent in arch section as provided in portal sections. Also, the number of rock bolts can be increased by reducing the spacing between the each rock bolt, which helps to reduce the amount of settlement and increase the loading carrying capacity of the poor rock.

References

1. Mandal G, Singh AK (2018) Finite element analysis of lining of tunnel. In: Indian geotechnical conference
2. Huo W, Shifengxue Z, Zongzhi G, Zhiyer MS (2021) Numerical simulation and analysis of the three-step excavation of and extra-large cross-section and a low flat-ratio railway tunnel. Res Square
3. Elsyed AA (2011) Study of rock-lining Interaction for circular tunnel using finite element analysis. Jordan J Civil Eng 50–64
4. Gao X-W, Davies TG (2002) Boundary element programming in mechanics. Cambridge University Press, Cambridge
5. Zhu B, Kou W, Xi J, Shen Y (2016) Numerical simulation on research of construction method for shallow buried large section tunnel. Open Civil Eng J 597–601
6. Sarkar C, Mukherjee S, Pitchumani NK (2020) Effect of over burden pressure on tunnel induced ground settlement. In: Proceeding of Indian geotechnical conference, pp 163–175

Pavement Geotechnics

Waste Plastic as Fiber Reinforcement in Pavement Quality Concrete

Roopa Kolan, G. Narendra Goud, T. Naveen Kumar, B. Alekhya, and K. Srujana

1 Introduction

1.1 Plastic Pollution Present Scenario and Treatment Need

In today's culture, plastic is one of the most widely used and practical materials. However, because of its detrimental effects on the environment, biodiversity, and public health, plastic use is becoming more and more controversial. Due to the inability of plastic garbage to decompose, the issue became more problematic. Authorities, society, and companies are looking for effective and affordable products to limit the use of plastic as awareness of the harm caused by plastic items grows. Instead, plastic disintegrates into tiny fragments that can damage animals and marine life and have a detrimental effect on human health by releasing toxins into the food chain. It is estimated that over 8300 + million tonnes of virgin plastic have been produced to date. Nevertheless, it is estimated that India generates 9.46 million metric tonnes of plastic waste annually. Nearly 40% of this waste remains uncollected, as per the environment ministry. Unfortunately, most cities and towns are unable to efficiently implement the plastic waste management rules.

A Central Pollution Control Board (CPCB) report (2020–21) [1] states that the total annual plastic waste generation in India is humungous at 3.4 million metric tonnes per year. Recycling and co-processing account for less than 50% of the generated plastic. Plastic waste is becoming extremely threatening to the environment due to the high quantities generated, which pose serious harm to both the environment and its inhabitants. According to worldwide material balance research on plastics, 79% of all plastics produced globally end up in the environment as garbage. Only 9% of the world's total plastic garbage is recycled [2]. Treatment through incineration

Present Address:
R. Kolan · G. N. Goud (✉) · T. N. Kumar · B. Alekhya · K. Srujana
CED-Maturi Venkata Subbarao Engineering College, Hyderabad, Telangana 501510, India
e-mail: gnarendragoud@gmail.com

S. Kolathayar et al. (eds.), *Best Practices in Geotechnical and Pavement Engineering*,
Lecture Notes in Civil Engineering 449, https://doi.org/10.1007/978-981-99-8505-0_33

will release toxic gases like dioxin that could be dangerous to human health. One of the potential means to address the problem is to recycle the plastic and use it in the construction industry [3–8]. Waste plastic fibers could be employed to make concrete, which would be a sustainable way to manage plastic garbage. This strategy minimises the amount of plastic that is disposed of or burned, decreasing pollution and resource depletion. The mechanical characteristics of concrete can be enhanced by the addition of used plastic fibers. The inclusion of plastic fibers in concrete can aid in crack control and stop cracks from spreading. It was concluded that the use of plastic waste for construction applications will improve the sustainability of the environment significantly, and also serve as a reliable source of materials for construction purposes.

2 Usage of Plastic in Construction Practice

Research by Guptha et al. [9] showed that polyester fiber reinforced concrete (PFRC) exhibited improved flexural and compressive strength, abrasion resistance, and reduced drying shrinkage over that of plain cement concrete (PCC). It was concluded that polyester fibers were alkali resistant, and PFRC can be used in pavement quality concrete (PQC), and as overlays, with no adverse effect on concrete. A study by Torres et al. [10] on the use of synthetic fibers of terephthalate polyethylene (PET) and polypropylene (PP) (0–4%) in PQC concluded that the residual stress increases as a function of the content of synthetic fiber compared with conventional concrete. The presence of fiber contributed to the reduction of contraction cracks generated in the first 24 h of concrete hardening and increased ductility after rupture. The results of the study by Chandrakar and Qureshi [11] indicated an improvement in the mechanical properties of concrete as well as the concrete bond strength and reduced cracks in the concrete with the use of plastic fiber in the range of 0.5–2.0%.

2.1 *Motivation*

Plastic is non-biodegradable and is becoming a big hazard to the environment. Applying sustainable practices in the construction sector is the need of the hour. Because of the harmful effects of plastic on the environment, human health, and biodiversity, reduce, reuse and recycle the plastic has become a trending sustainability practice. The primary factor in opting for the current work was reusing the PET plastic bottles instead of using costly fiber in the mix. The secondary factor was the limited research on the application of plastic fiber in PQC. Moreover, the use of waste plastic as fiber in PQC can consume a lot of plastic.

2.2 *Objectives*

The main objectives of the present work were to investigate the performance of concrete in compression and flexure when waste plastic was used as fiber in PQC and to appraise and compare the flexural strength in the following cases: (a) conventional mix with no fiber; (b) fiber reinforced concrete (FRC) mix with macro-synthetic fiber; and (c) FRC mix with waste plastic (PET) as fiber. Finally, to propose the mix constituents with waste plastic that provide enhanced or equivalent flexural strength to that of the traditional mix.

3 Materials

Materials used in the present study include two types of fibers (macro-synthetic fiber and cut pieces of waste plastic PET bottles), cement, aggregate, and super-plasticizer. The detailed descriptions of individual materials incorporated in the formulated mix are as follows:

3.1 *Fiber*

The macro-synthetic fiber brand used was Tuf Strand SF as one of the fibers. The cost of this fiber was 700 Rs./Kg. Figure 1 shows the macro-synthetic fiber Tuf Strand used in the mix. A literature review indicates that the use of fiber content could be in the range of 0–4%; however, IRC: SP: 46–2013 [12] suggests it could be less than 1.5% to avoid the strain hardening effect. In the current work, experiments were conducted with a fiber content of 1.2 and 2.5% by weight of cement to assess the flexural performance of FRC. Cut pieces of waste PET bottles were used as an alternate fiber. Figure 2 shows the PET fiber cut manually from waste plastic bottles proposed to be used in the concrete mixture. Table 1 provides the characteristics of the two types of fibers used in the concrete mix.

3.2 *Cement*

Ordinary Portland cement of 53 grade was used with a standard consistency of 29.1%, initial setting time of 110 min, and final setting time of 195 min. The compressive strength of cement was found to be 38.4 kg/sq.cm, 52 kg/sq.cm, and 68 kg/sq.cm at 3 days, 7 days, and 28 days, respectively. The fineness of cement was 380 sq.m/

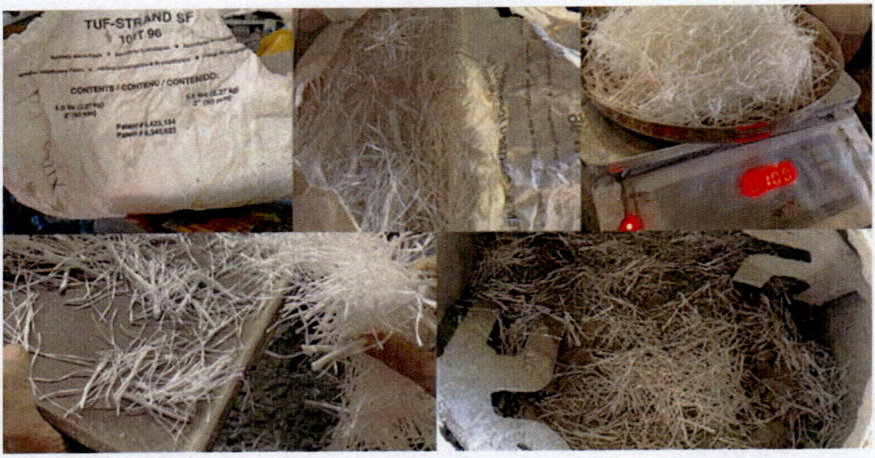

Fig. 1 Macro-synthetic fiber Tuf Strand

Fig. 2 PET fiber (waste plastic bottles fiber)

Table 1 Properties of fiber used in the mix

S. No.	Parameter	Macro-synthetic fiber	PET bottles	IRC specification
1	Aspect ratio	74	250	200–2000
2	Size, mm	51	45–50	6–48
3	Specific gravity	0.92	1.38	Above 1
4	Dosage	1.2	1.2, 2.5	0.2–0.6% by weight of cement
5	Tensile strength, MPa	600–650	170	–
6	Yield strength, Mpa	–	40	–

Table 2 Aggregate properties

S. No.	Property	Coarse aggregate (20 mm)	Coarse aggregate (10 mm)	Fine aggregate
1	Specific gravity	2.640	2.630	2.510
2	Absolute specific gravity	2.650	2.640	2.694
3	Water absorption, %	0.30	0.31	2.4
4	Los Angeles abrasion value, %	18 (Max. 35)		
5	Flakiness and elongation index	26 (Max.35)		

kg and the specific gravity was 3.12. Cement was found to be sound (Le Chatelier method). The tests on cement were performed in accordance with various parts of IS: 4031.

3.3 *Aggregate*

Crushed stone aggregate of granite was used in the concrete mix. The properties of 20 mm aggregate, 10 mm aggregate, and fine aggregate are presented in Table 2. The ratio of coarse aggregate to fine aggregate was 1.27 (0.56:0.44), and the ratio of 20 mm aggregate to 10 mm aggregate was 1 (0.5:0.5). Figure 3 shows the gradation of all-in aggregate used in the concrete mixture. The tests on coarse and fine aggregate were performed in accordance with various parts of IS: 2386.

3.4 *Admixture*

CONPLAST Super Plasticizer-430 was used as an admixture for better workability at a low water cement ratio. The maximum free water cement ratio prescribed by IRC [13] is 0.45 when OPC is used and 0.5 when blended cement is used in concrete mix. The water cement ratio adopted in the present work was 0.43.

4 Methodology

Materials selected were mixed in accordance with the design mix. The experiments were conducted to determine the mechanical properties of concrete, viz., compressive and flexural strength of the M30 grade design mix. Four types of mix formulations

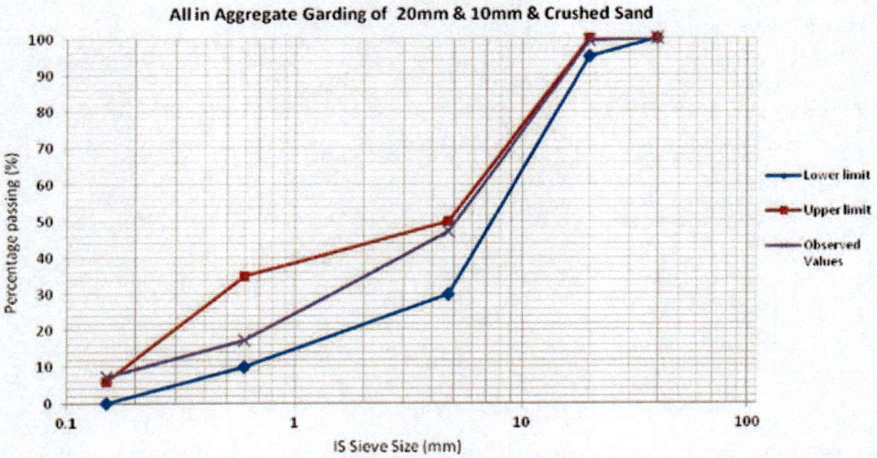

Fig. 3 All-in aggregate gradation

were adopted, viz., 1. conventional cement concrete mix without fiber, 2. cement concrete mix with macro-synthetic fiber Tuf Strand, 3. cement concrete mix with PET fiber (waste plastic fiber) 1.2%, and 4. cement concrete mix with PET fiber (waste plastic fiber) 2.5%.

A concrete drum mixer was employed for mixing. In the case of conventional concrete mix, the material addition sequence was course aggregate, fine aggregate, cement, water, and admixture; in the case of fiber mix, the material addition sequence was course aggregate, fine aggregate, cement, fiber, water, and admixture. A uniform and cohesive mix was ensured before checking workability and casting specimens for testing. Table 3 provides the details of various trial mixes, dosages of fiber, and admixture. The workability of the mix was measured using a slump cone, and slump values are presented in Table 3. As per IRC for PQC, the slump value shall be 25 ± 15 mm when fiber is used in the mix. Inclusion of fiber in the form of cut pieces of PET bottles or macro-synthetic fiber into the concrete, provides high plasticity and strength but reduces workability. An increase in fiber content reduces workability.

4.1 Design Mix

The cement content adopted was 360 kg/cu.m. The concrete mix was designed in accordance with IS: 10262 [14], and the proportion employed for cement, fine aggregate, and coarse aggregate was 1:2.27:2.93.

The temperature of the lab was maintained at 27 ± 2°C. After casting, the specimens were placed on a firm, vibration-free platform. After 24 h, the specimens were de-moulded and given identification by a permanent marker. Then the specimens

Table 3 Details of fiber and admixture dosage

Trial mix	Fiber used	Percentage of fiber used (% by weight of cement)	Admixture dose (% by weight of cement)	Slump in mm
TRIAL 1	–	0.0	0.60	45
TRIAL 2	Macro-synthetic Fiber Tuf Strand	1.2	1.00	30
TRIAL 3	PET fiber (waste plastic fiber)	1.2	0.70	45
TRIAL 4	PET fiber (waste plastic fiber)	2.5	0.85	40

were placed in the water tank for curing. Potable water free from impurities and salt was used for casting and curing of the concrete specimen.

4.2 Cubes for Compressive Strength

For each trial, nine cubes of 150 mm × 150 mm × 150 mm size were cast and tested for compressive strength at 3, 7, and 28-day time periods in accordance with IS: 516-1999 [15]. Figure 4 shows the processing of concrete cubes. The cured specimens were placed on the lower platen of the compression testing machine, and proper alignment and centering of the specimen were ensured. Continuously increasing compressive force was applied at a constant rate (140 kg/sq.cm/min) until failure occurred. The greatest load sustained by the specimen was reported.

4.3 Beams for Flexural Strength

Three beams of 150 mm × 150 mm × 700 mm size for each trial were cast and tested for 7, 14, and 28 days of strength in accordance with IS: 516-1999 [15]. Figure 5 shows the casting of beams and casted beams. The cured beam specimen was mounted on two supports with a span length of 500 mm. A progressively increasing load of 400 kg/min at the centre (two points of loading 200 mm apart) of the beam was applied until failure occurred. The greatest load sustained by the specimen was reported.

Fig. 4 Concrete cubes being cast for M30 grade

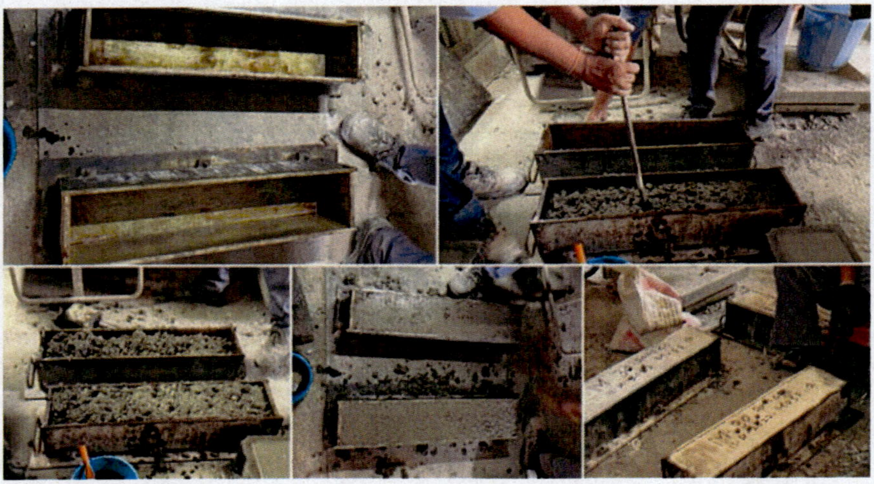

Fig. 5 Concrete beams being cast for M30 grade

5 Results and Discussion

All the mixes achieved 45 MPa and above compressive strength at 28 days; however, the target strength was 38.5 MPa. At the age of 7 days, the compressive strengths of all the mixes were in the range of 32–35 MPa. Figure 6 shows the compressive strength of different mixes with and without fiber. Conventional concrete without fiber exhibited marginally (2–4 MPa) high compressive strength at an early age and also at 28 days in comparison with FRC mixes. All the FRC mixes and conventional mixes have shown similar compressive strengths at 28 days.

As per IRC SP 46-2013 [12], the usual flexural strength of FRC ranges from 5 to 8 MPa, and the minimum shall not be less than 4.5 MPa. Figure 7 shows the variation in flexural strength of the various mixes under consideration. Mixes with fiber showed marginally higher strength in flexure in comparison with conventional mixes without fiber. A concrete mixture with a macro-synthetic fiber content of 1.2% exhibited higher strength at the early stage and also at 28 days in comparison with other mixes. This result was due to the higher tensile strength of synthetic fiber in comparison with that of PET fiber.

5.1 Challenges

Compatibility and uniform distribution of fiber: Ensuring good compatibility between waste plastic fibers and the concrete mixture is vital. *Long-term performance and ageing*: The long-term performance and ageing features of waste plastic fiber-reinforced concrete need more research and assessment. *Standardisation and design*

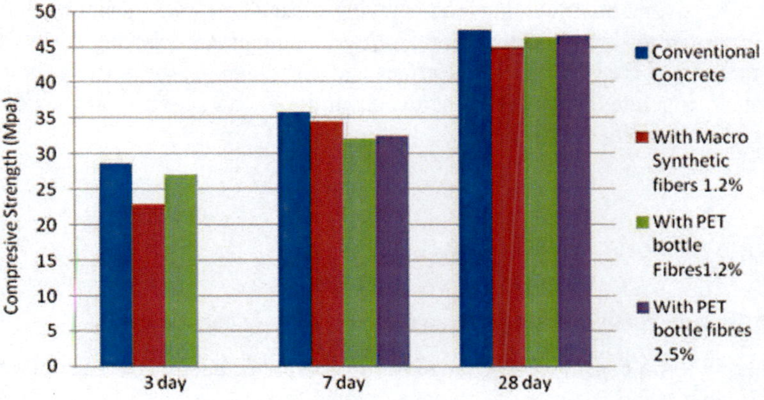

Fig. 6 Compressive strength of mixes with and without fiber

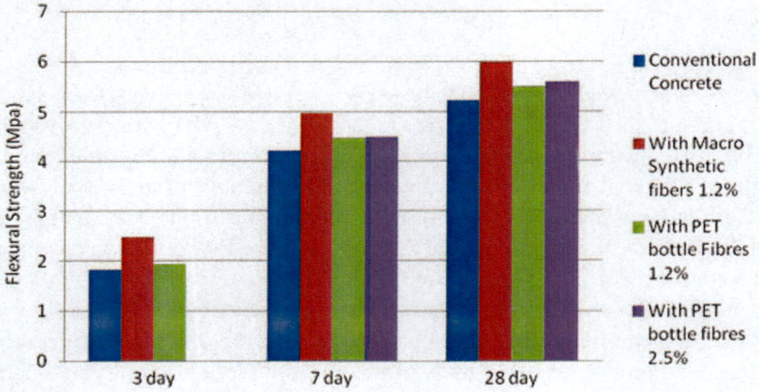

Fig. 7 Variation of flexural strength of various mixes

guidelines: The integration of waste plastic fibers in pavement quality concrete necessitates the creation of standardised testing procedures, design guidelines, and standards.

5.2 *Future Studies*

Durability studies: Long-term durability studies are required to evaluate the resilience of waste plastic fiber-reinforced concrete against conditions such as freeze–thaw cycles, chemical exposure, and ageing impacts. *Structural performance assessment*: Detailed investigations on the structural performance of waste plastic fiber-reinforced pavements, including load-carrying capacity, fatigue resistance, and deformation properties are important. *Life cycle assessment*: Conducting lifecycle analyses to analyse the environmental effect and sustainability of waste plastic fiber-reinforced concrete in comparison with conventional concrete is vital for a thorough understanding of its overall benefits.

6 Conclusions

The following conclusions were drawn from the study carried out:

- There was no considerable change in the compression strength of concrete with PET fiber when compared with conventional mixes.
- The concrete with macro-synthetic fiber showed a 14% increase in flexural strength, and this may be due to the very fine width of the fiber as well as the rough surface.

- An increase in the flexural strength of about 5% was observed in the PET fiber-reinforced concrete of 1.2% when compared with the conventional mix.
- Even though there was an increase in fiber percentage dosage of 2.5%, not much difference in flexural strength was observed compared with the mix with 1.2% fiber
- The mode of failure was bond failure, indicating that by increasing the bond strength of concrete, better flexural strength can be achieved by PET fiber.

Inclusion of PET fiber in concrete for pavement applications can be an effective way of using waste plastic bottles as it increases flexural strength. This can prevent plastic waste from being an environmental hazard. The importance of employing waste plastic as a fiber reinforcement in PQC lies in its potential to address waste management challenges, improve durability, control cracking, achieve cost savings, minimise the carbon footprint, encourage sustainability, and contribute to a circular economy. Embracing this sustainable approach can contribute to a greener and more sustainable construction industry.

Acknowledgements We gratefully acknowledge Stedrant Pvt. Ltd., for allowing us to perform our project in their laboratory. Our sincere thanks to the Management-Matrusri Educational Society, Principal, Vice-Principal, and Head-CED, Maturi Venkata Subba Rao Engineering College for the support.

References

1. Central Pollution Control Board (2020) Annual Report 2020–2021 [Internet]. (December):2–2. Available from: https://www.pvh.com/-/media/Files/pvh/investor-relations/PVH-Annual-Report-2020.pdf
2. Centre for Science and Environment (2020) Managing plastic waste in India [Internet]. Available from https://www.cseindia.org/managing-plastic-waste-in-india-10352
3. Suresha SN, Varghese G, Udya A, Shankar R (2019) Properties of cellulose fibres and waste plastic modified porous friction course mixes. Baltic J Road Bridge Eng 2010(5):156–163. https://doi.org/10.3846/bjrbe.2010.22
4. Sojobi AO (2016) Recycling of polyethylene terephthalate (PET) plastic bottle wastes in bituminous asphaltic concrete. Cogent Eng [Internet] 11(1):1–28. Available from https://doi.org/10.1080/23311916.2015.1133480
5. Tiwari AV, Rao YRM (2017) Study of plastic waste bituminous concrete using dry process of mixing for road construction. Trans Log [Internet] 17(43):47–56. Available from http://search.ebscohost.com/login.aspx?direct=true&db=a9h&AN=129744712&site=ehost-live
6. Kotresh KM, YB Kebede, Bhavya R, Vageesh HP (2015) A study on use of plastic wastes in road pavement construction. Int J Innov Res Sci Eng Tech 4(11):10387–10398
7. Nouali M, Derriche Z, Ghorbel E, Chuanqiang L (2019) Plastic bag waste modified bitumen a possible solution to the Algerian road pavements. Road Mater Pavement Design [Internet]. 0(0):1–13. Available from 14680629.2018.1560355
8. Appiah JK, Berko-Boateng VN, Tagbor TA (2017) Use of waste plastic materials for road construction in Ghana. Case Stud Constr Mater [Internet] 6:1–7. Available from https://doi.org/10.1016/j.cscm.2016.11.001
9. Gupta S, Sengupta J, Rao VVLK (2008) Evaluation of polyester fiber reinforced concrete for use in cement concrete pavement works. Road Mater Pavement Desig. 9(3):441–461

10. Torres DA, Bastidas JG, Ruge Cardenas JC (2018) Reinforced concrete with synthetic fibers (PET+PP) for rigid pavement structures. In: 2018 Congreso internacional de innovacion y tendencias en ingenieria, CONIITI 2018—proceedings
11. Chandrakar M, Qureshi AA (2019) Study on utilization of waste plastic as fibres in fibre reinforced concrete (6):1522–1526
12. Indian Roads Congress (IRC) SP 46 (2013) Guidelines for design and construction of fibre reinforced concrete pavements, p 48. Indian Roads Congress
13. Indian Roads Congress (IRC) (2013) Specifications for road and bridge works, p 906. New Delhi: MoRTH IRC
14. IS-10262 (2004) Recommended guidelines for concrete mix design Bureau of Indian standards
15. IS: 516. Methods of tests for strength of concrete (1999)

"Waste Material-Based Fillers for Patching Potholes Using Warm Mix Asphalt: A Study"

H. S. Bhanusuresh, H. J. Pallavi, K. S. Navya, and M. Harish Sagar

1 Introduction

In 2011, a study conducted by Ramizraja et al. [1] focused on warm mixed design on bitumen concrete using modified binders, specifically polymer modified binder (PMB) and crumb rubber modified binder (CRMB). Sergin et al. [2] and Goh et al. [3] investigated the use of limestone (LS) as a mineral filler in different proportions (4, 5, 6, and 7%) in the production of four different serial asphalt concrete samples. Arabani et al. [4] recognized the growing concern of producing large volumes of agricultural and industrial waste and highlighted the importance of studies focusing on pavement sustainability to reduce environmental hazards and conserve natural resources. The sustainability benefits of warm mix asphalt were concluded by Tutu et al. [5] in 2016. Khurram et al. [6] emphasized the impact of potholes on road and highway performance. Mistry et al. [7] investigated the effect of using fly ash (FA) as a replacement for common fillers in asphalt mixtures. Meghana et al. [8] highlighted the high level of emissions generated by conventional bituminous pavements and emphasized the need for sustainable solutions. Albayati et al. [9] discussed sustainable pavements that meet present generation requirements without compromising the ability of future generations. Shenghua et al. [10] documented the long-term field performance of water-containing and water-based foaming warm mix asphalt (WMA) technologies. Vispute et al. [11] addressed the problem of road deterioration in India, with a particular focus on potholes and the need for engineering solutions to mitigate this issue.

H. S. Bhanusuresh (✉) · M. H. Sagar
School of Civil Engineering, REVA University, Bengaluru, India
e-mail: bhanusuresh.hs@reva.edu.in

H. J. Pallavi
Department of Civil Engineering, GAT Bengaluru, Bengaluru, India

K. S. Navya
Department of Civil Engineering, ACSCE Bengaluru, Bengaluru, India

S. Kolathayar et al. (eds.), *Best Practices in Geotechnical and Pavement Engineering*, Lecture Notes in Civil Engineering 449, https://doi.org/10.1007/978-981-99-8505-0_34

1.1 Objectives of the Study and Methodology

In this project, the aim is to explore the use of waste materials as fillers in the development of sustainable warm mix asphalt using warm mix technology. Rice husk ash, waste brick powder, and waste glass powder are selected as waste material fillers, which have shown to have filler properties in previous research. The primary objectives of this study are to design and develop warm mix asphalt using standard marshal mix design method in the laboratory and to conduct Marshall Stability (MS) test on the developed mix. The study also focuses on evaluating the performance of waste materials as fillers by replacing conventional stone dust with different percentages of these waste materials.

The Marshall Stability test is conducted to determine the stability characteristics of the developed mixture. The main goal of the project is to recommend the most suitable WMA mixture with the filler to patch potholes effectively, which will ultimately help in reducing the environmental impact and enhancing the sustainability of road infrastructure. By utilizing these waste materials, this project aims to contribute towards a more sustainable and eco-friendly approach to road construction in India.

Each step followed in the present investigation is explained in the paragraph with the first step being collection of materials. The most important thing is to collect the materials, such as aggregates, binder, and fillers followed by conduction of laboratory tests on materials. Using the test results, Marshall Mix design was done further conventional and filler-replaced specimens will be fabricated for the obtained OBC. In the final step, Marshal Stability test was conducted on the above said specimens. Test results will be analyzed and best mix will be recommended for potholes as per IRC: 116-2014 code specifications.

2 Experimental Procedure

Waste materials, such as rice husk ash, waste brick powder, and waste glass powder are used in the project rice husk which is one of the main agricultural residues obtained from the outer covering of rice grains during the milling process. The husk is incinerated and converted to ash which contains some of the filler characteristics. Likewise, from the various research works the WBP and WGP also have filler properties. Hence these wastes are planned to implement in this project and all the materials incorporated in the present investigation were tested as per the standard guidelines given in codes. Various properties of materials tested are listed in the form of tables given below. The aggregate was tested for specific gravity using pycnometer. Obtained result was 2.69, it was found to be well within the limit, according to codal provisions as specified in Table 1.

Table 1 Tests on aggregates

Sl. No.	Property	Code
1	Specific gravity	IS: 2386-(Part III)-1963
2	Sieve analysis	IS: 2386-(Part I)-1963
3	Aggregate impact value	IS: 2386-(Part IV)-1963
4	Aggregate crushing value	IS: 2386-(Part IV)-1963
5	Los Angeles abrasion	IS: 2386-(Part IV)-1963
6	Water absorption	IS: 2386-(Part IV)-1963

The sieve analysis test was conducted by taking 1200 grams of aggregate and arranging them in descending order according to IS Sieves. The aggregate was also tested for Impact value using an Impact testing machine, which resulted in 27.725%, well within the limit. The compressive testing machine was used to test the aggregate for crushing value, resulting in 25.96%, which was also within the limit. The LA test was used to determine the abrasion value, which was found to be 32.80%. In addition, the water absorption test was conducted, and the result obtained was 1.5%, which is within the safe permissible limit of less than or equal to 2%. The permissible range of the VG-30 test is between 50 and 70 mm, and the result was found to be within the permissible limit. The softening point was determined to be 48 °C, and the ductility was measured to be 55 cm. For this investigation, three fillers were used: brick powder, glass powder, and rice husk ash (RHA). The asphalt mixture in which recycled brick powder was used as a replacement for filler exhibited properties similar to the mixture with the traditional filler according to Marshall. Additionally, the stability of a mixture increased with an increase in bitumen content.

The average specific gravity of brick powder, glass powder, stone dust, and rice husk ash were determined to be 3.61, 2.166, 2.129, and 2.5, respectively. For this study, the aggregate gradation was selected from the IRC: SP: 116-2014 code for Bituminous Macadam (BM) layer. Mid-value percentage passing values were incorporated for the study.

2.1 *Proportion of Fillers*

The following proportions of fillers are used in the project as given in Table 2.

2.2 *Binder Content*

Following the relevant code, a range of trial binder contents between 4 and 6% was selected for this project, with a constant interval of 0.5%. After selecting the gradation and trial binder contents, specimens with 100% stone dust filler were fabricated using

Table 2 Proportions of fillers

Sl. No.	Stone dust (per cent) (%)	Rice husk ash (percent)	Waste brick powder (percent)	Waste glass powder (per cent)
1	0	33.33	33.33	33.33
2	25	25	25	25
3	50	16.67	16.67	16.67
4	75	8.33	8.33	8.33
5	100	0	0	0

the following procedure. First, 1200 g of well-graded aggregate was weighed and heated to 110 °C. Meanwhile, the binder was heated to 125–130 °C. The aggregate was then mixed with the first binder content, which was 4% by the total weight of aggregate, at a temperature of around 130 °C. The mixture was thoroughly mixed and transferred to the compaction mould placed on the compaction pedestal. 75 blows were given on the top and bottom sides of the specimen using a standard hammer (45 cm, 4.86 kg). The moulds were then kept at room temperature for 24 h before being air-cured for about 3 h and burnt in an oven for one hour at 40 °C before conducting the Marshall Stability test. To determine the optimum binder content (OBC), the maximum stability, maximum unit weight, and 4% air voids were considered. The standard values were calculated as per the Marshall Mix design formulae, and the Marshall Stability test was conducted on the specimen to record the corresponding Marshall Stability value, Marshall Flow value, and density of the specimen. The same procedure was repeated for other percentages of binder, and the corresponding MS values and flow values were recorded. By plotting standard graphs, the optimum binder content (OBC) was determined. The remaining specimens with different percentages of fillers were cast and tested with the obtained OBC value. The Marshall specimens were prepared according to standard specifications with dimensions of 63.5 mm diameter and 100 mm thickness, while maintaining standard temperatures during the process. The casted specimens are as shown in the Fig. 1. With that the filler provisions used in the project are as given in Table 3.

Fig. 1 Casted specimen

Table 3 Number of specimens to be casted

Filler percentages				Number of specimen
SD	RHA	WBP	WGP	
100	0	0	0	15
0	33.33	33.33	33.33	15
25	25	25	25	15
50	16.67	16.67	16.67	15
75	8.33	8.33	8.33	15
Total				75

The fabricated samples were cured as said in the preparation procedure and were kept in the standard temperature before conduction of Marshal Stability test.

Marshall Stability test was conducted in order to obtain the OBC value for further proceedings of the project and the results are as follows. Also standard graphs were plotted in connection with the results so as to obtain the optimum binder content. The standard graphs are as follows.

i. Marshall Stability versus binder content
ii. Marshall Flow versus binder content
iii. Air voids versus binder content
iv. Bulk specific gravity versus binder content
v. Voids filled with bitumen versus binder content.

Among the above graphs, the following three graphs may be considered as reference for determining the optimum binder content of the bituminous mixture. They are as follows,

a. Binder content corresponding to maximum stability
b. Binder content corresponding to maximum bulk specific gravity
c. Binder content corresponding to the median of designed limits of per cent air voids (Vv) in the total mix (i.e. 4%).

After noting down the values, average of three may be taken for obtaining the final optimum binder content (OBC) in percentage. Marshall Stability test results for 100% stone dust is as given in Table 4.

Binder content versus voids filled with bitumen for 100% stone dust from the above table, it was observed that for a conventional WBM with 100% stone dust the optimum binder content obtained was 5% and it yielded higher MS value with 3.98% air voids. The mix showed initial increase in the stability but after it reached the 10.58 kN value of stability due to the increase in binder content, majority of the bitumen would have been absorbed by filler. Hence, there was a decrease in stability after 10.58 kN Marshal Stability value. Marshall Stability test results for 25% stone dust is as given in Table 5.

Table 4 Marshall Stability test results for 100% stone dust

Sl. No.	Binder content%	Average Marshall Stability (kN)	Average Marshall Flow (mm)	Bulk specific gravity	Air voids Vv (%)	VMA (%)	VFB (%)
1	4	6.41	3.63	2.38	4.23	28.21	37.84
2	4.5	7.19	3.37	2.44	4.02	26.18	39.62
3	5	10.58	2.98	2.73	3.98	25.31	43.08
4	5.5	9.78	2.67	2.68	3.67	23.53	45.11
5	6	9.07	2.45	2.61	3.46	21.35	46.70

Table 5 Marshall Stability test results for 25% stone dust

Sl. No.	Binder content %	Average Marshall Stability (kN)	Average Marshall Flow (mm)	Bulk specific gravity	Air voids Vv (%)	VMA (%)	VFB (%)
1	4	8.74	3.27	2.57	4.02	25.97	35.75
2	4.5	9.93	3.41	2.63	3.98	25.71	37.25
3	5	11.08	2.79	2.68	3.71	25.16	41.86
4	5.5	13.56	2.52	2.92	3.54	24.39	44.29
5	6	12.18	2.33	2.77	3.26	22.78	45.73

The Table 5 presented above displays the results of the Marshall Stability test for a filler content consisting of 25% stone dust and 75% waste materials, which were rice husk ash, waste glass powder, and waste brick powder, each accounting for 2% of the total filler content. The optimum binder content (OBC) value obtained for this proportion of fillers was 5.5%, which was found to be the highest stability value of 13.56 kN. Each filler contributed equally to the strength of the mix, although they possessed different filler characteristics. The minimal flow value suggested a lower likelihood of proximity towards viscoelastic failure.

The Table 6 above as given specifies the results of Marshal Stability test for a warm bituminous mix with 50% stone dust filler and other additional fillers. The OBC value obtained was 5% with an MS value of 11.96 kN. This represents a significant increase in strength, approximately 13%, compared with conventional WBM with 100% stone dust as filler. However, it should be noted that the flow values may not be as reliable as those obtained for WBM with 25% stone dust filler. Overall, the use of these alternative fillers has shown promise in improving the strength of the warm bituminous mix.

Above representation in Table 7 indicates the results of MS test conducted on WBM with 75% stone dust specimens. The MS value obtained was again 10.85 kN for a binder content of 5%. Compared with the strength values of conventional WBM, there was only an increment of 2.74% in strength. This was because the stone dust

Table 6 Marshall Stability test results for 50% stone dust

Sl. No.	Binder content %	Average Marshall Stability (kN)	Average Marshall Flow (mm)	Bulk specific gravity	Air voids Vv (%)	VMA (%)	VFB (%)
1	4	9.57	3.75	3.07	3.96	26.43	36.22
2	4.5	10.64	3.46	2.92	3.79	25.77	36.89
3	5	11.96	2.69	2.86	3.52	25.43	39.78
4	5.5	11.63	2.36	2.53	3.36	24.77	42.56
5	6	11.02	2.21	2.17	3.08	23.80	44.73

percentage acquires majority of the air voids again, it would be mostly filled with bitumen which decelerates the strength parameter. However it showed a good result compared with conventional WBM.

Based on the Table 8, presented above, it can be inferred that the warm bituminous mix containing 0% stone dust and 100% other fillers, such as RHA, WGP and WBP showed promising results. The mix had an OBC value of 5.5% with a MS strength value of 11.56 kN, which was the second-best result achieved after the 25% stone dust WBM. However, it is important to note that the Marshal Flow value was higher in this mix compared with the 25% stone dust WBM. Despite this, the mix can still be considered reliable and capable of delivering good strength values, given its high MS strength value. These results highlight the potential of using alternative fillers in warm bituminous mixes and their contribution to improving the overall strength and performance of the mix.

Upon analyzing the consolidated table from Table 9 above, it is evident that the warm bituminous mixture with 25% stone dust and other fillers in equal proportions (25% RHA, 25% WGP, and 25% WBP) demonstrated the highest Marshal Stability value of 13.43 kN with an optimum binder content of 5.5%. This mix may be considered as the optimum strength mix for patching potholes. In comparison with the conventional WBM, which uses 100% stone dust as filler, there was a significant increase of 26.93% in this mix's Marshal Stability value as shown in Fig. 2, making it

Table 7 Marshall Stability test results for 75% stone dust

Sl. No.	Binder content %	Average Marshall Stability (kN)	Average Marshall Flow (mm)	Bulk specific gravity	Air voids Vv (%)	VMA (%)	VFB (%)
1	4	9.12	3.22	3.76	3.61	25.34	37.43
2	4.5	9.78	3.07	3.29	3.54	25.18	38.82
3	5	10.85	2.73	3.02	3.25	25.03	39.30
4	5.5	10.05	2.64	2.83	3.17	24.86	41.68
5	6	9.20	2.39	2.56	2.96	24.61	44.17

Table 8 Marshall Stability test results for 0% stone dust

Sl. No.	Binder content %	Average Marshall Stability (kN)	Average Marshall Flow (mm)	Bulk specific gravity	Air voids Vv (%)	VMA (%)	VFB (%)
1	4	7.47	4.09	2.46	3.86	26.48	36.84
2	4.5	8.68	3.87	2.51	3.61	26.21	37.88
3	5	9.45	3.65	2.58	3.53	25.96	40.63
4	5.5	11.56	3.17	2.63	3.37	24.73	42.30
5	6	10.96	2.73	2.69	3.08	23.34	44.06

a highly stable option. Additionally, as the availability of conventional stone dust has decreased, this mix's combination of fillers would serve the need for higher stability criteria. Moreover, utilizing these waste products for pavement construction with an innovative approach would be an effective method of waste management.

Table 9 Consolidated results of OBC, MS, and filler proportions

Sl. No.	Filler proportions (stone dust percentage)	Optimum binder content (%)	Marshal Stability value (kN)
1	0	5.5	11.56
2	25	5.5	13.43
3	50	5	11.96
4	75	5	10.85
5	100	5	10.58

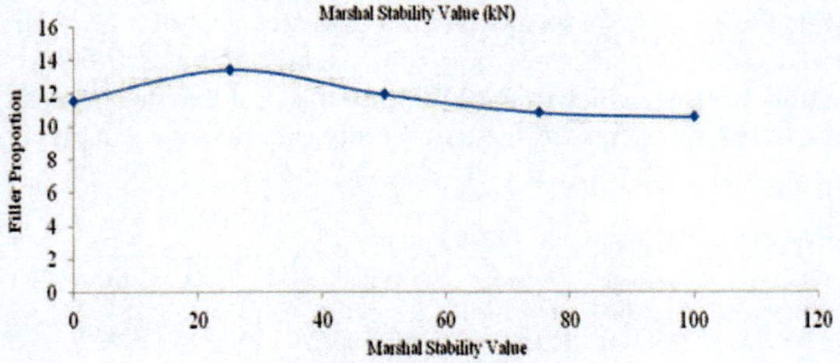

Fig. 2 Consolidated results of MS and filler proportions

3 Conclusions and Future Scope of Work

Based on the detailed study conducted on warm mix asphalt with different fillers, several reliable conclusions were drawn in line with standard specifications. The WMS was designed according to standard Marshal Mix design, and all fabricated samples were subjected to Marshal Stability test. The conventional WMA (stone dust-100%) showed an OBC value of 5% with an MS value of 10.58 kN, while the filler-replaced WMA exhibited optimum Marshal Stability with OBC 5.5% for the filler proportion of 25% stone dust and 25% each of RHA, WGP, and WGP. The WMA with 25% of all the fillers (SD: RHA: WGP: WBP) showed a Marshal Stability 26.93% higher than that of conventional WMA, indicating a promising option for sustainable warm mix technology for pothole patching. In terms of future work, the present investigation can be extended to include detailed performance tests to study the moisture susceptibility characteristics using different types of waste material fillers. This study can serve as a supporting reference for future research in the warm mix asphalt sector. Further research can also explore the potential of other waste materials as fillers in warm mix asphalt and their effectiveness in enhancing pavement performance.

References

1. Ramizraja M (2011) A study on warm mix asphalt technology on bituminous mixes using rediset-warm mix. Int Organiz Sci Res J 13:245–278
2. Sergin S, Saltan M, Morova N (2013) Evaluation of rice husk ash as filler in hot mix asphalt. J Constr Build Mater 48:390–397
3. Goh BH (2014) Review paper on use of rice husk ash as mineral filler in mastic asphalt. Int Res J Eng Res 05
4. Arabani M, Tahami SA (2016) Laboratory Investigation of hot mix asphalt containing waste materials. J Road Mater Pavement design 18(3):713–729
5. Tutu K (2016) Influence of recycled filler on asphalt mix properties. Int J Eng Res Sci Tech 76:227–232. ISSN: 2395-0056
6. Kamal K (2016) Introduction to potholes conjuration in order to visualize 90. ISSN: 2345-0048
7. Mistry R (2016) Ageing and performance of warm mix asphalt pavements. In: Laboratory for road engineering/sealing components, swiss federal laboratories for materials science and technology (EMPA), vol 8, pp 307–309
8. Meghana A (2017) Review of warm mix asphalt in the US. J Basic Appl Eng Res 3(4):309–314
9. Albayati AH (2018) A review on different types of wastes used as fillers in bituminous mix 9(9):289–300
10. Shenghua W (2019) Warm mix asphalt pavements: technology of the future 2:235–280
11. Vispute G (2019) Evaluation of rice husk ash as filler in hot mix asphalt. J Constr Build Mater 48:390–397
12. Celauro C, Airey GD, Collop AC (2016) Assessment of warm mix asphalt for pavement surfacing. In: Proceedings of the institution of civil engineers-transport, vol 169(no 6). Thomas Telford Ltd, pp 379–389

Influence of Jute Reinforcement on the Stiffness Capacity of Cohesionless Pavement Geomaterials

Pradeep Kumar, Yakshansh Kumar, and Ashutosh Trivedi

1 Introduction

The use of natural fibre composites in road construction has become popular for their eco-friendliness and cost-effectiveness. Unpaved roads often suffer from ruts and deformations, but geosynthetic reinforcement has shown promise in addressing this issue in limited lab studies [1]. The observations lead to the conclusion that displacement behaviour, equivalent stress, and principle stress distribution more realistically explain the basis of rutting in pavement subjected to moving load [2]. The study investigates the impact of replacing some of the aggregates with waste rubber tyre particles and replacing a portion of the bitumen with crumb waste rubber tyres, along with the addition of wax content in the bituminous samples [3]. Low strain GeoGauge stiffness tests on compacted silts show the link between stiffness and other variables, including water content, dry unit weight, saturation level, volume change after soaking, shear strength, and soil plasticity [4]. Few researchers have studied geomaterial reinforced with jute, geogrid, geosynthetic, etc. [5–8]. To evaluate the stiffness and strength properties of a compacted subgrade, a cross-hole dynamic cone penetrometer (DCP) was utilized [9]. A study uses stiffness measurements to estimate the compactness of granular geomaterials used in road sub-base and base courses [10]. CASM-n is a unified critical state model for bonded geomaterials that extends an

P. Kumar (✉) · Y. Kumar · A. Trivedi
Delhi Technological University, Shahbad Daulatpur, Main Bawana Road, Delhi 110042, India
e-mail: pradeep.237cv006@nitk.edu.in

Y. Kumar
e-mail: yakshanshkumar_2k20phdce502@dtu.ac.in

A. Trivedi
e-mail: atrivedi@dce.ac.in

P. Kumar
Department of Civil Engineering, National Institute of Technology, Mangalore 575025, India

S. Kolathayar et al. (eds.), *Best Practices in Geotechnical and Pavement Engineering*,
Lecture Notes in Civil Engineering 449, https://doi.org/10.1007/978-981-99-8505-0_35

existing model for reconstituted geomaterials (CASM). CASM-n takes into consideration pre-yield higher strength and stiffness, as well as the cohesive-frictional shearing mode in the post-yield zone, to more correctly characterize bonded geomaterial [11]. The proposed paper suggests optimizing stiffness and damping of structural systems simultaneously by minimizing the sum of mean square responses to stationary random excitations and ensuring constraints on total stiffness and damper capacity. The best design for a constant total stiffness and damper capacity is found in the first phase of a two-step optimization approach, and a number of best designs for changing total stiffness and damper capacity are found in the second step [12, 13]. The study investigates deflections driven by cyclic strength from structural vibrations caused by high-velocity movement loads on rigid pavement. A new finite-element cyclic response model is introduced, using velocity-induced stress ratio, for designing rigid pavement. This model considers characteristic strength and load cycles. It measures central deflection for various loading sources, velocities, and pavement thicknesses [14]. This paper reviews existing literature on the impact of jute reinforcement on the stiffness capacity of geomaterials. Stiffness capacity measures a soil specimen's resistance to deformation when subjected to external forces such as the weight of a building or pavement. It is a fundamental parameter that determines the soil's ability to support applied loads without excessive settlement or deformation. Factors like composition, density, moisture content, and reinforcement materials affect soil stiffness capacity. A higher stiffness capacity indicates stronger and more stable soil that can bear greater loads and stresses without significant deformation. Even many researchers have reported their studies on the stiffness capacity of the soil.

2 Materials and Test Procedure

2.1 Soil

In this research, a soil sample was collected from nearby local places and subjected to sieve analysis test to determine the particle size distribution curve as shown in Fig. 1a. Based on the test results, the soil was categorized as SP-SM (poorly graded sand with silt) in accordance with the IS: 2720 Part-4 (1985) standard. A soil's liquid limit (LL) refers to the moisture content on which the soil exhibits characteristics similar to that of a liquid yet displays minimal shear strength. This can be determined by using Casagrande's liquid limit device which involves closing a groove in the soil sample by repeatedly striking it with a standard sized cup. To obtain the liquid limit of the soil, a semi-log plot is created where the logarithm of the number of blows is plotted against water content. The liquid limit is then determined as the moisture content corresponding to 25 no. of blows as achieved from the plot as shown in Fig. 1b. To determine the maximum dry density (MDD) and optimum moisture content (OMC) of the soil, a standard proctor test (light compaction) was performed according to IS: 2720 Part-7 (1980). The soil sample was compacted in three equal layers in a 1000 cc mould by applying 25 numbers of blows to each layer using a 2.6 kg rammer and a free fall height of 31 cm. The MDD and OMC achieved from this test were 18.63 kN/m^3 and 12.94%, respectively, as shown in Fig. 1c. In addition, other geotechnical characteristics of the soil, including Atterberg limits and specific gravity, were examined in the lab and are shown in Table 1.

2.2 *Jute Geotextile*

A woven jute fibre sheets are used in this research as shown in Fig. 2. The jute geotextile is a sustainable and cost-effective solution for a variety of geotechnical and environmental applications offering superior performance and durability compared to other natural fibre-based materials. Jute fibre is a versatile natural fibre with a number of notable physical and chemical properties which are shown in Table 2.

2.3 *Test Procedure*

To assess the California bearing ratio (CBR) on samples of both reinforced and unreinforced soils, standard procedures were followed. The first step was to take the required amount of oven-dried soil and mix it thoroughly with water until it reached its optimum moisture content (OMC). This mixture was then placed in a CBR mould that had a diameter of 15 cm and a depth of 17.5 cm along with a base plate that can be detached and has perforations as per IS: 2720 Part-16 (1987). To achieve the maximum dry density, a laboratory standard proctor test (light compaction) was conducted and the soil was compacted accordingly. Filter paper and a perforated metallic disc were placed over the specimen to prepare the soil samples and a spacer disc was inserted into the mould reducing the effective height to 12.7 cm with a net capacity of 2250 cm^3. After the soil samples were prepared, the CBR mould containing the unsoaked soil sample was subjected to testing using a CBR testing machine shown in Fig. 3. The CBR values of both the unreinforced and reinforced soil samples were then evaluated based on the plunger penetration of 2.5 and 5 mm. The testing was conducted by applying a load on the sample's top surface through the plunger at a constant rate of penetration (1.25 mm/min). The load and corresponding penetrations were recorded and the CBR values were calculated as the ratio of the load required to penetrate the soil sample by the plunger at a depth of 2.5 or 5 mm to the standard loads. This testing process helps to determine the strength and stiffness capacity of the soil which is an important factor in various engineering applications.

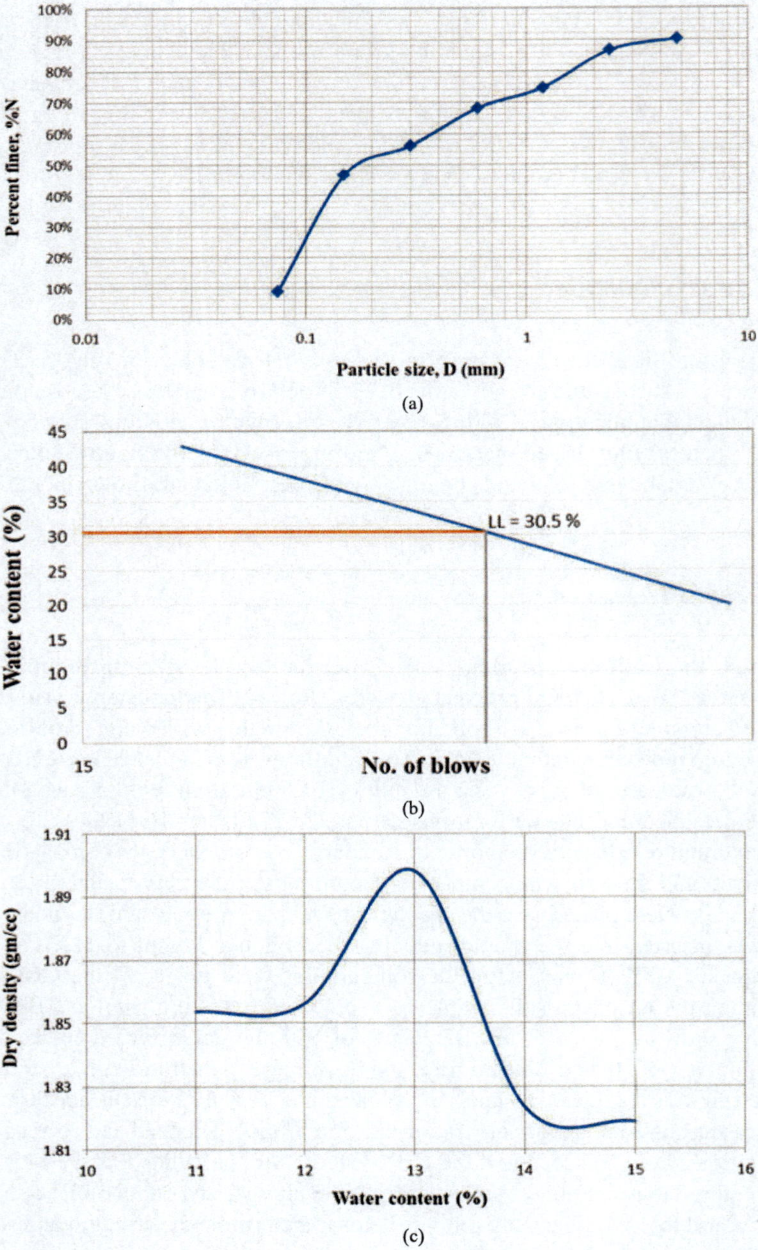

Fig. 1 **a** Grain size distribution curve of soil, **b** Atterberg limits graph of soil and **c** standard proctor test compaction curve for soil

Table 1 Soil properties

Serial no	Property	Notations	Values	Units
1	Gravel fraction	–	9.7	%
2	Sand fraction	–	81.3	%
3	Silt and clay fraction	–	9	%
4	Specific gravity	G	2.63	–
5	Liquid limit	LL	30.5	%
6	Plastic limit	PL	25	%
7	Soil classification	SP-SM	–	–
8	Maximum dry density	γ_{dmax}.	18.63	kN/m^3
9	Optimum moisture content	OMC	12.94	%

Fig. 2 Woven-type jute sheet

Table 2 Properties of jute fibre

Jute fibre	Values	Units
Physical properties		
Density	1.3	g/cm^3
Elongation at break	1.5–1.8	%
Tensile strength	393–773	MPa
Young's modulus	26.5	GPa
Colour	Light brown to grey	–
Texture	Coarse, rough, and stiff	–
Chemical properties		
Lignin	12–15	%
Cellulose	65–70	%
Hemicellulose	12–14	%
Pectin	0.5–1	%
Reaction with acids	Decomposes	–

3 Results and Discussions

3.1 CBR Testing

The test involves mixing an unreinforced soil sample with single, double, and triple layers of jute fibre at various depths to determine the optimal depth of the layers. From the data of CBR results, evaluate the spring constant (K) values and stiffness capacity ($K/K_{\text{max.}}$) of soil which is shown in Table 3 as well as in the equation forms. The reinforcement of single, double, and triple layers of jute fibre in an unreinforced soil sample results in the highest stiffness capacity at a zero penetration factor value. This can be attributed to the significantly higher compaction level at the top surface. However, as the penetration factor increases, both the reinforced and unreinforced soils exhibit a decrease in stiffness capacity. Interestingly, Table 3 reveals that at a depth of $D/4$ (3.175 cm), there is a minimal decrement in stiffness capacity compared to other layers and the unreinforced soil. Moreover, this depth demonstrates the highest stiffness capacity at different penetration factors. Consequently, the optimum depth at $D/4$ represents higher stiffness capacity and contributes to an increased strength of the pavement when compared to the unreinforced soil specimen. Figure 4 illustrates the stiffness capacity versus penetration factor plot curves for single, double, and triple layers of jute fibre-reinforced soil as well as a comparison between reinforced and unreinforced soil. The plot shows that the reinforced soil has a significantly higher stiffness capacity than the unreinforced soil at the same penetration factor.

$$k = \frac{F}{\delta} \rightarrow \frac{Q(t, L_f, P_r)}{\delta(t, d_{gr})} \tag{1}$$

Fig. 3 CBR testing machine

Equation 1 relates the spring constant (k) of soil to the applied force (F) divided by the amount of deflection (δ) that it experiences under the load.

$$Q = f\left(t, L_f, P_r\right) \tag{2}$$

Equation 2 represents the applied force (Q) on a sample of soil as a function of time (t), load factor (L_f), and proving ring reading (P_r). The equation suggests that the applied force is influenced by several factors including the time over which the load is applied, the size of the soil, and the output signals of measuring devices used to measure the load.

$$\delta = f\left(t, d_{\mathrm{gr}}\right) \tag{3}$$

Equation 3 relates the deflection (δ) of soil to the variables time (t) and dial gauge reading (d_{gr}). The dial gauge reading refers to the displacement or deflection of the material under the applied load. The equation suggests that the amount of deflection of soil is influenced by several factors incorporating the time over which the load is applied and the displacement of the soil specimen as measured by a dial gauge.

$$P_f = \frac{\delta}{\delta_{\mathrm{max.}}} \tag{4}$$

where P_f is the penetration factor, δ is the deflection of the specimen and δ_{max}. is the maximum deflection.

4 Conclusions

Based on the information provided in this research paper, it can be concluded that incorporation of jute fibre in unreinforced soil sample enhances the stiffness capacity of soil specimens at different penetration factor values particularly at the optimal depth of *D*/4 (3.175 cm) which resulting in improvement in strength of the pavement. The stiffness capacity has increased from 0.378 to 0.682 at maximum penetration factor shown in Table 3 which increases 80.42% of the capacity of stiffness when compared to unreinforced soil. The results obtained from the stiffness capacity versus penetration factor plot curves which are shown in Sect. 3 demonstrate that the stiffness capacity increases with the number of jute fibre layers which indicating that the addition of jute fibre reinforcement can significantly enhance the soil's stiffness. The single-layered jute fibre-reinforced soil shows the highest stiffness capacity among all the reinforced soil configurations tested. Moreover, the plot curves highlight that the reinforced soil's stiffness capacity is significantly higher than the unreinforced soil at the same penetration factor indicating that jute fibre reinforcement can effectively improve the soil's load-carrying capacity thereby increases the strength of the pavement. Therefore, the findings of this study highlight the potential of jute

Table 3 Comparison of stiffness capacity values at different penetration factor between unreinforced and reinforced soils ($D = 12.7$ cm)

Penetration factor (P_f)	K/K_{max} (unreinforced soil)	K/K_{max} (jute fibre embedded in single layer)			K/K_{max} (jute fibre embedded in double layers)			K/K_{max} (jute fibre embedded in triple layers)
		D/2 (6.35 cm)	*D*/3 (4.23 cm)	*D*/4 (3.175 cm)	*D*/2 (6.35 cm) and *D*/3 (4.23 cm)	*D*/3 (4.23 cm) and *D*/4 (3.175 cm)	*D*/2 (6.35 cm) and *D*/4 (3.175 cm)	*D*/2 (6.35 cm) and *D*/3 (4.23 cm) and *D*/4 (3.175 cm)
0	1	1	1	1	1	1	1	1
0.2	0.643	0.718	0.7	0.727	0.667	0.674	0.656	0.675
0.3	0.548	0.604	0.6	0.636	0.533	0.579	0.521	0.516
0.4	0.4911	0.546	0.562	0.613	0.449	0.533	0.468	0.437
0.5	0.457	0.506	0.54	0.618	0.406	0.495	0.425	0.4
0.8	0.402	0.484	0.493	0.631	0.35	0.516	0.398	0.35
1	0.378	0.484	0.51	0.682	0.346	0.543	0.387	0.379

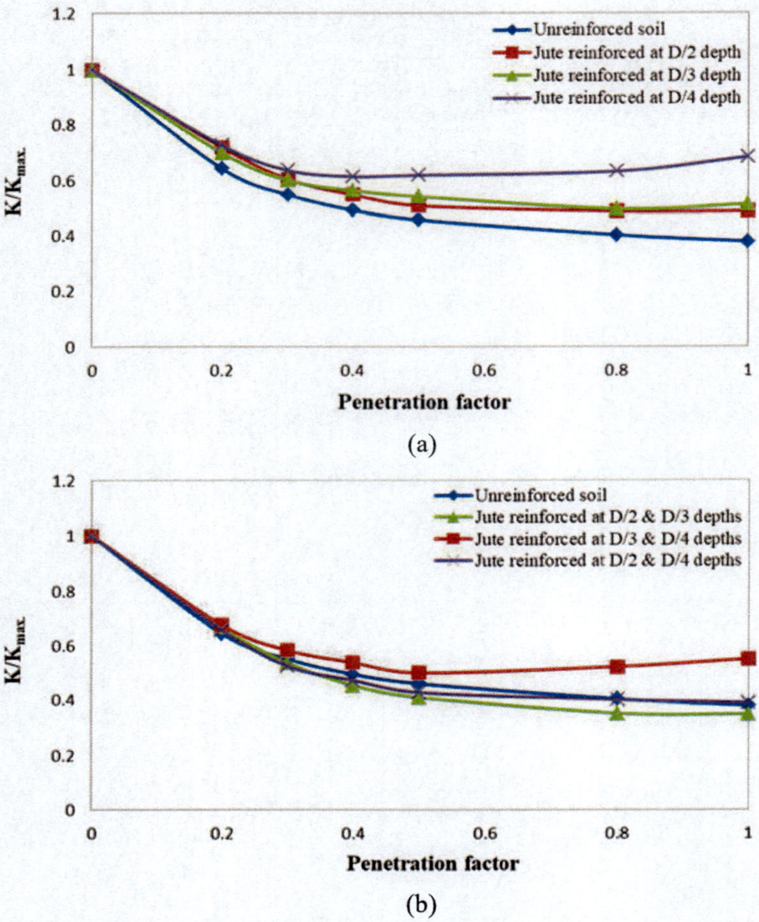

Fig. 4 Stiffness capacity versus penetration factor plot for **a** soil reinforced with single, **b** double, **c** triple, and **d** optimum depths of jute fibre among all layers

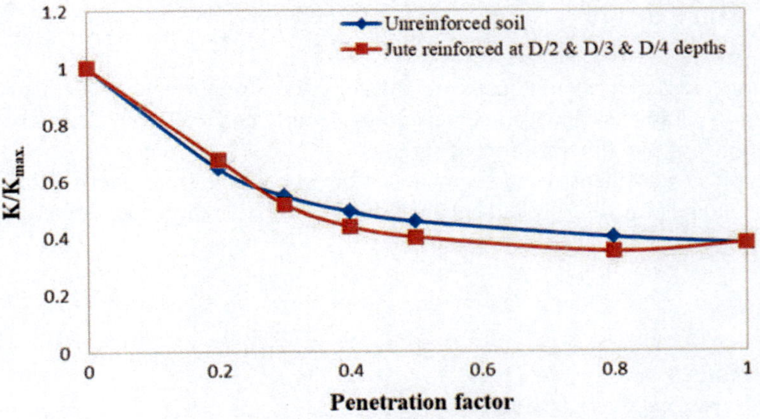

(c)

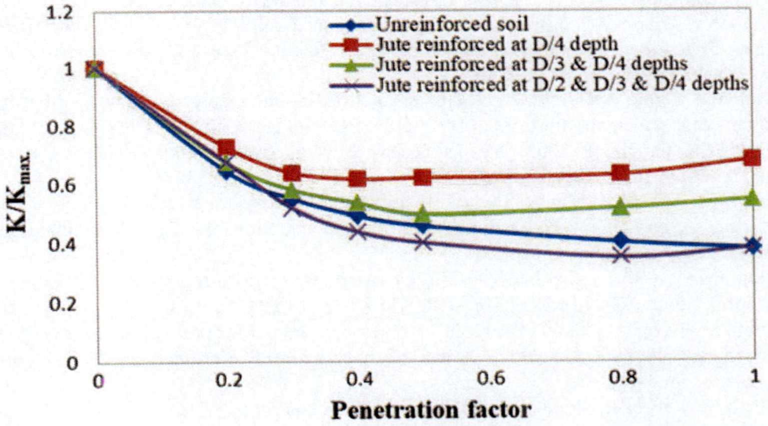

(d)

Fig. 4 (continued)

fibre reinforcement as a viable alternative to conventional reinforcement materials in geotechnical engineering, providing an economical and sustainable solution for enhancing the capacity of stiffness and load-carrying capacity of soil in various geotechnical applications.

5 Future Scope

- Future research should focus on the long-term performance, optimization of design parameters, environmental considerations, and exploring alternative applications for jute fibre-reinforced soil.
- These advancements would provide a more comprehensive understanding of the technique's capabilities and enhance its cost-effective implementation in various geotechnical engineering projects.

References

1. Singh M, Trivedi A, Shukla SK (2022) Evaluation of geosynthetic reinforcement in unpaved road using moving wheel load test. Geotext Geomem 50(4):655–667
2. Kumar Y, Trivedi A, Shukla SK (2023) Damage evaluation in pavement-geomaterial system using finite element-scaled accelerated pavement testing. Transp Infrastruct Geotech. https://doi.org/10.1007/s40515-023-00309-y
3. Mohan R, Gupta A, Gaur k (2021) Utilization of bitumen, aggregate and wax with rubber tyre in a flexible pavement. Int Conf Adv Civil Eng 1197(1):012017
4. Phillip SK, Jianping P (2003) Use of stiffness for evaluating compactness of cohesive pavement geomaterials. In: Transp Res Rec, J Transp Res Board 1849(1):11–19
5. Singh M, Trivedi A, Shukla SK (2020) Influence of geosynthetic reinforcement on unpaved roads based on CBR, and static and dynamic cone penetration tests. Int J Geosynthetics Ground Eng 6(2):13
6. Singh M, Trivedi A, Shukla SK (2020) Unpaved test sections reinforced with geotextile and geogrid. Mater Today Proc 32:706–711
7. Singh M, Trivedi A, Shukla SK (2020) Fuzzy-based model for predicting strength of geogrid-reinforced subgrade soil with optimal depth of geogrid reinforcement. Transp Infrastruct Geotech 7:664–683
8. Singh M, Trivedi A, Shukla SK (2019) Strength enhancement of the subgrade soil of unpaved road with geosynthetic reinforcement layers. Transp Geotech 19:54–60
9. Lee J-S, Tutumluer E, Hong WT (2021) Stiffness evaluation of compacted geo-materials using crosshole-type dynamic cone penetrometer (CDP), rPLT, and LFWD. Constr Build Mater 303:124015
10. Kazemi Darabadi B, Khavandi Khiavi A, Ouria A (2018) Evaluation of the compactness of subbase and base geomaterials by using stiffness. Sādhanā 43:195
11. Yu HS, Tan SM, Schnaid F (2007) A critical state framework for modelling bonded geomaterials. Int J Geomech Geoeng 2(1):61–74
12. Takewaki I (1999) Displacement-acceleration control via stiffness-damping collaboration. Earthquake Eng Struct Dynam 28(12):1567–1585
13. Takewaki I (2000) An approach to stiffness-damping simultaneous optimization. Comput Methods Appl Mech Eng 189(2):641–650
14. Kumar Y, Trivedi A, Shukla SK (2023) Deflections governed by the cyclic strength of rigid pavement subjected to structural vibration due to high-velocity moving loads. J Vibr Eng Tech. https://doi.org/10.1007/s42417-023-01063-8

Stability Analysis of a Road Vertical Cut

Priyanjali A. Bhonsle and Sumitra S. Kandolkar

1 Introduction

The downward and outward movement of soil mass is referred to as slope failure. The pores of soil mass filled with water exert pressure on the soil grain decreasing the frictional resistance; also the type of soil and geometry affects the stability of slope. Factor of safety < 1 indicates marginally unstable slope that requires attention, monitoring and needs to be stabilize to increase the safety factor and to reduce the slope movement. The main purpose for analysing the stability of slopes is to assess the equilibrium conditions for the excavated slopes and investigating the possibility of failure. If the critical failure surface of the slope is identified then the factor of safety can be accurately determined. Limit equilibrium analysis and finite element analysis are the method used for the analysis of slope failure. Researchers have focused on the slope failure due to the presence of cohesive soil and improper drainage of rainwater in soil [1–7]. Authors [8–13] presented the results of factor of safety which were numerically analysed using different slope stabilization techniques on week soil. The present study makes an attempt to provide an extensive review on the slope which is likely to cause sudden failure in monsoons due to the change in geometry of slope done by the residence of the area. The stability analyses is done by using Bishop's limit equilibrium method and using FEM-based software PLAXIS 2D. The study presents improvement in factor of safety with the variations in geometric profile, surcharge values, and soil properties. Habtemariam et al. [14], Taher et al. [15] conducted study on the effects due to change in profile of the slope.

P. A. Bhonsle · S. S. Kandolkar (✉)
Civil Engineering Department, Goa Engineering College, Farmagudi, Goa 403401, India
e-mail: sumitra@gec.ac.in

S. Kolathayar et al. (eds.), *Best Practices in Geotechnical and Pavement Engineering*, Lecture Notes in Civil Engineering 449, https://doi.org/10.1007/978-981-99-8505-0_36

2 Material and Methods

The site is situated in Moira village of Bardez taluka, North Goa District. The geographical co-ordinates of site are 15°35′56″N and 73°50′22″E. The site condition is depicted in Fig. 1. The soil encountered on the site was classified as SM based on the various laboratory tests conducted on the soil which are presented in Table 1 (Sample 1). The soil shear strength properties indicate cohesion in the soil. Hence the soil having frictional properties was considered. Author [16] found embankment to be stable with the use of lateritic soil. The properties of this soil are exhibited in Table 1 (Sample 2).

2.1 Analyses Method

The main purpose for analysing the stability of slopes is to assess the equilibrium conditions for the excavated slopes and investigating the possibility of failure. If the critical failure surface of the slope is identified then the factor of safety can be accurately determined. Here the stability of slope/cut shown in Fig. 2 was calculated using Bishop's method of slices [17]. Method uses graphical trial and error method where the angle of slope is assumed to be 85°. The height of the slope is 10 m. The factor of safety was calculated using Eq. (1). The theoretical results of the analysis

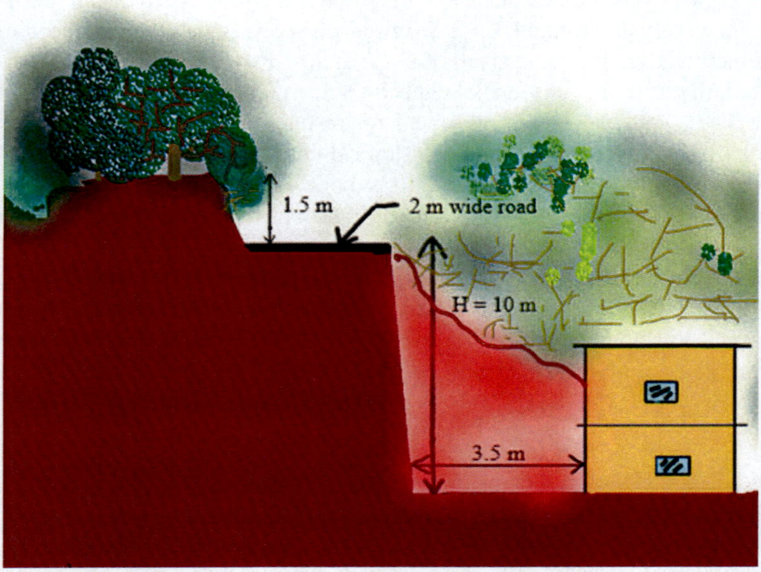

Fig. 1 A view of vertical cut at Moira

Table 1 Input parameters considered for analysis

Parameters	Sample 1	Sample 2
Bulk unit weight, γ (kN/m^3)	19	18
Saturated unit weight γs_{at} (kN/m^3)	20	19
Young's modulus, E	47,000	14,000
Poison's ratio μ	0.35	0.3
Internal friction, ϕ(°)	5.56	30
Cohesion (kN/m^2)	14.32	15
Material model	Mohr coulomb	Mohr coulomb
Behavior type	Drained	Drained

were further verified using PLAXIS 2D V 2023.1.0 software. PLAXIS 2D is a finite element software used to analyse the vertical cut and determine the factor of safety of the slope using the strength reduction method. The vertical cut was checked for stability in the existing condition and its factor of safety was found to be < 1 hence the analyses was carried out as presented in Table 2 for different values of height of vertical cut, base widths based on site restrictions. The vertical cut was analysed by considering batter. All the cases were analysed (i) with surcharge (ii) without surcharge conditions.

$$\text{FOS} = \frac{\text{CL} + (w\cos\alpha - \mu l)\tan\phi}{w\sin\alpha} \quad (1)$$

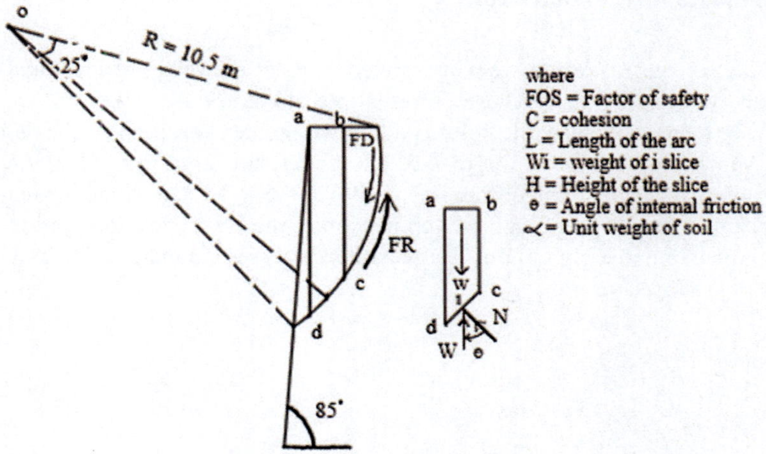

Fig. 2 Slope stability analysis using Bishop's method

Table 2 Different cases considered for the analysis

Case	Height (m)	Base (m)	Angle, β (°)	Surcharge load, q (kN/m^2)
1	10	1.1	85	50
2	10	3.5, 3, 2	71, 73, 79	0
	8	3.5, 3, 2	66, 69, 76	0
	6	3.5, 3, 2	60, 63, 72	0
	4	3.5, 3, 2	49, 53,63	0
3	10	3.5, 3, 2	71, 73, 79	50
	8	3.5, 3, 2	66, 69, 76	50
	6	3.5, 3, 2	60, 63, 72	50
	4	3.5, 3, 2	49, 53,63	50

These analyses were adopted to understand the effect of each geometric profile variation on the stability condition of the vertical cut under study. The vertical cut was further modelled using PLAXIS 2D software. The problem is treated as plane strain case and the soil is discretized into 15 node triangular element. The unit of length, force, and time are taken as m, kN, and day. The x–y plane is used to generate the model. The standard fixities are assigned to the model which appears as two parallel lines perpendicular to the fixed direction. The fine mesh was generated. The laboratory test results were used to evaluate the parameters needed for analyses with the Mohr–Coulomb material model under drained condition.

3 Results and Discussion

PLAXIS 2D analysis results were interpreted based on the values of stress and deformations for existing case. These deformation and stress are displayed in Fig. 3. Figure 3b helps to identify the failure pattern within the slope model. The deformed mesh shows slope deformation of 0.053 H under surcharge load 50 kN/m^2. The values of displacement range from 4.8×10^{-3} to 5.6×10^{-3} m which indicates that displacements are higher along the top portion of the cut. Hence top portion of the cut is vulnerable to early failure. The maximum stresses are indicated by red colour in the PLAXIS 2D results.

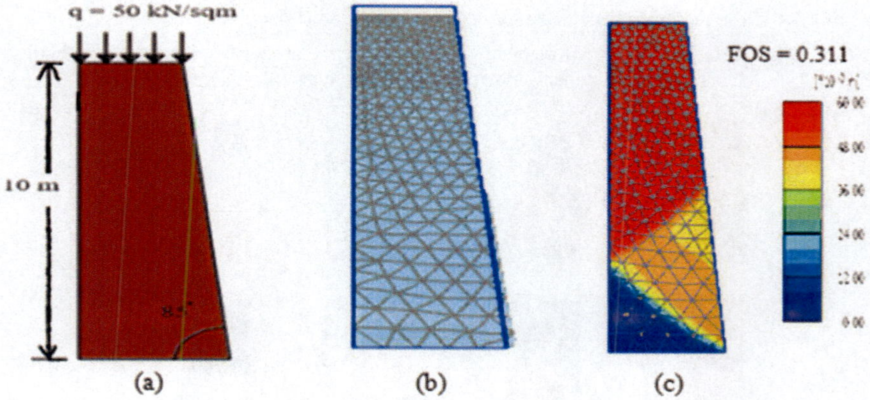

Fig. 3 Results of PLAXIS 2D analysis (**a**) Cuts used for analysis (**b**) Deformation of the vertical cut model (**c**) Total stress of slope

Table 3 Safety factor calculated using Bishop's method and PLAXIS 2D

Sr. no	Safety factor (Bishop's method)	Description	Safety factor (PLAXIS 2D)	Description
1	0.309	Unsafe	0.311	Unsafe

The theoretical analysis using Bishop's method and the numerical analysis using PLAXIS 2D generates a value of factor of safety < 1 which is highlighted in Table 3. Hence it can be stated that the current status of the vertical cut is unstable and it is likely to undergo failure at any time.

3.1 Effect of Geometric Profile on Factor of Safety (Without Surcharge)

As an attempt to stabilize the cut under existing state, an attempt was made to study the effect of variations in geometric profile with respect to height of cut, base slope, base width of cut, angle of the cut, and the surcharge load. A total of twelve numerical analysis were performed to study the effect of varying geometrical profile on the factor of safety and the stability of the vertical cut as whole. Table 4 below shows the displacement pattern of the cut. The results shown in Fig. 4 indicate improvement in the stability condition prominently with the reduction in height of the cut, while the increase in base width does not improve the stability of the cut appreciably.

Table 4 Below shows displacements for different cases under study

H, B, β	Total displacement	Horizontal displacement	Vertical displacement	Effective stress	FOS
10, 3.5, 71	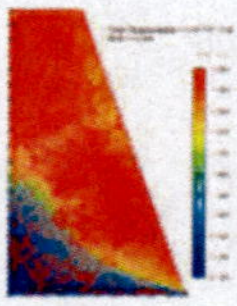	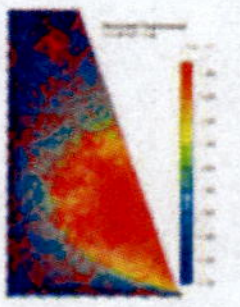	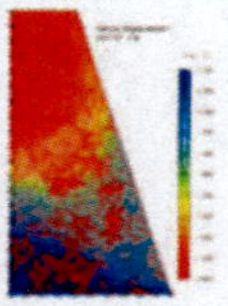	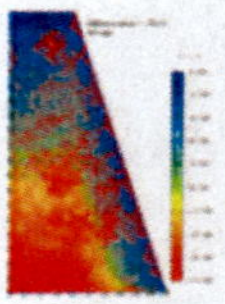	0.404
8, 3.5, 66	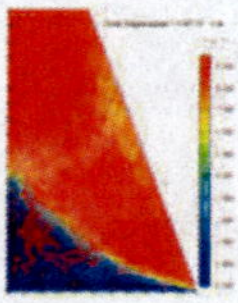	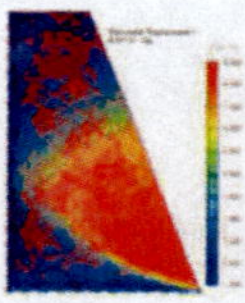	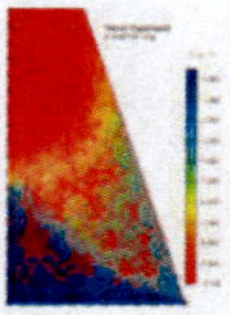	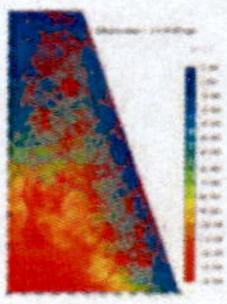	0.515
6, 3.5, 60	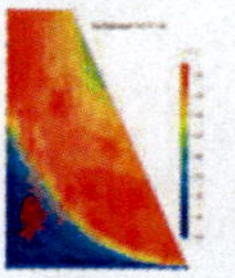	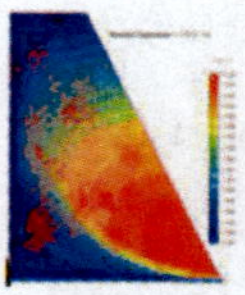	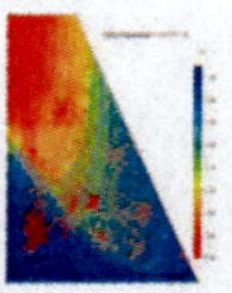	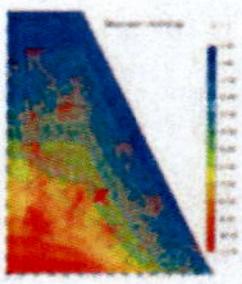	0.741
4, 3.5, 49	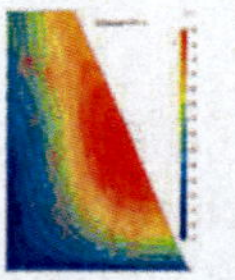	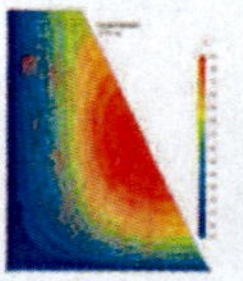	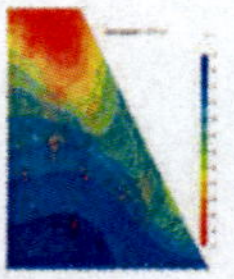	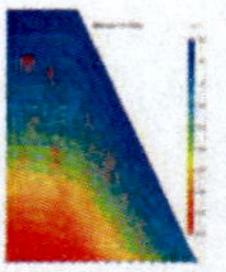	0.99

Height = H, Base = B, Angle = β

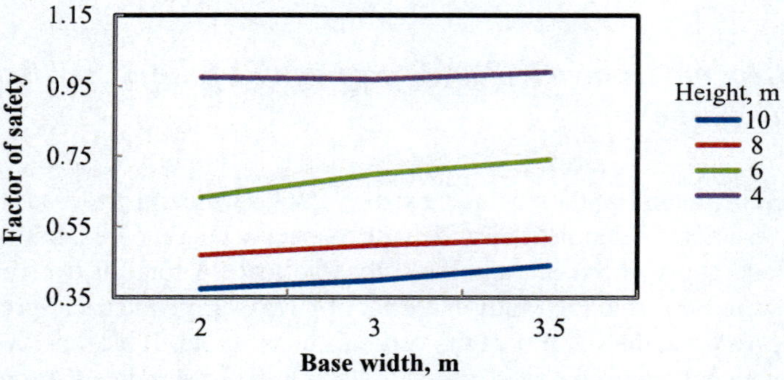

Fig. 4 Effect of varying geometric profile on FOS (without surcharge)

3.2 *Effect of Geometric Profile of the Cut on Factor of Safety (with Surcharge)*

The vertical cut under study is a part of a village road, hence analyses was performed with a surcharge value of 50 kN/m^2 to account for traffic loads. Twelve cases were analysed again with varying geometric profiles with surcharge effect. The village road cut under study is presently not subjected too much traffic load, but anticipating future situation of the road to be used for vehicular traffic loads, the stability of the cut was studied with surcharge. Table 5 below shows the displacement cut profile obtained for various cases taken in consideration. The results presented in Fig. 5, shows the impact of surcharge on the factor of safety values. The graph indicates that the FOS is affected adversely and the stability of cut is at stake. Adopting a specific solution like gabion wall or soil nailing for this case requires additional land acquisition, hence a new soil altogether with better shear strength parameters was considered and the stability analysis were repeated using PLAXIS 2D.

Table 5 Below shows displacement obtained for different cases

H, B, β	Total displacement	Horizontal displacement	Vertical displacement	Effective stress	FOS
10, 3.5, 71	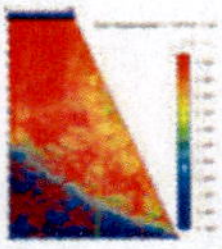	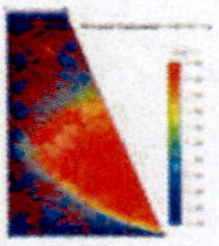	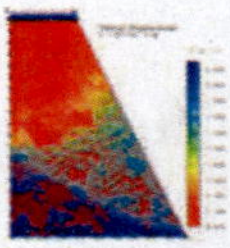	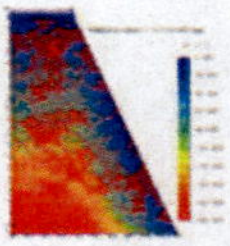	0.337
8, 3.5, 66	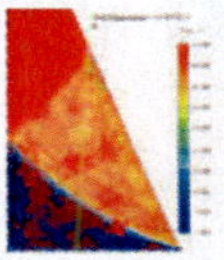	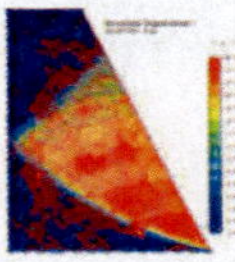	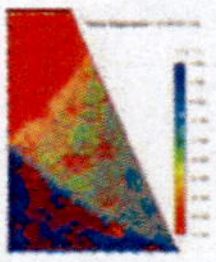	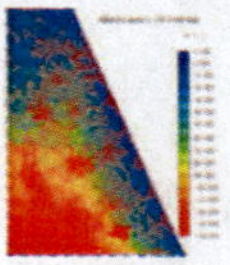	0.401
6, 3.5, 60	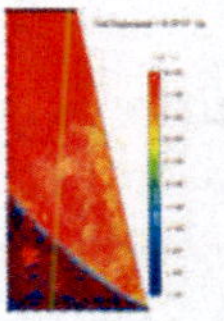	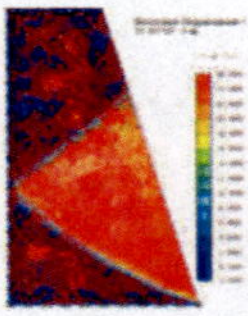	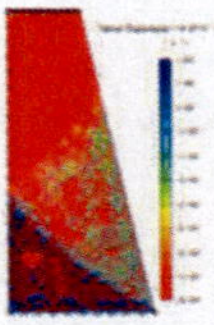	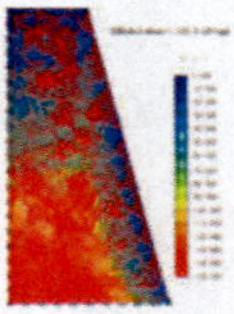	0.501
4, 3.5, 49	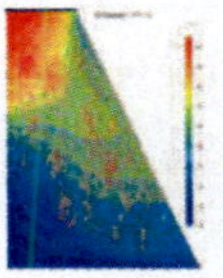	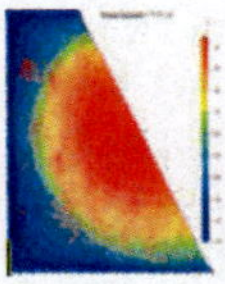	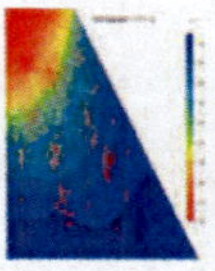	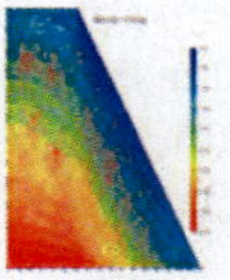	0.675

Height = H, Base = B, Angle = β

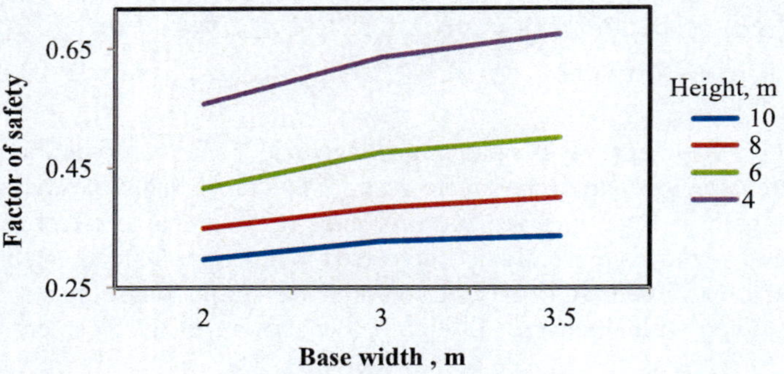

Fig. 5 Variation of FOS of the cut for various cut heights with, $q = 50$ kN/m^2

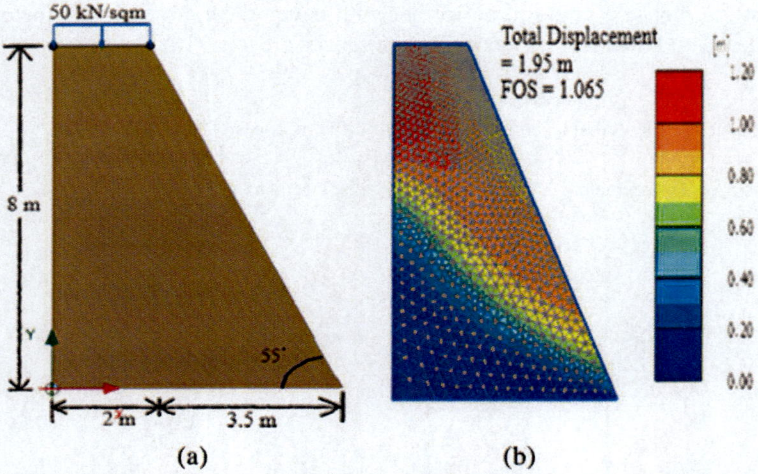

Fig. 6 Results of analysis (**a**) Model of the vertical cut (**b**) Total displacement of the slope

3.3 *Effect of Improved Soil on FOS (with Surcharge = 50 kN/m^2)*

To arrive at a stable condition of the cut, the numerical analysis was carried out by changing the soil forming the vertical cut. A suitable soil with improved friction angle was used in the analysis along with a reduced height of the vertical cut to 8 m which depicted in Fig. 7. Figure 6 shows the results of the analysis. The factor of safety obtained was > 1 (Fig. 7).

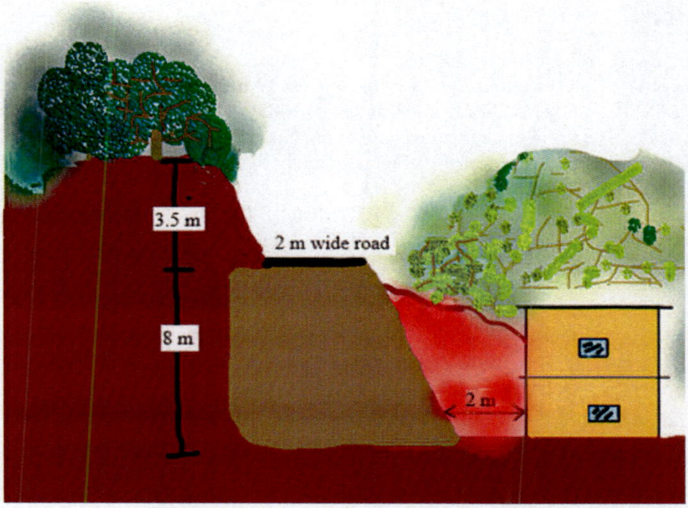

Fig. 7 A view of vertical cut after stabilization

4 Conclusion

This study was aimed at providing a viable solution for stabilizing the vertical cut, which is found to be on the verge of failure as the factor of safety in the existing condition is < 1.

1. The soil visual inspection and laboratory test indicate slope cut is made of silty sand with traces of clay and organic matter.
2. The slope vertical cut in current situation is unstable and unsafe. This is indicated through the results of factor of safety less than 1 as obtained from theoretical analysis and PLAXIS 2D analysis.
3. The variations in geometric profile of slope with respect to height and angle with vertical were considered. Twelve different profiles were considered for analyses in PLAXIS 2D with and without surcharge and factor of safety for all twelve cases is < 1.
4. The slope/vertical cut is found to be stable after performing the analysis with different soil having properties favourable for improving the stability of the vertical cut.

References

1. Heyman J (1973) The stability of a vertical cut. Int J Mevh 15:845–854
2. Zimmie TF, Pamuk A, Adalier K, Mahmud MB (2005) Retrofit-reinforcement of cohesive soil slopes using high strength geotextiles with drainage capability. Geotech Geol Eng 23:447–459
3. Hu J, Ma F (2017) Failure investigation at a collapsed deep open cut slope excavation in soft clay. Geotech Geol Eng
4. Bushira KM, Gebregiorgis YB, Verma RK, Sheng Z (2018) Cut soil slope stability analysis along national highway at Wozeka—Gidole Road, Ethiopia. Model Earth Syst Environ
5. Ramkrishnan R, Karthik V, Sruthy MR, Sharma A (2019) A soil reinforcement and slope stabilization using natural jute fibres. In: Springer International Publishing AG, part of Springer Nature
6. Yanamandra G, Oh WT (2020) Estimation of stability of an unsupported deep vertical cut in clay. ResearchGate
7. Kumar P, Thakur M, Singh TN (2022) Slope stability analysis of road cut slopes along NH-58 in Alaknanda Valley from Dhari Devi to Rudraprayag, Uttarakhand. India. J Earth Syst Sci 131:82
8. Al-Homoud AS, Tubeileh TK (1997) Analysis and remedies of landslides of cut slopes due to the presence of weak cohesive layers within stronger formations. Environ Geol 33(4):3138
9. Woodward PK (1999) Stability of slopes with berms on rigid foundations. Geotech Geol Eng 16:309–320
10. Vekli M, Ayteki M, Ikizler SB, Calik U (2012) Experimental and numerical investigation of slope stabilization by stone columns. Nat Hazards 64(64):797–820
11. Komadja GC, Pradhan SP, Roul AR, Adebato B, Habinshuti JB, Glodji LA, Onwualu AP (2020) Assessment of stability of a Himalayan road cut slope with varying degrees of weathering: a finite element model based approach. In: Heliyon
12. Perumalsamy K, Ranganathan S (2022) Single pile in cohesionless soil in sloping ground under lateral loading. Int J Geo-Eng
13. Singh D, Singh PK, Kainthola A, Pandey HK, Kumar S, Singh TN (2022) Analysis of failure pattern in cut slopes of bedded sandstone: a case study. Environ Earth Sci
14. Habtemariam BG, Shirago KB, Dirate DD (2022) Effects of soil properties and slope angle on deformation and stability of cut slopes. In: Hindawi advances in civil engineering, p 10
15. Taher NR, Gor M, Aksoy HS, Awlla HA (2022) Numerical investigation of the effect of slope angle and height on the stability of a slope composed of sandy soil. GUFBD/GUJS 12(2):664–675
16. Kumar A, George V, Marathe S (2016) Stability analysis of lateritic soil Embankment sub-grade using plaxis-2D. Int J Res Civil Eng
17. Sungkar M, Munirwansyah M, Munirwan RP, Safrina D (2020) Slope stability analysis using Bishop's and finite element method. Mater Sci Eng 14:933

A Study on Flexural Fatigue Performance of Cement-Treated Base in Flexible Pavements Due to Repetitive Loading

B. V. Kiran Kumar and N. Shiva Prasad

1 Introduction

Flexible pavement has become one of the most preferred pavement system for road construction across the globe. Flexible pavements are constructed using rock aggregates as a multi-layered structure made up of the top bituminous bound aggregate layer, the intermediate aggregate base, and sub-base layers placed over subgrade soil. Construction of pavements involves excessive use of non-renewable resources such as natural aggregates and fossil fuels which accounts for high embodied energy of the pavement structure. This happens to create a dearth of the world's non-renewable resources and hence it becomes the fundamental responsibility for all levels of government to consider pavement structures on a sustainability basis rather than just on initial cost basis [1]. Sustainable development should be considered vital because it helps to reduce investments, meet societal needs, save natural resources, maintain synchronization with natural resources and people, and retain the availability of natural resources for impending generations [2]. Every developing country's monetary growth is essentially reliant on expanding and modernizing its infrastructure. Over the past two decades, researchers have been especially interested in exploring alternative means for durable and sustainable construction materials [3]. The construction industry and institutions have kept on with the research activities for improving the use of lower quality materials by enhancing their strength and durability aspects which is factual with the pavement industry as well.

B. V. K. Kumar
Department of Civil Engineering, Goverment Sri Krishnarajendra Silver Jubilee Technological Institute, Bengaluru, Karnataka 560001, India
e-mail: dr.kirankumar@ka.gov.in

N. S. Prasad (✉)
Department of Civil Engineering, B. M. S. College of Engineering, Bengaluru, Karnataka 560019, India
e-mail: shivaprasad.civ85@gmail.com

S. Kolathayar et al. (eds.), *Best Practices in Geotechnical and Pavement Engineering*, Lecture Notes in Civil Engineering 449, https://doi.org/10.1007/978-981-99-8505-0_37

The use of cement stabilization technique along with additional cementitious materials, such as ceramic powder, fly ash, Kota stone slurry waste, metakaolin, silica fume, will have a substantial influence on construction costs and carbon footprint reduction [4]. The use of marginal and waste materials reduces the use of virgin aggregates, which have become a scarce resource in recent decades. Subgrade soils, aggregate bases, and marginal materials can now be strengthened and made more durable by stabilization techniques [5]. The base courses used in pavement construction come in a wide range of classifications, including recycled pavement base, asphalt-stabilized base, permeable aggregate base, and cement-stabilized base [6].

Cement-stabilized aggregate base, or in general, cement-treated base (CTB) forms the easiest and economical methods of pavement construction on long run. CTB is widely used to balance between high worth with low costs [6]. In its simplest form, CTB is a blend of graded aggregate mixture with cement and optimal water content that is pulverized to an ideal level to form cement stabilized mixture. In comparison to unbound granular layers, CTB is a layer of pavement structure that significantly reduces stresses before they reach the subgrade [7]. Flexible pavements with cement-bound base primarily consist of CTB layer sandwiched between a prepared sub-base/ subgrade at bottom and an overlying surfacing of bituminous materials [8]. The type of surfacing and its thickness are governed by factors such as traffic volume, material availability, climatic conditions, construction cost, and indigenous practices [9]. CTB contain enough cemented granular components—typically between 3 and 6% of the dry mass of granular material, to provide a pavement base layer with noteworthy flexural strength [10]. After a number of repeated applications of stresses greater than some limiting value, CTB material become prone to fatigue failure. The stiffness of base layer, which may have an effect on tensile strains above it and the compressive strains over the subgrade layer, is greatly influenced by the material quality of base layer [11]. Flexural fatigue is of primary interest because of its impact on pavement cracking which overtime reflects on the pavement surface as reflection cracks [9]. Fatigue is the gradual, localized, and irreversible structural change that occurs in a material under repeated loads. Repeated loading at modest loads can cause fatigue cracks to form and perhaps lead to failure [12]. Wheel loads, temperature changes, and moisture degradation are typically attributed for these cracks. The progression of material deterioration, crack initiation, crack propagation, and finally complete failure are the phases of fatigue cracking [13].

In accordance with testing protocols, the beam-fatigue test can be carried under strain-controlled (constant strain) and stress-controlled (constant stress) testing conditions [14]. In strain-controlled test, the applied load is reduced during the test to maintain strain; in the stress-controlled test, stress is maintained while the strain increases. Because of its high stiffness, CTB material is extremely sensitive to changes in strain parameters. It is challenging to regulate strains per minute with current instrumentation for the reason that CTB is a brittle material with a 200 μm fracture strain [15]. Although no evident damage could be seen from the CTB surface during the initial phase of low load action, the interior of the material would have suffered minor damages. These minor damages constantly add up due to the repetitive load cycle and result in the material to suffer from sudden damage known as fatigue damage [16]. In contrast to the material's modulus, breaking strain has a substantial impact on the fatigue relationship for cemented materials [17].

The design guidelines for flexible pavement in India, IRC-37, third revision (2012) and fourth revision (2018) promotes the use of CTB in flexible pavement construction [18, 19]. Although it is past a decade, the nation at present is heavily invested in the construction of flexible pavements using traditional unbound base and sub-base layers. The fundamental cause of this is due to lack of trust in planners and contractors on how pavements made with cement-bound base materials would behave. Primarily at the laboratory level, one might consider a rigorous process to design and assess the performance of CTB material. The current study is mainly focussed on a methodical approach to the design and laboratory evaluation of CTB material. This article discusses the laboratory testing of CTB material using beam specimens that were subjected to flexural fatigue analysis by repetitive loading under fatigue testing equipment at various stress ratios over 7 and 28 days curing periods. Additionally, in the current study, stress is recognized as one of the primary controlled markers for evaluating fatigue. The validation of the laboratory results were done by comparing the strains under repetitive loading with IIT pave strain output and IRC-37 fatigue model which was mainly based on the Australian experience [20]. To provide an appropriate methodology to use CTB as a base layer material, the objective of the presented research is envisioned to study the effect of repetitive loading and fatigue damage on cement-treated base for applications in flexible pavements.

2 Materials and Tests

2.1 Material Characterization

Ordinary Portland cement (OPC) Type I—53 grade was used in the preparation of CTB specimens. The preliminary tests on cement were conducted in accordance to Bureau of Indian Standard (BIS) code and the corresponding test results are presented in Table 1. The coarse aggregates and fine aggregates (M-Sand) used for the current research were sourced from a quarry located in Bidadi, near Bengaluru. The aggregates are geologically classified as igneous granite. The aggregates were subjected to preliminary examination to explore the basic properties as per IS 2386 Part I to V [21–23]. Table 1 presents test results of coarse aggregates and fine aggregates. The aggregate gradation for CTB was chosen in compliance with IRC: SP:89-2010 [24]. Accordingly aggregates of sizes varying from 26.5 to 0.075 mm were blended to match the desired gradation using Rothfutch's mix proportioning method. The obtained gradation for CTB is shown in Fig. 1.

Table 1 Material characterization

Material	Test/parameter	Results	Specified limits as per BIS/MoRTH
Cement	Specific gravity	3.15	3.1–3.16
	Normal consistency (%)	28	26–35
	Initial setting time (minutes)	40	> 30
	Final setting time (minutes)	620	> 600
	Fineness	252	> 225
	Grade	53	–
Coarse aggregate	Aggregate origin/type	Igneous/granite	–
	Specific gravity	2.68	–
	Water absorption (%)	0.27	2
	Impact value (%)	15.23	30
	Crushing value (%)	18.5	30
	Abrasion value (%)	29	35
	Elongation index (%)	12.34	35
	Flakiness index (%)	9.62	35
	Angularity number	7	–
Fine aggregate	Aggregate type/Origin	M–sand/igneous/granite	–
	Specific gravity	2.68	–
	Water absorption (%)	1.04	–
	Bulk density (g/cm^3)	1.72	–
	Bulking (%)	8	–

2.2 *Specimen Preparation*

Determination of optimum moisture content for CTB. After obtaining the desired gradation for CTB, modified compaction test was carried out on CTB material in accord with IS-2720 Part-8 [25]. The cement contents in the range of 2.5% to 5% of total weight of aggregates was chosen to evaluate the influence of cement content on dry density and moisture content relationship of CTB. The results of modified compaction tests are graphically presented in Fig. 2. Data in figure depict that the variation in cement content has a greater impact on maximum dry density of CTB.

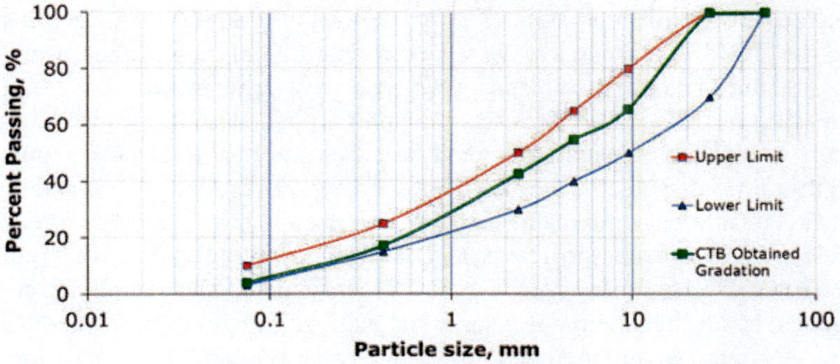

Fig. 1 Gradation requirements for cement-treated base layers as per IRC: SP-89:2010

The maximum dry density (MDD) in gm/cm^3, for 2.5, 3, 3.5, 4, 4.5, and 5% cement content was found to be 2.07, 2.12, 2.14, 2.24, 2.25, and 2.28, respectively. Further, the corresponding optimum moisture content (OMC) was 4.7, 5.85, 4, 6, 5, and 5.5%, respectively.

Preparation of CTB cube specimen. The cement-treated base mixture was prepared in laboratory using a pan type rotating drum mixer. The mixture was mixed well for 3 min using optimum moisture content obtained from modified compaction test. Cubes of dimension 150 × 150 × 150 mm were casted and were compacted well to achieve MDD. For this a specially fabricated cube mould of 150 × 150 × 150 mm dimension, fitted with a collar was used. Following the compaction energy of heavy compaction (2710.5 kJ/m^3), the CTB material was compacted using a rammer weighing 4.5 kg in three layers, with 55 blows being applied on each layer to reach the maximum dry density possible. Cement content was adjusted from 2.0 to 5.0% in increments of 0.5% to prepare cubes. The cast specimens were kept for 7-day curing period at a temperature of 22 ± 3°C in the lab.

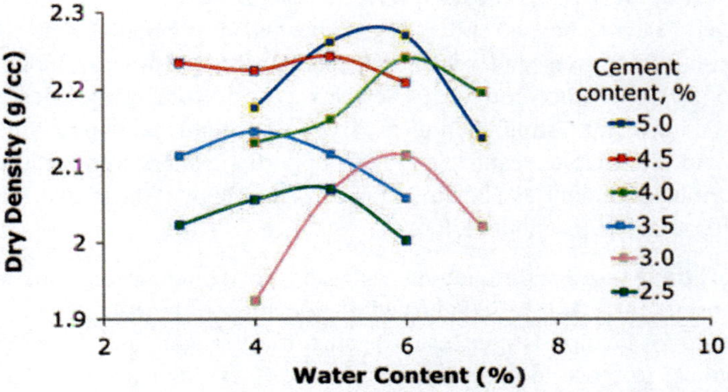

Fig. 2 Dry density—moisture content relationship of CTB for different cement contents

Preparation of beam specimen. To examine flexural strength and fatigue characteristics, the CTB beam specimens were casted within steel moulds measuring 100 × 100 × 500 mm. The test specimens were casted in two batches, each batch consisting of three beams. The CTB mixes were mixed in a pan-style drum mixer using 5% cement content which was determined based on the results of cube compressive strength test (as discussed under Sect. 3.1). The water content at optimum moisture level was considered based on the modified compaction test. Rodding was employed to first compact the beam specimens, and then a table vibrator with a 3600 rpm vibration rate was used to further compact the CTB specimens. The overall density of the compacted beam specimen was ensured for MDD of modified compaction test. The beam specimens were demoulded 36–48 h after casting and then cured for 7 days (3 beams) and 28 days (3 beams) by covering with wet cloth. The beam specimens were covered with thin polythene covers to prevent the loss of moisture. At a lab temperature of 25 ± 3°C, the cured beam specimens were left in the air for roughly 5–6 h before subjecting them to flexure test.

2.3 Tests on CTB Specimens

Cube compressive strength test. For the evaluation of stabilized bases of flexible pavements, several standards, and codes of practice frequently employ the cube compressive strength. The cube compressive strength was determined in accordance with 'Bureau of Indian Standard code IS-516: 1959' procedure [26]. The specimens were subjected to compressive strength check using a compression testing machine after 7-day curing period at a temperature of 22 ± 3°C in the lab. The cubes were gradually loaded at a rate of 0.5 ± 0.05 N/mm^2 per second until the cube failed. The ratio of ultimate load, in newton, to cross-sectional area of the cube specimen, in mm^2 is represented as the cube compressive strength of CTB specimen and shall be expressed to the nearest N/ mm^2.

Flexure test on beams. The beam specimens (of size 100 × 100 × 500 mm) which had attained 7 days (3 beams) and 28 days (3 beams) curing period were subjected to static flexural strength test under four-point loading as per 'Bureau of Indian Standard code IS-516: 1959' procedure [26]. The test was performed using a designated test fixture on a universal testing equipment. Two loading pins were positioned equally apart around the sample's centre, and two supporting rollers were spaced apart by a predetermined amount as shown in Fig. 3. From above, these two loadings are descended steadily till sample failure.

Flexural fatigue test. Flexural fatigue test was carried out on beam specimens (with dimensions of 100 × 100 × 500 mm) with 5% cement. Figure 4. depicts the experimental setup for flexural fatigue test. The span was 450 mm, and the loading points were identical to those in the static flexure test (four-point loading). The ratio of maximum fatigue stress to static flexural stress used for flexural fatigue tests ranged from 0.65 to 0.85. Throughout the experiment, the fatigue stress ratio (ratio of

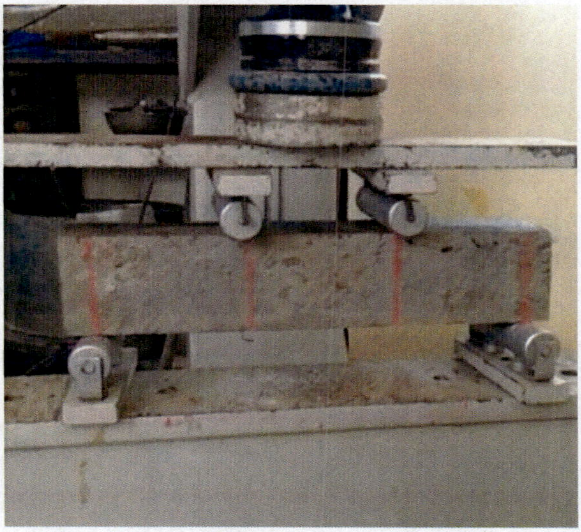

Fig. 3 Static flexural strength test four-point loading

maximum fatigue stress to lowest fatigue stress) was held constant. During every phase of the test, sinusoidal load was applied at a frequency of 5 Hz with a constant amplitude. When the beam reached the failure point or after achieving an average vertical deformation of 2 mm, the flexural fatigue test was terminated.

3 Results and Discussion

3.1 Cube Compressive Strength of CTB

The results of 7-day compressive strength test on CTB cube specimens with varying cement contents are shown in Table 2. A graphical representation of the change in compressive strength of CTB cube specimens for various cement contents is displayed in Fig. 5. The graph shows that the 7-day compressive strength of CTB increases linearly as cement content increases. According to IRC: SP: 89-2010, the cube compressive strength of the CTB material should be between 4.5 and 7 MPa. The typical cement stabilized material should achieve this required strength in 7 days, and the average laboratory strength value should be 1.5 times the required minimum field strength [24]. From Table 2, it is observed that, optimum cement content for CTB material is 5.0% (> 1.5 times the minimum value of 4.5 MPa). This value is well within the range of 3–6% of cement content as observed in reviewed literature [10].

Fig. 4 Flexural fatigue test on CTB beam

Table 2 Average compressive strength of CTB specimens at different cement content

Cement content (%)	OMC (%)	MDD (g/cc)	7 days' average compressive strength (MPa)
2.5	4.7	2.07	1.92
3.0	5.85	2.12	2.61
3.5	4	2.14	3.72
4.0	6	2.24	4.6
4.5	5	2.25	5.76
5.0	5.5	2.28	6.84

3.2 *Static Flexural Strength of CTB*

Four-point flexure test is adopted mainly for brittle materials like CTB and bituminous mix specimens. The four-point flexure test provides values of material's bending modulus of elasticity, flexural stress, flexural strain, and flexural stress–strain response. To analyse fatigue response behaviour of CTB beam specimens, the maximum and minimum stress levels are required as input parameters that are selected based on the ultimate flexural load of beam specimens at failure. In contrast to three-point bending, which only stresses the material directly beneath central bearing, four-point flexural test stresses the entire length of the beam between two

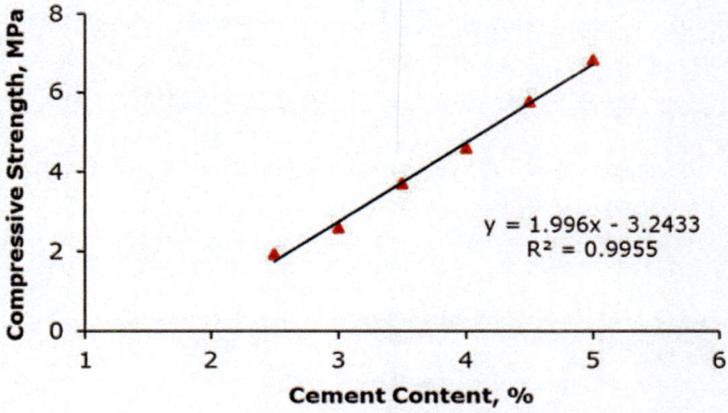

Fig. 5 Graphical representation of cement content versus 7 days' compressive strength of CTB

loading points. The region between two loading pins is not subjected to shearing stresses during four-point bending flexure test. As brittle materials are less able to endure direct shear loads, four-point bending test is especially well suited for CTB material. The average flexural strength of three beam specimens obtained for 7 and 28 days curing period respectively was found to be 1.03 and 1.29 MPa. The obtained result is well within the limiting flexural strength (maximum) value of 1.40 MPa for cementitious stabilized aggregates as per IRC-37: 2018 [19].

3.3 Fatigue Test and S–N Relationships

Flexural fatigue test was performed at different stress levels between 0.85 and 0.55. The findings of the flexural fatigue test of beam specimens cured for a duration of 7 and 28 days' time period in terms of load-to-failure repetitions under various stress levels are outlined in Table 3. The relationship between applied maximum stress (S) and number of load cycles (N) until a specimen fails is visually represented by the S–N curves. The established S–N models of the present research in Fig. 6(a) and 6(b) predicted the variation of stiffness in CTB beam specimens cured for different time periods, which is clearly visible with the shift in load repetitions sustained by 28 days cured specimens as compared with those of 7 days cured specimens.

3.4 Validation of Results

The 'Guidelines for the design of flexible pavements'—2018 revised version from Indian Roads Congress (IRC) incorporates 'Fatigue performance model' based on the Australian experience [10]. According to Austroads, tensile strain at the base of

Table 3 Variation of fatigue behaviour of CTB beams for different stress ratio

Stress ratio	7 days cured beam specimens		28 days cured beam specimens	
	No. of repetitions (N)	Log (N)	No. of repetitions (N)	Log (N)
0.85	2800	3.447	45,000	4.653
0.75	12,800	4.107	70,000	4.845
0.65	35,800	4.554	90,000	4.954
0.55	62,000	4.792	105,000	5.021

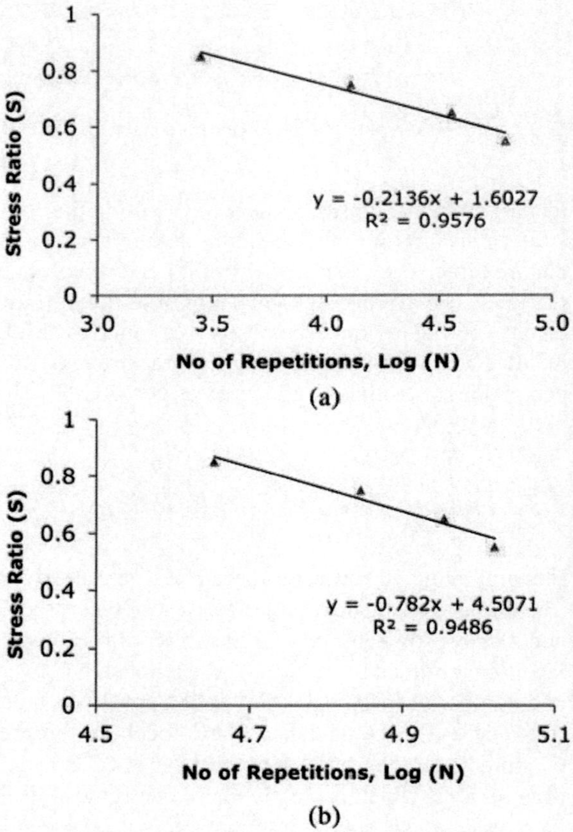

Fig. 6 S–N Curves; (**a**) Beams cured for 7 days; (**b**) Beams cured for 28 days

cement-treated base layer is converted to number of standard axle load repetitions and contrasted with intended number of standard axle repetitions. Equation (1) represents fatigue relationship of CTB proposed for structural analysis and design of flexible pavements.

$$N = RF\left[\frac{\left(\frac{113{,}000}{E^{0.804}} + 191\right)}{et}\right]^{12} \quad (1)$$

where N = number of wheel load repetitions in standard axles which CTB material can sustain; RF = reliability factor of cementitious materials for fatigue failure; E = 5000 MPa, elastic modulus of CTB material; e_t = tensile strain at the bottom of the CTB layer in micro strains.

To find tensile strains (e_t) at the bottom of CTB layer, pavement analysis can be done using IITPAVE with input parameters E = 5000 MPa, Poisson's ratio = 0.25 for, and contact stress = 0.8 MPa on the pavement surface, in accordance with IRC-37:2018 standards [19]. When the estimated tensile strain value is substituted along with other inputs to arrive at Eq. (1), the number of standard axle loads (wheel load repetition) must be less than the intended traffic.

On the other hand, for any given number of wheel load repetition 'N', Eq. (1) can be conversely used to determine tensile strain (e_t) value at the bottom of the CTB layer. Further, strain value corresponding to the maximum number of wheel load repetition 'N', that causes distress to the CTB layer shall be considered as the allowable strain (e_{tmax}) at the bottom of CTB layer. Accordingly, Eq. (1) takes the form of Eq. (2) as shown below. In the present study, Eq. (2) is used to determine allowable strains for a given number of load repetition 'N'. The graphical representation of measured strains during the flexural fatigue test by repetitive load method, versus number of load repetition 'N' is depicted in Fig. 7.

$$etmax = \left[\frac{81.803 \times 10^{28}}{N}\right]^{\frac{1}{12}} \quad (2)$$

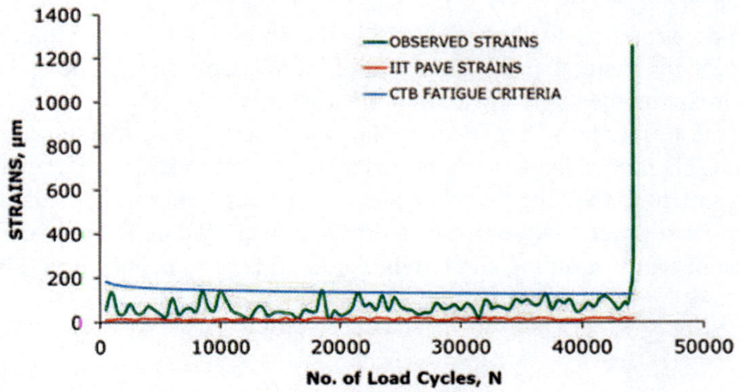

Fig. 7 Comparison of measured strains with IIT PAVE and CTB 'Fatigue performance model—IRC-37: 2018'

Further a comparison is made with the strains determined from IIT PAVE software and the strains calculated from CTB fatigue criteria using Eq. (2). From Fig. 7, it is seen that the measured strains are within the allowable strains but are greater than the strains output from IIT PAVE software. During flexural fatigue test, failure point of the beam specimen is remarked by a drastic rise in the strain value much above the allowable value witnessing the ultimate fracture point of the test specimen. The results are consistent with reviewed literature, which states that CTB beam specimens fail at fracture strains > 200 μm [15].

4 Conclusions

The proposed methodology of designing the mix for cement-treated base and its laboratory evaluation of flexural fatigue performance is found to be systematic which is evident from the results with reasonable correlation. Compressive strength of CTB materials depend essentially on the percentage of cement added. The compression test results are found to be satisfactory with linear fit, indicating a good correlation between cement content and average 7 days' compressive strength. The static flexural strength test conducted on beam specimen under four-point loading, with optimal cement content of 5%, has gained a flexural strength of 1.29 MPa upon 28 days curing period. The obtained result is well within the limiting flexural strength (maximum) value of 1.40 MPa for cementitious stabilized aggregates as per IRC-37: 2018.

The S–N curves generated using the mean fatigue data of three beam specimens for different stress ratios of 0.85, 0.75, 0.65, and 0.55 have shown linear relationship with number of wheel load repetitions indicating a good correlation for both 7 and 28 days of curing period. A graphical comparison with measured strains of flexural fatigue test, output strains from IIT PAVE software, and the calculated strains from 'Fatigue Performance model' for CTB from IRC-37:2018 has revealed that the measured strains under repetitive loading vary within the allowable limiting strains up to the failure point and there after the strain values immediately shoots above the limiting strains witnessing the breaking point of the test specimen.

To the best of authors' knowledge, this paper has demonstrated a general guide to design CTB mix in laboratory and assertively recommend for its application in field as a structural base for flexible pavements through appropriate quality checks. However, a comprehensive research can further be carried on the field performance evaluation in terms of rutting and fatigue for flexible pavements with CTB layers in future scope.

References

1. Smith T Jolly R (2005) Concrete pavement a sustainable choice. In: Proceedings of the 8th international conference on concrete pavements colorado Springs CO international society of concrete pavements, pp 585–606
2. Kumar P, Shukla S (2022) Flexible pavement construction using different waste materials: a review. Mater Today Proc
3. Gautam L, Kalla P, Jain JK, Choudhary R, Jain A (2022) Robustness of self-compacting concrete incorporating bone china ceramic waste powder along with granite cutting waste for sustainable development. J Clean Prod 367:132969
4. Gautam L, Jain JK, Jain A, Kalla P (2022) Recycling of bone china ceramic waste as cement replacement to produce sustainable self-compacting concrete. Structures. Elsevier, Amsterdam, pp 364–378
5. Berthelot C, Podborochynski D, Marjerison B, Saarenketo T (2010) Mechanistic characterization of cement stabilized marginal granular base materials for road construction. Can J Civ Eng 37(12):1613–1620
6. Su N, Xiao F, Wang J, Amirkhanian S (2017) Characterizations of base and subbase layers for mechanistic-empirical pavement design. Constr Build Mater 152:731–745
7. Sounthararajah A, Bui HH, Nguyen N, Jitsangiam P, Kodikara J (2018) Early-age fatigue damage assessment of cement-treated bases under repetitive heavy traffic loading. J Mater Civ Eng 30(6):04018079
8. Halsted GE, Luhr DR, Adaska WS (2006) Guide to cement-treated base (CTB). trid.trb.org
9. George KP (1990) Characterization and structural design of cement-treated base. Transp Res Rec 1288:78–87
10. Austroads P (2006) 4D: stabilised materials. In: Guide pavement technology. Sidney
11. Francois A, Ali A, Mehta Y (2019) Evaluating the impact of different types of stabilised bases on the overall performance of flexible pavements. Int J Pavement Eng 20(8):938–946
12. López MAC, Fedrigo W, Kleinert TR, Matuella MF, Núñez WP, Ceratti JAP (2018) Flexural fatigue evaluation of cement-treated mixtures of reclaimed asphalt pavement and crushed aggregates. Constr Build Mater 158:320–325
13. Zhao X, Dong Q, Chen X, Xiao Y, Zheng D (2021) Fatigue damage numerical simulation of cement-treated base materials by discrete element method. Constr Build Mater 276:122142
14. Jitsangiam P, Nusit K, Chummuneerat S, Chindaprasirt P, Pichayapan P (2016) Fatigue assessment of cement-treated base for roads: an examination of beam-fatigue tests. J Mater Civ Eng 28(10):04016095
15. Dong S, Hao P, Wang J, Bi J, Zhang M (2021) A novel method for testing the fatigue performance of cement stabilized base field coring samples. Constr Build Mater 274:122065
16. Lv S et al (2021) Strength and fatigue performance for cement-treated aggregate base materials. Int J Pavement Eng 22(6):690–699
17. González A, Jameson G, de Carteret R, Yeo R (2013) Laboratory fatigue life of cemented materials in Australia. Road Mater Pavement Des 14(3):518–536
18. IR Congress (2012) Guidelines for the design of flexible pavements IRC: 37-2012, 3rd revision. In: Indian roads congress, New Delhi
19. IR Congress (2018) IRC: 37-guidelines for the design of flexible pavements (fourth Revision). In: Indian roads congress, New Delhi
20. Austroads (2012) Guide to pavement technology part 2: pavement structural design. Austroads Sydney
21. IS 2386 (1963) Methods of test for aggregates for concrete, Part I, pp 11–14
22. BIS (1963) IS 2386: methods of test for aggregates for concrete, part 3: specific gravity, density, voids, absorption and bulking. BIS New Delhi
23. IS: 2386 (1963) Method of test for aggregates for concrete, mechanical properties, Part-4. Bureau of Indian Standards New Delhi
24. Indian Roads Congress (2010) IRC SP 89: guidelines for soil and granular material stabilization using cement, lime and fly ash. IRC SP

25. IS: 2720 Part 8 (1983) Laboratory determination of water content—dry density relation using heavy compaction. Indian Standard methods of test for soils. Bureau of Indian Standards New Delhi
26. IS: 516–1959 (1979) Indian standard code of practice-methods of test for strength of concrete. Bureau of Indian Standards New Delhi

Characterization of Flexible Concrete Pavements Using Demolition Waste and Laterite

N. T. Chethan Kumar, K. E. Prakash, and Rajani V. Akki

1 Introduction

1.1 *Flexible Concrete Pavements*

Rigid pavements subjected to thermal expansion and contraction that depends on site conditions, seasonal variations, etc. To avoid this issue, a new type of pavement is suggested as flexible concrete pavements. Here rectangular/hexagonal blocks will be made using waste plastic. Prepared concrete mix will be filled inside the plastic blocks and allowed to set as per standards. After proper curing process, the pavement can be open to traffic. Here the advantage of this type of pavement is, it avoids provision of joints of thermal expansion. Total quantity of aggregates is less compared to normal concrete pavement works. This type of pavement is also known as cell-filled pavement/block-filled pavement. Maintenance cost is less compared to normal concrete pavements. Hence these types of pavements are gaining importance in recent periods.

N. T. C. Kumar (✉)
Malnad College of Engineering, Hassan, Karnataka, India
e-mail: chethancv.mce@gmail.com

K. E. Prakash
Shree Devi Institute of Technology, Mangalore, Karnataka, India

R. V. Akki
East Point College of Engineering and Technology, Bangalore, Karnataka, India

S. Kolathayar et al. (eds.), *Best Practices in Geotechnical and Pavement Engineering*, Lecture Notes in Civil Engineering 449, https://doi.org/10.1007/978-981-99-8505-0_38

1.2 Cement Concrete

Concrete is amongst the oldest and most widely used construction materials in the world, employed in various structures across civil engineering projects, from infrastructure to both low and high-rise buildings. It primarily consists of fine and coarse aggregates. These aggregates can either be naturally sourced or artificially manufactured. The growing demand for concrete has led to increased costs for the binding material (cement) and the depletion of natural sources of fine aggregate, resulting in higher concrete production expenses. This research project explores the outcomes of concrete mixes that incorporate partial substitutions of fine aggregate with Laterite soil and coarse aggregate with recycled materials. Such replacements impact the engineering properties of concrete. Concrete has gradually supplanted many alternative construction materials due to its versatility and widespread availability. Architects and structural engineers favor it because of its durability, long lifespan and minimal maintenance requirements. In recent years, rigid pavements have gained prominence over flexible pavements due to their lower maintenance needs, superior performance, and extended durability throughout their designed lifespan.

1.3 Aggregates

Concrete relies on a strong matrix to bind together aggregate particles, effectively filling the gaps between them. These aggregates can have their properties enhanced using suitable admixtures. Concrete serves as a fundamental material for various construction applications, including building foundations, retaining walls, pavement, highways, runways, parking structures, dams, reservoirs, and pipelines, among numerous architectural and civil engineering projects. However, due to the factors mentioned above, there is a growing need for alternative materials to partially or completely replace Portland cement, fine aggregate, or coarse aggregate in concrete mixtures to sustain construction work without compromising critical concrete attributes like strength, workability, and durability. The strength of concrete varies based on the type of coarse aggregate used and the proportion of aggregate in recycled aggregate concrete (RCA). Therefore, it is crucial to examine the aggregate properties of recycled aggregate since coarse aggregate plays a central role in concrete composition, and there is a compelling rationale for reusing it instead of disposing of it in landfills. RCA contributes significantly to sustainable development efforts by conserving natural resources and reducing the disposal of demolition waste from used concrete structures. In-depth studies have been conducted on the properties of recycled aggregate concrete, involving various comparisons and comprehensive testing of their mechanical and physical characteristics. Recycled aggregate is derived from aged concrete that has been demolished and removed from foundations, pavements, bridges, or buildings. This concrete is then crushed and processed into different size fractions. Recycled aggregates are obtained through the processing of debris generated during the demolition of concrete structures and other construction waste, such as discarded precast concrete components, broken masonry, concrete roadbeds, and asphalt pavement.

2 Objectives

The main aim of this research work is to reduce the demand on natural resources and use of alternate materials effectively.

Main objectives of the project are as follows:

- To determine the physical properties of recycled aggregate and Laterite aggregate.
- To determine the quantity of materials required as per IS 10262-2009.
- To study the effect of recycled aggregate on fresh and hardened properties of concrete.
- To determine optimum content of recycled aggregate and Laterite.

3 Materials and Methodology

Following are the materials used for the research work.

3.1 Cement

Ordinary Portland cement of 53 grade was used throughout the research work. The cement was tested as per the relevant standard codes. And obtained results are satisfactory. Birla super 53 grade cement with initial setting time 60 min. and the final setting time of 330 min. recorded. Standard consistency obtained as 31%. Specific gravity is obtained as 3.12.

3.2 Fine Aggregate

Sand is commonly used as the fine aggregate in cement concrete. Artificial sand or *M* sand is used for this purpose. Here also physical properties of the fine aggregate were determined as per IS codes and obtained results are satisfactory. Specific gravity test is conducted using Pycnometer with a result of 2.58. A fineness modulus of 3.58, moisture content as 2.6% was obtained.

3.3 Coarse Aggregate

It is obtained by natural stone quarries, here also natural aggregates physical properties were tested, and satisfactory results were obtained. Specific gravity of 2.45, fineness modulus of 6.5 and moisture content of 1.2% is obtained. Aggregate crushing value is 24.2%.

3.4 Water

The water used in concrete plays an important role in mixing. The strength of concrete directly depends on the quality and quantity of water used in the mix. Portable drinking water was used for the work.

3.5 Recycled Aggregates

These are produced from aged concrete that has been demolished and removed from foundations, pavements, bridges, or buildings. This material is collected from nearby sites and segregated as per our requirements. Obtained recycled aggregate was converted to as per our requirement of 20 mm downsize. The results of specific gravity 2.26, fineness modulus of 5.6, and moisture content of 2.6% was obtained. Aggregate crushing value obtained as 34.3%.

3.6 Laterite

As the Laterite is rich in iron oxide and used in a very limited way in the agriculture field. Laterite is widely available in western parts of Karnataka. For this project work, Laterite is obtained from Kundapura taluk and tested for the physical properties. Specific gravity of 2.2, fineness modulus of 5.6, and moisture content of 13.9% is recorded.

Recycled aggregates caters the strength requirements, hence it can be used as a replacement of coarse aggregates. Laterite also used as fine aggregate in concrete [1]. As per relevant code provisions M25 grade of mix proportion is considered for the proposed work [2]. To ascertain the strength parameter of recycled aggregates, in the first trial coarse aggregates replacement was done and the experimental results are represented in Table 1. From the table it can be observed that up to 20% of replacement can be done without altering strength parameter. In case of Laterite also, replacement was considered as fine aggregate [3]. The results are mentioned in Table 2. Here also up to 20% of replacement can be suggested.

Table 1 Comparison of test results using recycled concrete aggregate

Sl. No	Types of coarse aggregates used, natural aggregate (NA), recycled concrete aggregate (RCA)	Compressive strength, N/mm^2		
		7 days	14 days	28 days
1	100% NA + 0% RCA	19.44	22.04	25.44
2	90% NA + 10% RCA	19.51	23.47	25.97
3	80% NA + 20% RCA	17.92	21.98	24.77
4	70% NA + 30% RCA	17.01	21.11	23.88

Table 2 Comparison of test results using Laterite

Sl. No	Types of fine aggregates used, natural fine aggregate (NFA), laterite aggregate (LA)	Compressive strength, N/mm^2		
		7 days	14 days	28 days
1	100% NFA + 0% LA	18.59	22.10	24.61
2	90% NFA + 10% LA	18.42	22.19	24.82
3	80% NFA + 20% LA	17.49	21.42	24.50
4	70% NFA + 30% LA	17.17	20.58	24.41

From the above tables, it is evident that 20% of replacement of coarse and fine aggregate can be done without affecting strength parameter. Figures 1 and 2 represents the graphical representation of test results.

After obtaining optimum content of recycled concrete aggregate and Laterite of 20%, an effort has been made to use the materials together and the test results are represented in Table 3. Graphical representation is represented in Fig. 3.

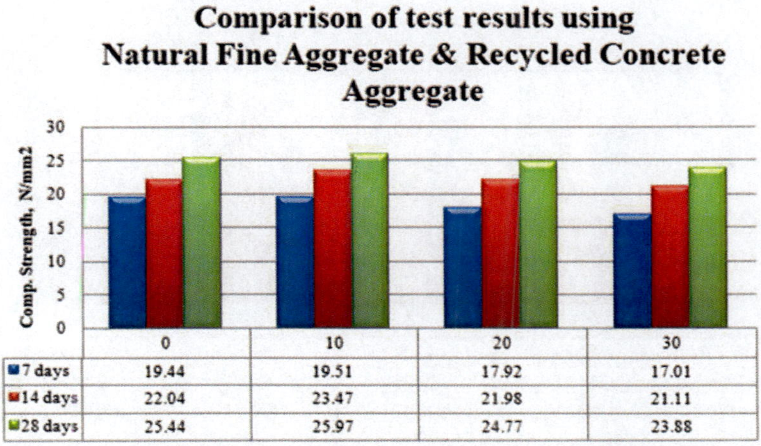

Fig. 1 Comparison of test results using recycled concrete aggregate

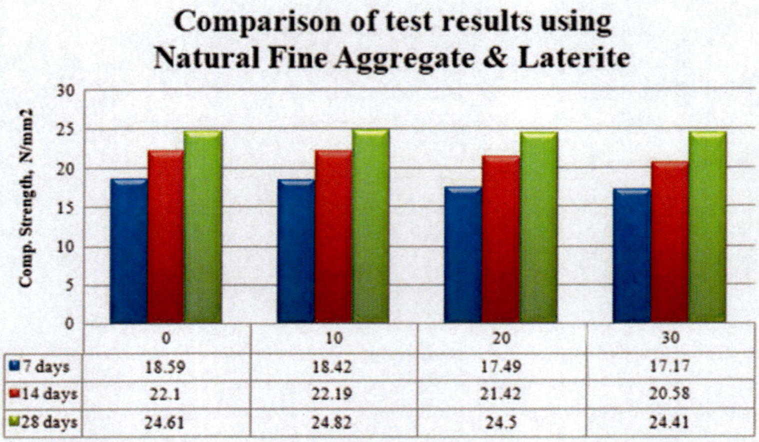

Fig. 2 Comparison of test results using recycled concrete aggregate

Table 3 Comparison of test results using recycled concrete aggregate and laterite

Sl. No	Types of aggregates used, natural aggregate (NA), recycled concrete aggregate (RCA) natural fine aggregate (NFA), laterite aggregate (LA)	Compressive strength, N/mm^2		
		7 days	14 days	28 days
1	100% NCA + 100% NFA + 0% RCA + 0% LA	19.23	23.51	24.31
2	90% NCA + 90% NFA + 10% RCA + 10% LA	19.56	23.33	25.17
3	80% NCA + 80% NFA + 20% RCA + 20% LA	17.62	22.61	24.71
4	70% NCA + 70% NFA + 30% RCA + 30% LA	16.15	22.68	23.14

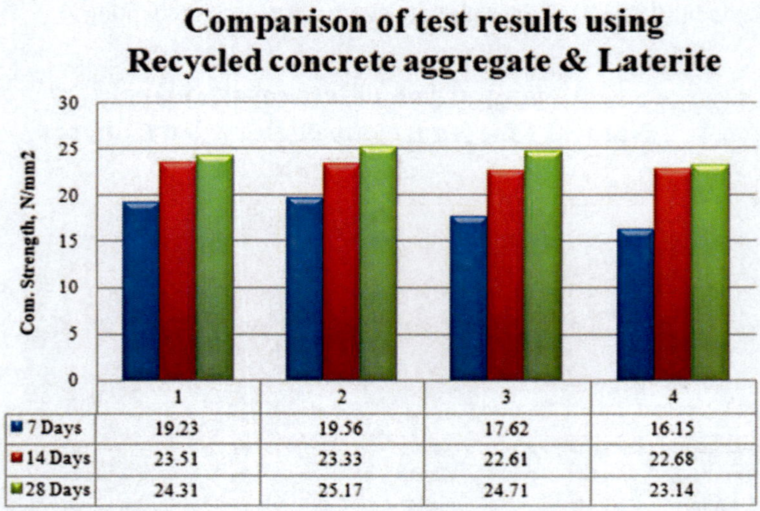

Fig. 3 Comparison of test results using recycled concrete aggregate and laterite together

4 Results and Conclusions

As stated in the introduction, waste materials are creating problems in landfill issues, and associated problems on surrounding environment. Demolition waste is a material which we must use effectively by segregation. Quality materials can be expected from the demolition wastes. Recycled concrete aggregate can be used as a coarse aggregate or fine aggregate depending on the requirement [4]. As stated in the literature review, powder form can also used as replacement of cement. In this regard, research work is essential to assess the use in structural concrete works. Hence, in this work proportions of demolition waste were used as a replacement of coarse aggregates. And 20% of the demolition waste can be used in concrete work.

In the case of Laterite also, various proportions were used to replace the fine aggregate. 20% of the replacement was found satisfactory along with the demolition waste materials [5]. By doing so, land fill effects by demolition waste can be reduced drastically. At the same time, demand for natural resources can be reduced. Laterite can also be effectively used in concrete work as a replacement of fine aggregate. Due to scarcity of natural river sand, we are shifting toward manufactured sand or M sand. Proper use of Laterite can reduce the problem of river sand as well as m sand. In the case of flexible pavement, it demands periodic maintenance with less initial cost of construction. In case of failure, a particular failure spot can be repaired in the initial stage. In the case of rigid pavement, failure rates are very rare, and the initial cost of construction is high. The type of pavement is similar to interlock pavement. This is also known as flexible concrete pavements. In this case, particular failure section/block can be repaired easily. This type of pavement can be provided over low traffic roads/rural roads/parking lots and in petrol bunks. Locally available materials usage in the concrete work reduces the cost of construction as well as the cost of maintenance.

References

1. Shekhawat P, Sharma G, Singh RM (2023) Strength characteristics of hazardous wastes flyash and eggshell powder-based geopolymer-stabilized soft soil cured at ambient temperature. Arab J Geosci 16:127. https://doi.org/10.1007/s12517-023-11234-2
2. IS 10262 (2019) Guidelines for concrete mix proportioning
3. Shekhawat P, Sharma G, Singh RM (2023) Morphology and microstructure of waste material-based geopolymer with flyash, eggshell powder, and soft soil. Mater Lett 334:133621. https://doi.org/10.1016/j.matlet.2022.133621
4. Murali A (2021) Experimental investigation on partial replacement of coarse aggregates by demolished concrete. Int J Eng Res Technol 09(03):128
5. Novakova I (2016) Properties of concrete with partial replacement of natural aggregate by recycled concrete aggregates from precast production. In: International conference on ecology and new building materials and products, ICEBMP 2016. https://doi.org/10.1016/j.proeng.2016.07.387
6. Srivastava V (2013) Demolition waste as cement replacement in concrete. J Environ Nanotechnol 02, 17–20. https://doi.org/10.13074/jent2013.09.132026

Experimental Investigation on Cold Mix for Road Construction Using Granite Aggregates and Rubber Waste

Vedant N. Nawale, Rutuja A. Girme, Tejal V. Sonawane, Shubham A. Wagh, and Atteshamuddin S. Sayyad

1 Introduction

Road construction using hot mix technique (HMT) has a number of negative effects on the environment, human health, and the national economy. The drawbacks of hot mix technologies include high levels of noise and air pollution, the release of greenhouse gases, high energy consumption, safety issues for the maintenance staff, and a compromise in the bitumen's durability owing to ageing during heating. This article primarily focuses on the advancement of cold mix technology (CMT) in road construction as a means of resolving these issues. Unheated aggregate, emulsion or cutback, and binders make up the cold mix. Since there is no heating procedure involved, the primary distinction between HMT and CMT is the use of natural room temperature to mix the aggregates and emulsions. (1) It does not heat the aggregate or the binder. (2) It benefits the environment and saves electricity. Cold mix pavement can provide energy savings of more than 50% when compared to hot mix. It can be thought of as the green bituminous mixture used to construct roads. (3) It is simple to prepare or make utilizing a minimal set-up locally. For small-scale projects, it can also be created manually. (4) This paving material is suitable for constructing roads in distant and inaccessible areas of a nation where HMT that was made in a facility would have hardened before arriving at the construction place. (5) CMT can be laid in situations that are rainy or humid. (6) It is cost-effective and allows for large production with little outlay of capital. Currently, practically all of the road 2 IRC: SP:100-2014 construction and maintenance standards may be satisfied employing bitumen emulsion as a binder using warm and cold mix technologies. Careful selection and deployment of these technologies can produce significant cost savings, environmental benefits, and energy security in the construction and maintenance of

V. N. Nawale · R. A. Girme · T. V. Sonawane · S. A. Wagh · A. S. Sayyad (✉)
Department of Structural Engineering, Sanjivani College of Engineering, Savitribai Phule Pune University, Kopargaon, Maharashtra 423603, India
e-mail: hodstcoe@sanjivani.org.in

S. Kolathayar et al. (eds.), *Best Practices in Geotechnical and Pavement Engineering*, Lecture Notes in Civil Engineering 449, https://doi.org/10.1007/978-981-99-8505-0_39

roads. Bituminous macadam (BM) is an open-graded bituminous mixture that is used to strengthen flexible pavements and construct bituminous foundation courses. It is appropriate for highways with light traffic. Cold-mixed bituminous macadam (CMBM) is the construction of one or more courses of a compacted mixture formed with bitumen emulsion and mineral aggregate, applied immediately after mixing to the right grade and camber using suitable machinery.

Silva et al. [1] provided a set of case studies utilizing recycled aggregates in unbound, hydraulically bound, bitumen-bound, and non-structural concrete applications for building and road construction. Dash et al. [2] reviewed research works in terms of the design methodologies used, mix parameters, laboratory, and field performances of cold mix asphalts. Along with the future research agenda to broaden cold mix asphalts applications, some criteria go into successful cold mix asphalts applications. Nageim et al. [3] studied that hot mix asphalt offers equivalent mechanical performance to typical HMA combinations when used as surface pavement layers. The innovative cold mix asphalts shown in this work have better mechanical performance. Compared to HMA, it is more cost-effective and environmentally friendly since it can accept wastewater sludge ashes that are frequently dumped in landfills. Modified bitumen was introduced by Sahu and Joshi [4] and was blended using crumb rubber at specific temperatures. Marshall's mix design was implemented by varying the modified bitumen content while maintaining the ideal rubber content. Subsequent tests were conducted to identify the various mix design features for traditional bitumen as well. The use of waste rubber from tyre in the construction of bituminous roads was presented by Baraiya [5]. The viability of using waste tyre rubber as a blending component in bitumen, which is used in road construction, was investigated by Mane et al. [6]. A review on the utilization of waste/recycled plastic materials and waste rubber from tyres as road-building materials that strengthen the sub-base course of the pavement was presented by Chhabra and Marik [7]. The reuse of granite by-products for road construction was discussed by Rombi et al. [8]. Wulandari and Tjandra [9] studied the properties of cold mix asphalt. The Marshall Properties, indirect tensile strength, and fatigue behaviour of bituminous concrete mixes prepared with polymer-modified bitumen were examined in relation to mixing and compacting temperatures.

The following objectives are framed based on the aforementioned literature. (1) To determine whether granite is appropriate for the cold mix. (2) To calculate the quantity of binder needed for the mixture. (3) To ascertain the cold mix's water-resisting qualities.

2 Methodology

2.1 *Material*

(1) Granite: Granite is a name for a pale-coloured granular platonic rock made of feldspar, quartz, and trace amounts of other minerals such as biotite, hornblende, pyroxene, and iron oxides. India is blessed with an abundance of granite resources, including more than 200 different hues. Reserves comprise 94% coloured granite, and the remaining 6% is black granite. With shares of 42%, 30%, and 26%, respectively, Rajasthan, Odisha, and Karnataka are home to about 98% of the world's reserves. Compressive strength, tensile strength, density, *p*-wave velocity, and other attributes of granite are typically rated for use. Table 1 shows the BIS requirements for granite: IS: 3316-1974.

(2) **Emulsion**: A mixture of two or more liquids that are typically immiscible (UN-blendable or UN-mixable) is known as an emulsion. Emulsions fall within the broader category of colloids, which are two-phase systems of matter. Although the terms colloid and emulsion are occasionally used synonymously, emulsion is the preferred term when the dispersed and continuous phases are both liquids. Emulsions come in a variety of varieties, according to BIS classification in IS 8887-2004. Rapid Setting-1 (RS-1), Rapid Setting-2 (RS-2), Medium Setting (MS), Slow Setting-1 (SS-1), and Slow Setting-2 (SS-2) are the different types of emulsions. When the organic and aqueous phases separate into two distinct layers, or when the dispersion disappears, an emulsion is said to break. The

Table 1 Physical properties of granite as per IS: 14223-Part 1

Sr. No	Characteristics	Requirements (Multi-coloured and grey granite)
1	Moisture content (%) (max)	0.15
2	Dry density (m/v)	2.60–2.68
3	Apparent specific gravity (min)	2.75
4	Water absorption (%) (max)	0.50
5	Porosity (%)	1–2
6	Compressive strength (kg/cm^2)	1300–2200
7	Tensile strength (kg/cm^2) (min)	90
8	Shear strength (kg/cm^2)	300–540
9	Hardness (Mohs' scale)	6–7
10	Hardness (Schmidt No.)	85–110
11	Hardness (Shore No.)	46–61
12	Ultrasonic pulse velocities	5000
13	Resistance to wear	Not > 2 mm

breaking time varies depending on the type of emulsion, but it is mostly influenced by environmental factors like temperature, humidity, wind speed, etc. Anionic, cationic, clay-based, and non-ionic bitumen emulsions are categorized into these four groups. Road construction and maintenance frequently involve the usage of cationic and anionic emulsions.

(3) **Anti-striping agent**: The major factor affecting the integrity of the pavement, the bonding between the asphalt and aggregate, is of particular significance. The asphalt and aggregate must first make physical touch for this bonding to form, and it must remain throughout the lifespan of the pavement. Lower performance is the result of lost bonding. One of the key functional attributes ensuring the longevity of bituminous mixes is adhesion at the bitumen and aggregate contact. Stripping can result in fretting and eventual ravelling of the wearing course and is caused by the displacement of the bitumen layer from the aggregate surface. The two most frequent causes of stripping are moisture and dust surrounding aggregate surfaces before coating. According to numerous studies, stripping is influenced by the bitumen's chemistry, the mineral and chemical makeup of the aggregate, the surface texture (roughness), and the surface tension of the aggregate. The term "anti-striping agent" refers to the substance used to make up for issues with an aggregate's striping value.

2.2 *Preliminary Tests on Material*

(1) Aggregate—The following tests are important for deciding suitability aggregates in road construction.

   A. Aggregate impact value test
   B. Aggregate crushing value test
   C. Aggregate abrasion test
   D. Aggregate specific gravity and water absorption test.

(A) Impact value of aggregate: The overall impact value in the current investigation was determined to be 17.47%. The IS: 2386-1963 (Part-IV) states that a material's toughness quality is tough/strong if the aggregate impact value is between 10 and 20%.

(B) Crushing value of aggregate: The aggregate crushing value of granite was found to be 23.43% in the current investigation.

(C) Abrasion value of aggregate: According to the current analysis, granite has a LOS Angeles abrasion value of 28.51%, making it suitable for surfacing.

(D) Specific gravity (SG) of aggregate and water absorption (WA) of aggregate: The SG and WA in the current study were determined to be 2.8 and 0%, respectively. Table 2 shows the comparison of granite and basalt aggregate properties.

(2) Bitumen emulsion.

Material: Cationic bitumen emulsion for roads as per IS: 8887.2002.

Table 2 Comparison of granite aggregate with basalt aggregate

Sr. No	Type of aggregate	Test on aggregate					Remark
		Impact	Crushing	Abrasion	Specific gravity	Water absorption	
1	Granite aggregate	17.47	23.43	28.51	2.85	0.0%	Strong
2	Basalt aggregate	17.08	23.67	26.005	2.34	0.0%	Strong

Emulsion Type: Rapid Setting-1 (RS-1).
Table 3 highlights the testing report of bitumen emulsion.

(3) **Anti-striping Agent**

Description of Product: TIKI anti-striping agent.
Procure From: Tiki Tar Industries (Baroda) Limited.
Test Method: MORT and H.
Table 4 shows the test properties of the anti-striping agent.

Table 3 Test report for bitumen emulsion

Sr. No	Name of test	Test method	Result	Unit	Std. requirements of IS:8887.2004
1	Residue on 600 (IS Sieve)	IS:8887.2004	0.02	% by mass	0.05 (Max)
2	Viscosity at 50C	IS:3118.2004	35	Sec	20–100
3	Coagulation of emulsion at low temperature	IS:8887.2004	Nil	-	Nil
4	Storage stability after 24 h	IS:8887.2004	1.2	%	2.0 (Max)
5	Miscibility with water	IS:8887.2004	No coagulation	-	No Coagulation
6	Particle charges	IS:8887.2004	Positive	-	Positive
7	Test on Residues				
8	Residue by evaporation	IS:8887.2004	64.2	%	60 (Min)
9	Penetration 25C/ 100 g/5	IS:1208.1978	83	(1/10) mm	80–150
10	Ductility at 25C	IS:1208.1978	68.6	cm	50 (Min)
11	Solubility in trichloroethylene	IS:1208.1978	99.15	% by mass	98 (Min)

Table 4 Test report for anti-striping agent

Sr. No	Test parameters	Requirement	Results	Test method
1	Appearance	Liquid/solid	Thick liquid	Visual
2	Odour	Aggregate	Aggregate	Smelling
3	Pour Point	Max, 42C	27C	IS: 1448(P-10)
4	Specific gravity, 27C, gm/ml	0.86–1.03	0.93	IS: 101(P-1/S7)
5	Water content per. Vol. %	Max. 1.0	Nil	IS: 14,448(P-40)
6	Striping value with bitumen containing 1% Wt. anti-striping Comp	No striping		IS: 6241
7	Nitrogen content, 5 Wtmin	7.0	7.4	Element analyser
8	Boiling water test, % min. coating	95%	Above 95%	ASTM D 3625
9	Total base value mg, KOH/g	Min. 200	214,012	–

3 Design of Cold Mix Product

The design of cold-mixed BM or semi-dense bituminous concrete (SDBC) does not currently have a standardized design process. The cold mix design process includes optimizing both the water and bitumen emulsion content for the mix's aggregates. The aggregates are given a light water mist to moisten their surface, and then cationic bitumen emulsion is applied. A variety of aggregate sizes are blended in a variety of ratios to produce the best standard gradation of BM or SDBC. The procedures described below are necessary for designing cold-mixed BM or SDBC utilizing cationic bitumen emulsion.

(a) Testing of aggregate: In order to determine attributes like specific gravity, aggregate impact value, water absorption, and sand equivalent value, the supplied aggregates are put through sieve analysis and other physical tests.
(b) Testing of bitumen emulsion: The samples of the rapid setting bitumen emulsion (SS-2) that will be utilized to build the BM must pass the IS: 8887 test. Additionally, aggregates must pass the ASTM 2397 or AASHTOM 208 eating test.
(c) Determination of Emulsion content: The quantity of emulsion is first determined using the following equation given by Wulandari and Tjandra [9].

$$P = 0.05 \times A + 0.1 \times B + 0.5 \times C,$$

where P represents the quantity of bitumen emulsion in per cent, A represents the per cent aggregate retained on a 2.36 mm sieve, B represents the per cent aggregate passing 2.36 mm sieve and retained on the 90-micron sieve, and C represents the per cent aggregate passing on a 90-micron sieve.

The amount of bitumen emulsion is kept constant, while several samples of aggregates and emulsion mixes are generated with variable water contents. The binder then visually inspects the coated aggregates to determine the extent of the coating. The water concentration at which aggregate coating is greatest is known as the optimal water content. When creating the cold-mixed DM or cold-mixed SDBC, this water cement is used as a premixed water contest. It has been shown that a water content of 2–3% by weight of aggregate is ideal for BM or SDBC. However, this quality can change according on the weather (temperature, humidity, and wind speed), as well as the characteristics of the aggregates.

(a) Determination of emulsion percentage

For granitic aggregate: The quantity of emulsion is first determined for granitic aggregate using the following empirical formula.

$$P = 0.05A + 0.1B + 0.5C,$$

where P represents the quantity of bitumen emulsion in per cent, A represents the per cent aggregate retained on 2.36 mm sieve, B represents the per cent aggregate passing 2.36 mm sieve and retained on 10 micron sieve, and C represents the per cent aggregate passing on 90-micron sieve. Weight of sample is taken 1000 gm passing 4.75 mm.

(b) *Fixation gradation of aggregate*

The upkeep and development of roads will benefit from the usage of tested, sized, and graded aggregates. While there are other aspects to take into account as well that contribute to a good road, attention must be given in particular to well-prepared, sized, and graded aggregates. Establishing the ratios of bitumen, filler, fine and coarse aggregates is the aim of the bituminous mix design in order to produce a mix that is useful, sturdy, long-lasting, and inexpensive. Table 5 shows observation table of the Marshall Stability test.

3.1 Tests on Design Product

Numerous factors, including the source of the aggregate, the curing environment, the curing period, etc., affect the properties of cold mixes. Consequently, there is no well-established mix design methodology for cold mix. But the Marshall Method is frequently employed when creating emulsified mixtures.

Figures 1 and 2 show variations in flow value and stability value concerning percentage of rubber, respectively. From both the figures, it is observed that at 3%

Table 5 Observation table for Marshall Stability test

Sr. No	Material	Emulsion %	Rubber %	Max. proving ring reading	Flow dial reading in Div	Stability Value in KN (Max. proving ring reading × 5 × multiplication factor)	Flow Value, 0.25 mm, Units
S1	Granite	10	0	35	8	11.970	2.00
S2		10	1	31	4	10.602	1.00
S3		10	2	30	10	10.260	2.50
S4		10	3	23	9	7.866	2.25
S5		10	4	10	10	3.420	2.5
S6	Granite + Basalt	10.8	0	30	6	10.26	1.5
S7		10.8	1	25	15	8.55	3.7
S8		10.8	2	27	4	9.23	1.0
S9		10.8	3	29	27	9.91	6.7
S10		10.8	4	26	25	8.89	6.2
S11	Basalt	11.66	0	27.5	11	9.40	2.7
S12		11.66	1	34	6	11.62	1.5
S13		11.66	2	29	12	9.92	3.0
S14		11.66	3	23	22	7.87	5.5
S15		11.66	4	25	7	8.55	1.75

and 4% of rubber combination of aggregates (Granite + basalt) shows excellent results of flow value and stability value.

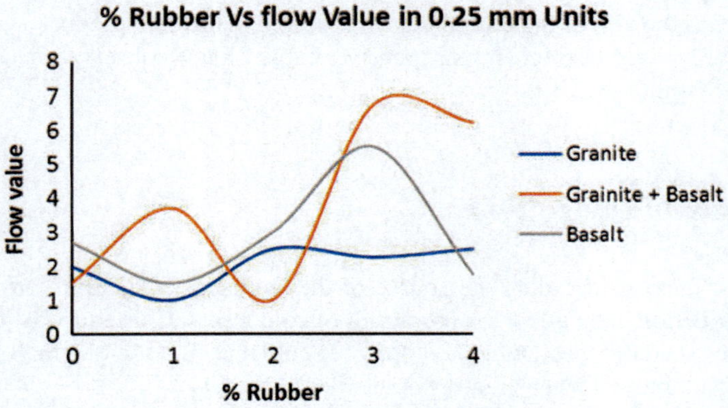

Fig. 1 Plot of flow value versus percentage of rubber

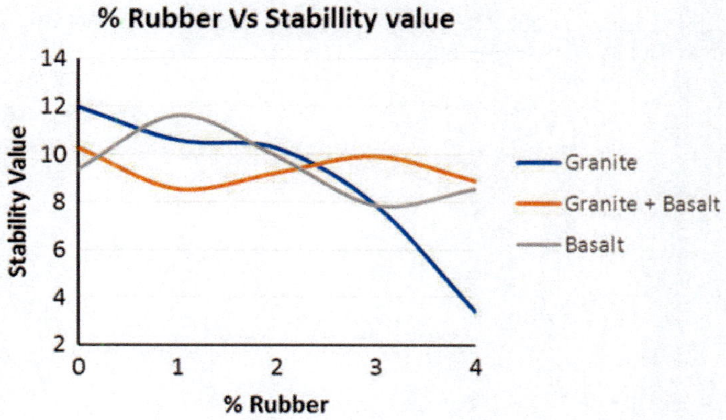

Fig. 2 Plot of stability value versus percentage of rubber

4 Conclusion

From the experimental investigation of the cold mix made by using rubber waste and granite aggregates as partial/full replacement of basalt aggregates, it is observed that the rubber waste may be used in cold mix for road construction with optimum dosage of 4%. Also, 50% use of granite aggregates replacing basalt aggregates gives excellent results of emulsion and the Marshall Stability value. Therefore, it is concluded that rubber waste as well as granite aggregates have a good potential to be used in the cold mix technology for the road construction.

References

1. Silva RV, de Brito J, Dhir RK (2019) Use of recycled aggregates arising from construction and demolition waste in new construction applications. J Clean Prod 236:117629
2. Dash SS, Chandrappa AK, Sahoo UC (2022) Design and performance of cold mix asphalt. Constr Build Mater 315:125687
3. Nageim HA, Dulaimi A, Al-Busaltan S, Kadhim MA, Al-Khuzaie A, Seton L, Croft J, Drake J (2023) The development of an eco-friendly cold mix asphalt using wastewater sludge ash. J Environ Manag 329:117015
4. Sahu R, Joshi R (2015) Economic construction of bituminous road by utilization of waste tyres rubber-CRMB. Int J Sci Technol Eng 1(12):35–40
5. Baraiya ND (2013) Use of waste rubber tyres in construction of bituminous road: an overview. Int J Appl Innov Eng Manag 2(7):12–18
6. Mane PA, Petkar DG, Bhosale SM (2013) Laboratory evaluation of usage of waste tyre rubber in bituminous concrete. Int J Sci Res Publ 3(9):1–7
7. Chhabra R, Marik S (2014) A review literature on the use of waste plastics and waste rubber tyres in pavement. Int J Core Eng Manag 1(1):1–5
8. Rombi J, Zedda V, Contu A, Etzi A, Portas S, Coni M (2013) Re-use of granite by-products for road construction. Int J Pavem Confer 15:169

9. Wulandari PS, Tjandra D (2023) Properties evaluation of cold mix asphalt based on compaction. IOP Confer Ser Earth Environ Sci 1195:0120249

Utilization of Construction and Demolition Waste Material in Low Volume Road Construction: Experimental Study

Snehal V. Waghchaure, Priyanka A. Rakshe, Rutika S. Jadhav, Rutuja A. Gavhane, and Bharti M. Shinde

1 Introduction

Concrete is a particle reinforced composite material in which the aggregates are embedded in a matrix of cement. It is a hard substance made from cement mixed with sand, water, aggregates, etc., combined in the proper ratios to create a paste. Concrete endure compression pressures after setting also, due to its strength, durability, and flexibility it can be shaped into a variety of shapes and textures. For added tensile strength and crack resistance, it can be strengthened with steel. Hence, the use of concrete in construction of various structures is increased. But, with the age the strength of structure is reduced due to various reasons like temperature changes, chemical attack, cracks developed, etc. Therefore to improve the service life of structure either it is altered or demolished. Demolition of structure generates a huge quantity of solid waste. As constructions like buildings, bridges, and highway pavement are demolished, construction and demolition (C&D) wastes are produced. Certainly, these industrial solid wastes have a negative impact on the environment. Cement concrete waste is one of the main types of C&D wastes, in fact, demolition waste is a source of solid wastes from the construction industry. As a result, since cement concrete is not a biodegradable material, it could be challenging to dispose of solid trash of it. The greatest way to lessen their negative environmental consequences is to dispose of these wastes sustainably by using them to create structural and non-structural concrete. The disposal of concrete has become a significant environmental concern because of increased environmental standards and the lack of space available. Due to the growing environmental issues associated with construction and demolition (C&D) waste, several nations are working to create

S. V. Waghchaure · P. A. Rakshe · R. S. Jadhav · R. A. Gavhane · B. M. Shinde (✉)
Department of Structural Engineering, Sanjivani College of Engineering, Savitribai Phule Pune University, Kopargaon, Maharashtra 423603, India
e-mail: shindebhartist@sanjivani.org.in

S. Kolathayar et al. (eds.), *Best Practices in Geotechnical and Pavement Engineering*, Lecture Notes in Civil Engineering 449, https://doi.org/10.1007/978-981-99-8505-0_40

new regulations and increase environmental protection awareness. Urbanization and urban renewal produce a significant amount of C&D waste. In terms of the world level, Muhammad et al. [1] presented that the urbanization rate was 54.3% in 2016 and 55% in 2018. By the year 2050, it is projected to reach 68% globally, which would result in an ongoing rise in C&D waste production. Among the nations that produce the most C&D waste are China and the USA. Today, science and technology must keep up with emerging trends that are economical and environmentally friendly. Old concrete structures that have been demolished can be repurposed to obtain recycled aggregate. Whenever RCA is used in place of natural aggregates, the mechanical qualities of the fabricated concrete suffer significantly, along with a discernible increase in drying shrinkage [2]. To reduce the negative impact of RCA on concrete qualities, researchers employed RCA as a partial substitute for natural one. The RCA is an adaptable construction product that may be used to build gutters, pavements (base layer and sub-base), and revetments. But due to lack of sufficient information and research application, the utilization of C&D waste and its application in roads is limited in India. When building roads, C&D waste replaces the need for traditional raw materials like gravel and rock which results in preserving the natural resources. Additionally, by minimizing their extraction and transportation, it also uses less energy and has a smaller carbon footprint. The addition of C&D waste to road development may strengthen and stabilize the infrastructure for roads, by providing stability and reinforcement through its materials. Building and demolition (C&D) waste use in road building can improve road infrastructure. When correctly handled, C&D waste can offer adequate materials for sub-base and base layers, enhancing the load-bearing capacity. This may result in more durable and long-lasting road surfaces, lowering maintenance costs and minimizing traffic delays. The effective utilization of C&D waste as a substitute material, thus, requires additional thorough research in this field. Hence, the feasibility analysis of using C&D trash in rural concrete roads is described in the current work.

1.1 Application of RCA

Crushing and separating of used concrete, masonry, and asphalt materials results in recycled coarse aggregate (RCA). The most popular RCA applications are as below:

1. **Road base and sub-base**

RCA is a suitable material for use in the base and sub-base of roads.

2. **Concrete manufacture**

RCA can be used to replace natural coarse aggregates in the production of fresh concrete. This decreases the requirement for virgin materials and the carbon footprint of the concrete industry.

3. **Drainage systems**

RCA can be used as a drainage material in applications like French drains and septic systems. RCA can be used as an ornamental landscaping material in applications such as roads, pathways, and garden paths.

4. **Erosion control**

RCA can be used in erosion control applications like rip rap to stabilize slopes and prevent erosion. Overall, the use of RCA can help reduce waste and save natural resources while still delivering a high-quality material for a wide range of construction applications.

2 Literature Review

Tavaakoli and Soroushin [2] have presented the study on, regular concrete's strength, aggregate content compared to RCA, and the proportion of fine to coarse particles in regular concrete, to determine the strength characteristics of RCAs. In his study, author has used Los Angeles abrasion loss and water absorption test to determine the strength characteristics of RCAs.

Limbachiya et al. [3] have presented a study on use of C&D waste in concrete. Authors have studied the effect of coarse RCA on ceiling strength, bulk engineering, and durability properties of concrete. They also concluded that up to 30% substitution with recycled aggregates had no negative impact on compressive strength.

Mandal et al. [4] have studied that the RCA concrete strength can be improved by changing the W/C ratio, while the concrete is being mixed. Based on the data, the author concluded that, for 28 days of design strength, specimens constructed using RCA concrete nearly equaled those constructed with normal aggregate concrete in terms of durability and technical performance.

Zaki et al. [5] have studied that the RCA may replace NA up to 50% and is suitable for non-structural applications such as sidewalks, curbs, and concrete driveways. The guidelines and aspirations for the future call for the addition of an admixture to strengthen recycled aggregate concrete used in structural applications. In addition, authors have discussed about the government incentive demolition contractors to build crushing and screening facilities to employ CDWs in concrete mixtures, for environmental and economic reasons.

Gumaster and Ramamurthy [6] have found that the RCA concrete's compressive strength is much lower and that this difference is solely due to the parent concrete's strength, which is made up of the acquired material.

Manasa et al. [7] have looked into the fact that crushed, inorganic particles from building and demolished materials constituents of recycled coarse aggregates (RCA). In order to replace natural aggregates in high strength concrete, this research examines the strength characteristics of recycled aggregates. It compares RCA concrete with natural aggregate concrete in terms of workability, compressive strength, tensile strength, elastic modulus, and flexural strength at 0, 25, 50, 75, and 100%. As the proportion of recycled aggregates increases, the parameters determined moderately decrease.

3 Experimental Study

3.1 Physical Characteristics of NA and RCA (Gravel)

3.1.1 Sieve Analysis

Sieve analysis is a technique for determining the particle size distribution of a material sample. Recycled aggregates are obtained from a demolished waste; therefore, it is very important to carry out the sieve analysis for particle size distribution as per the requirement.

3.1.2 Specific Gravity

Specific gravity is the weight ratio of a given volume of aggregate to a given volume of water.

3.1.3 Water Absorption

It is used to evaluate both the outside and interior concrete surfaces, water absorption rate (absorptive).When only one face of the specimen is exposed to water, the test comprises calculating the mass gain caused by water absorption as a function of time.

3.1.4 Bulk Density

The weight of a given volume of material is defined as bulk density. The weight of a known volume of material is measured in the test.

$$\text{Bulk density} = \text{substance weight}/\text{container volume}.$$

It is vital to note that a material's bulk density might change depending on parameters including moisture content, compaction, and particle size distribution.

Table 2 shows the basic properties of normal aggregate and recycled coarse aggregate used for present research work. Specific gravity is determined using basket method, and it is found in the range of standard specifications. Also, Table 1 shows that the water absorption of both NA and RCA is nearly same. Bulk density of RCA is slightly lower than that of NA.

For designing the concrete mix, the Bureau of Indian Standards [8] (IS10262:2019) is employed. Stipulation of proportioning according to [8] IS 10262:2009 codification and [9] IS 456:2000 (Grade of concrete = M25, Grade of cement = OPC 53 grade, Aggregate Size (Maximum) = 20 mm, Water to cement ratio = 0.5).

Step 1: Mean Target Strength

The mean target strength is expressed as

$$f_{ck'} = f_{ck} + 1.65 \times S, \quad (1)$$

$$\therefore \quad f'_{ck} = 25 + 1.65 \times 4 = 31.6\ \text{N/mm}^2.$$

Compressive strength of concrete at 28 days = 25 N/mm^2.

Step2: Selection of Water Content

$$\text{Max. water content} = 186\ \text{lit(for}\ 25 - 50\ \text{mm slump)}.$$

$$\therefore \quad \text{Corrected water content} = 186 + (186 \times 3)/100.$$

Table 1 Basics properties for NCA and RCA

Properties	Natural coarse aggregate (NA)	Recycle coarse aggregate (RCA)	Standard specifications
Specific gravity	2.7	2.5	2.40–2.90
Water absorption (%)	1.8	3.5	0.20–4.0% for NA 3.0–12.0% for RCA
Bulk density (kg/m^3)	1600	1423	1200–1750 kg/m^3

Table 2 Material quantities in various proportions for 1 m^3

Ingredient	Quantity	Proportion
W/c ratio	0.5	0.5
Cement	383 (kg)	1
Water	191.5 (L)	0.5
CA (20 mm)	737.50 (kg)	1.93
CA (< 20 mm)	491.66 (kg)	1.29
FA	669.94 (kg)	1.795

$$\therefore \quad W = 191.5 \text{ liters.}$$

Step 3: Determination of Cement Content

$$\text{Cement content per unit volume of concrete for W.C. ratio} = \frac{191.5}{0.5}$$
$$= 383.16 \text{ Kg/m}^3.$$

Step 4: Determination of Coarse Aggregate Content

$$\text{Aggregate Size} = 20 \text{ mm.}$$

Zone II

$$\text{W.C. ratio} = 0.5$$
$$\text{Volume of CA} = 0.62.$$

Step 5: Determination of Fine Aggregate Content

$$\text{Volume of FA} = 1 - \text{Volume of CA}$$
$$= 1 - 0.62$$
$$= 0.38.$$

Step 6: Mix Proportion Calculation

$$\text{Volume of Concrete} = 1 \text{ m}^3.$$

$$\text{Volume of cement} = (\text{mass of cement/specific gravity}) \times 1/1000$$
$$= (333.16/8.15) \times 1/1000$$
$$= 0.121 \text{ m}^3.$$

$$\text{Volume of water} = \text{mass of water}/(\text{specific gravity} \times 1000)$$
$$= 191.5/(1 \times 1000)$$
$$= 0.191 \text{ m}^3.$$

Volume of all in aggregate

$$e = [a - (b + c + d)]$$
$$= 1 - [0.121 + 0.193]$$
$$= 0.686 \text{ m}^3.$$

$$\begin{aligned}\text{Mass of CA} &= e \times \text{volume of CA} \times \text{Specific gravity of } CA \times 1000\\ &= 0.686 \times 0.62 \times 2.89 \times 1000\\ &= 1229.17\ \text{kg.}\end{aligned}$$

$$\begin{aligned}\text{Mass of FA} &= e \times \text{volume of FA} \times \text{Specific gravity of FA} \times 1000\\ &= 0.686 \times 0.62 \times 2.89 \times 1000\\ &= 669.94\ \text{kg.}\end{aligned}$$

Therefore, the final mix proportion is expressed as Cement: Fine aggregate: Coarse aggregate = 1:1.75:3.22. According to mix proportion, the quantities of materials for 1 m^3 of concrete for % replacement of RCA are as given in Table 2.

4 Results and Discussion

4.1 Properties of the Hardened Concrete

4.1.1 Slump Test

The slump test is a common method for determining the consistency and workability of fresh concrete. It is often used in the construction sector to ensure that the concrete has the desired consistency and will be easy to pour and compact once in place. Filling a metal cone with fresh concrete and then removing the cone to see how much the concrete slumps or settles is the test. Figure 1 shows the experimental set up for slump cone test. The results of workability of concrete using slump cone test are presented in Table 3.

4.1.2 Compression Test

A compression test is a common method for determining the compressive strength of a material such as concrete, masonry, or rock. A compressive load is applied to a sample of the material until it fails, and the load at failure is measured. The following are the steps for performing a compression test: Prepare a sample of the substance to be tested in accordance with the applicable standard standards. The results of compressive strength of concrete for various % replacement of RCA for and 28 days is given in Table 4. Figure 2 shows the variation of 7 and 28 days compressive strength of concrete for 0, 50, and 100% replacement of aggregate. From Table 4 and Fig. 2, it is observed that the 7 and 28 days strength of 50% replacement is more than that of 100% replacement. Also at the 50% replacement the results obtained are in close agreement with the 0% replacement.

Fig. 1 Concrete slump test for workability

Table 3 Slump values of concrete made up of various % replacement of RCA

% of RCA	RCA (0%)	RCA (50%)	RCA (100%)
Slump (mm)	100	90	82

Table 4 Results of average compressive strength of the concrete for RCA%

RCA percentage (%)	Average compression load (KN)	Compressive strength at 7 days (MPa)	Compressive strength at 28 days (MPa)
0.00	695	20.5	30.79
50.00	590	17.38	26.11
100.00	545	16.15	24.26

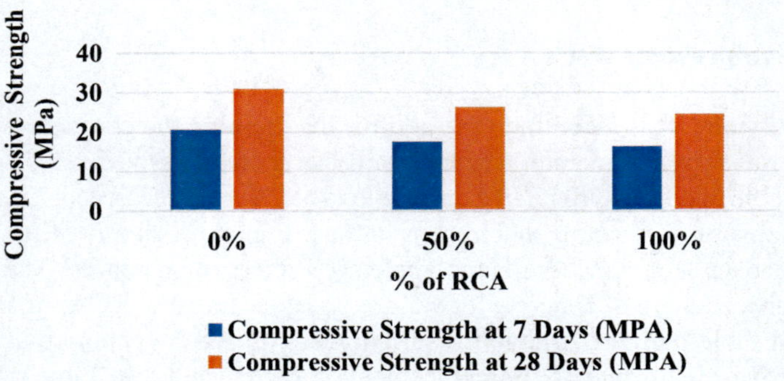

Fig. 2 Comparison of compressive strength of concrete for various proportions of RCA

5 Conclusions

In conclusion, using C&D waste in low volume road building has major positive effects on the environment, the economy, and society. It fosters the circular economy, lowers the use of natural resources, improves road performance, and helps with sustainable waste management. In the present study, the compressive strength of concrete is determined where natural coarse aggregate is replaced by RCA obtained from C&D waste. From the results obtained, it is concluded that the 28 days compressive strength of concrete at the 50% replacement of RCA is found to be in close agreement with the 0% replacement.

References

1. Muhammad SA, Beijja H, Lifeng C (2020) Review of construction and demolition waste management in China and USA. J Environ Manag 264:11045
2. Tovakoli M, Soroushin P (2022) Strength of recycled aggregate concrete made using field demolished concrete as aggregate. ACI Mater J 93:182–190
3. Limbachiya MC, Leelawat T, Dhier RK (2000) Usage of recycled aggregates in high strength concrete. Mater Struct Materiaux et Constr 33:574–580
4. Mandal S, Chakarborty S, Gupta A (2002) Some studies on durability of recycled aggregate concrete. Indian Concr J 76:385–388
5. Zaki A., Eldaly A., Abdelshakour M., Aboelkasim D.: Using construction demolition waste (CDW) as recycled aggregate in non-structural concrete elements. In: Proceedings of the 3rd international conference sustainable construction and project management integrated management for smart cities, pp 1–12
6. Rammurthy K, Gumaste KS (1998) Properties of recycled aggregate concrete. Indian Concr J 72:49–53
7. Manasa M, Bhaskar U, Naveen Kumar G (2019) Performance of recycled aggregate concrete for M25 grade concrete. Int J Eng Adv Technol 9(2):4694–4699
8. IS10262 (2019) Concrete mix proportioning
9. IS 456 (2000) Plain and reinforced concrete design